D0023884

COMPUTING FOR ENGINEERS AND SCIENTISTS WITH FORTRAN 77

COMPUTING FOR ENGINEERS AND SCIENTISTS WITH FORTRAN 77

Daniel D. McCracken

City College of New York

John Wiley & Sons

New York   Chichester   Brisbane   Toronto   Singapore

Copyright © 1984, by John Wiley & Sons, Inc.

All rights reserved. Published simultaneously in Canada.

Reproduction or translation of any part of
this work beyond that permitted by Sections
107 and 108 of the 1976 United States Copyright
Act without the permission of the copyright
owner is unlawful. Requests for permission
or further information should be addressed to
the Permissions Department, John Wiley & Sons.

Library of Congress Cataloging in Publication Data:

McCracken, Daniel D.
    Computing for engineers and scientists with Fortran 77.

    Includes index.
    1. Engineering—Data processing.   2. Science—Data
processing. 3. FORTRAN (Computer program language)
I. Title.
TA345.M395   1984   620'.0028'4   83-23473
ISBN 0-471-09701-2

Printed in the United States of America

10 9 8 7 6 5 4 3 2 1

Vaughan
TA345
.M395
1984
c.1

To Charles B. Stoll and Walker G. Stone
for their support and encouragement as my first editors

PREFACE

This book uses the study of programming in Fortran 77 to provide:

- A programming competence that an engineer or scientist can use in work or study.
- An understanding of the powers and limitations of computers.
- A basis for effective communication with programming experts when more difficult tasks are encountered.
- An appreciation for the increasingly common ways to use a computer as a tool *without* any conventional programming.

Since Fortran is the *lingua franca* of applications in engineering and science, it is the choice for this book. Fortran 77 is used, specifically, because it is the latest standard and because its newer features permit the writing of structured programs that are much easier to understand and maintain than the programs written in earlier versions of Fortran.

As with my previous books, there are many illustrative programs. The examples are drawn from engineering, science, and college mathematics, at a level consistent with the fact that a course of this type is often taught in the freshman year. A number of standard techniques in elementary numerical methods are introduced through programming examples, recognizing that many students will never take a formal course in numerical methods. Throughout the text there are warnings about traps for the unwary, such as the accumulation of roundoff errors, the effects of finite precision, and the foolishness of blind reliance on double precision.

Readers with the goals assumed here have no need to know all of the Fortran 77 language. Some of the older and less desirable features of the language (the arithmetic IF, the assigned GO TO, etc.) are not covered at all. Others are covered in part. List-directed input and output serve nicely while the reader is learning more fundamental matters, for example, and then formatted input and output are considered at an appropriate depth in Chapter 6.

Heavy emphasis is placed on good programming style. Logic structures are held to a bare minimum and are used in a consistent way. Statement numbers are never attached to statements other than CONTINUE and FORMAT. The GO TO statement is used only in simple Fortran 77 implementations of the WHILE and REPEAT structures. Pseudocode is introduced in the third chapter, and subprograms are introduced in the fourth—the latter much earlier than in most other Fortran texts. Subprograms are used consistently thereafter as a program organizing tool, and effective use of subroutine libraries is emphasized with several examples from the IMSL library. In Chapter 5, after the reader has enough background to understand the issues, there is a full treatment of program development, including modularization, top-down and stepwise development, and program testing. These issues are reinforced in later chapters with larger example programs.

There are many exercises, ranging from "fingerwork" to reinforce the syntax, to challenges that might serve for term problems. Answers to nearly half of the exercises are given at the end of the book.

The final chapter is a "horizon-broadener," with a wide-ranging sampling of ways to use computers without any conventional programming at all: mathematical and engineering packages, symbolic mathematics, bibliographic searching, NOMAD as an example of a fourth-generation language, and text editing and formatting.

In sum: The book is—among other things—a text on Fortran programming, and is intended for use in a conventional course on that subject. I, for one, have no doubts about the legitimacy of such a course for future engineers and scientists. But the reader will be left with a clear understanding that when the computer is viewed as a tool of the practicing engineer or scientist, there are increasingly attractive alternatives to actual programming for getting applications done. That statement is true even today, and will be an unremarkable commonplace in the working careers of students now in school. The book thus provides a solid core of skills for use in today's world of computer applications, and a bridge to tomorrow's.

An extensive Instructor's Manual gives answers to all exercises that do not have answers in the book. It also contains suggestions for newer teachers on how to teach programming, an annotated model syllabus, sample exams, and a summary of the very few programming changes that are needed when the book is used in the WATFIV environment.

Daniel D. McCracken

New York
October, 1983

ACKNOWLEDGMENTS _____

Like most programming, producing a book is a team effort. It is a pleasure to acknowledge the many and varied contributions of the following people.

Gregory P. Williams, Masstor Systems Corporation, an old friend from General Electric days, served as a faithful reviewer and provided early guidance that influenced the shape of the book in fundamental ways.

Charles L. Baker, Science Applications, Inc., supplied much good advice on programming style matters and on the way programming is really done.

Jeffrey R. Sampson, University of Alberta, and Paul P. Clement, Advanced Systems, Inc., were the most thorough and helpful reviewers I have ever worked with. (And that's saying something, because I value the contribution of reviewers very highly indeed, and have worked with many good ones.) Their contributions ranged from catching programming slips, to improving style, to suggesting major revisions in early drafts.

John L. Lowther, Michigan Technological University, also did a fine job of reviewing, and suggested the Fortran 77 implementation of the WHILE construct that I have used. He disclaims credit for inventing it, but I hadn't seen it and it is much better than what I had been planning to use.

Michael J. Clancy, University of California, Berkeley, and Leon Levine, University of California, Los Angeles, were particularly helpful in the early planning stages, drawing on their extensive experience with the kind of course for which this book is intended.

My editors at Wiley, first Gene A. Davenport and now Carol Beasley, capably assisted by Judith Watkins, were uniformly helpful and supportive. High on the long list of things they did to assist me was the provision of an astonishing number of highly talented reviewers, without whose services I simply would not know how to approach this kind of project. Most of these reviewers worked anonymously; I am delighted that all of them said "yes" when my request to give them their due credit here was passed along.

The Wiley reviewers: J. Mack Adams, New Mexico State University; Donald E. Burlingame, State University of New York, Potsdam; Donner A. Dowd, Jr., Lake Superior State College; Henry A. Etlinger, Rochester Institute of Technology; Elaine N. Frankowski, Honeywell Information Systems; Robert M. Graham, University of Massachusetts; Charles E. Hughes, University of Central Florida; Thomas E. Kurtz, Dartmouth College; Richard J. LeBlanc, Georgia Institute of Technology; Hans E. Lee, Michigan State University; George L. Miller, North Seattle Community College; Frederick J. Mowle, Purdue University; Steven S. Muchnick, University of California, Berkeley; James T. Perry, San Diego State University; Franklin Prosser, Indiana University; Woodrow E. Robbins, North Carolina State University; Bernard H. Rosman, Framingham State College; J. Denbigh Starkey, Washington State University; David B. Teague, Western Carolina University; John F. Wakerly, Stanford University;

Jerry M. Waxman, Queens College; Lloyd Weaver, Purdue University; R. A. Williams, University of Akron.

Chapter 9 contains several examples drawn from the subroutine library of IMSL, Inc. Thomas J. Aird and Granville Sewell were generous in their assistance. Dr. Sewell also created the input and produced the graph using IMSL's TWODEPEP in Chapter 10.

Chapter 10 includes a section on muMATH, a symbolic mathematics package from the Soft Warehouse, Honolulu. David R. Stoutemyer made several helpful suggestions in getting this material right.

Chapter 10 also contains a section on bibliographic searching based on the facilities of DIALOG, Inc. Charles T. Meadow and Charles D. Sullivan were most helpful in adapting this material so that it both showed what is possible and remained intelligible to the reader unfamiliar with such services.

All of the text and almost all the programs were developed using the facilities of Dun & Bradstreet Computing Services, formerly National CSS, Inc. (This name change came too late, unfortunately, to be able to change the text references to National CSS.) My closest contacts there in recent years have been Nicholas A. Rawlings and Christopher Grejtak, with much help on VS Fortran from Lloyd E. Fuller and Walter H. Horowitz. Mr. Rawlings also reviewed much of the manuscript and assisted in a variety of other ways, especially with the NOMAD material in Chapter 10. The letter reproduced on page 320 attempts to express my appreciation to some of the many others now or formerly at the company to whom I am indebted.

A few of the programs were run on the facilities of the City University of New York, and I wish to express my appreciation for that support. My own institution, the City College of New York, is a component of the City University, and it is a pleasure to express my appreciation for the support given to me by my colleagues and by my chairman, George G. Ross. My appreciation also goes to the director of the CCNY computer center, George W. Elder, and to Paul Fortoul, a systems programmer who has been a major factor in my education in the past two years.

Too little credit is given to the people who turn a manuscript into a book: the production staff, with their wide range of talents and responsibilities. Their usual thanks is that if they do their job right, nobody notices! The production people I worked with at Wiley were uniformly competent, cooperative, and effective. I am happy to name them: Deborah Herbert, Madelyn Lesure, Eugene Patti, Elaine Rauschal, and Ruth Sandweiss. I appreciate all of their efforts.

Finally, I wish to say "thank you" to my students at City College. I suppose it's been said before, but I think I have learned more from them than they have learned from me. They make teaching a pleasure.

D.D.M.

CONTENTS

COMPUTING FOR ENGINEERS AND
SCIENTISTS WITH FORTRAN 77

ONE

GETTING STARTED IN COMPUTING AND FORTRAN

The uses of computers

Electronic computers are widely used in the solution of the problems of science, engineering, business, and education. This use is based upon their ability to operate at great speed, to produce accurate results, to store large quantities of data, and to carry out long sequences of operations without human intervention.

Here are some examples of the kinds of applications that we assume to be of interest to readers of this book.

■ Designing a chemical plant requires calculations of capacities, operating conditions, and yields, under a variety of circumstances. Determining the optimum operating conditions, taking into account technical and economic factors, requires large amounts of computer time.

■ Weather prediction studies involve large amounts of data and the solution of equations that, although not inherently difficult, call for vast amounts of computation.

■ Statistical studies of the relationships among various factors that affect a person's learning ability often require computers. The computations may be as modest as a student research project, or as complex as a study following millions of people for many years.

■ The communications industry uses computers to store, process, and disseminate information. Telephone systems make intensive use of computers for billing, network management, and—in increasing numbers—within the telephones themselves. In a different kind of communication, all the text for this book was processed by computer, with drafts produced by one computer and the photocomposition of the final book by another. More and more computing involves information of a textual sort or digitized voice data, rather than what we ordinarily think of as "numbers."

■ The investigation of the possible structure of a complex organic compound could involve a combination of computations of binding energies, interatomic distances, and so on, with an elaborate computer program to present the results in a graphical form, often in color.

■ The design of an airplane uses thousands of hours of computer time to investigate the interrelated requirements of structures, aerodynamics, power plants, and control systems as they would operate under various flight conditions. After a prototype has been built, flight testing generates voluminous data that must be captured and analyzed, the latter often using statistical techniques. Then, during manufacturing, many applications that would usually be called "data processing" come into play: project control, materials requirements planning, inventory management, purchasing, and quality control, among many others. Finally, operating such a fleet of equipment requires computers to order spare parts and keep track of them, schedule crew assignments and aircraft maintenance, and plan flights.

The computer techniques needed to work with applications as diverse as these depend to a certain extent on the subject matter of the task. But the person using a computer in any of them would need to know something about how to specify the desired processing to a computer, which is essentially what this book is about. And while we are learning to "talk to" a computer, we shall also look at small but representative examples of a variety of applications that are typical of the way computers are used in engineering and science.

The steps in solving a problem with a computer

There is much more to solving a problem with a computer than the actual use of the computer. It is instructive to outline the complete process of setting up a technical problem for computer solution to see just what people do and what the computer does.

Problem identification and goal definition

A computer cannot decide for us what we want to do. *We* have to decide what the system under development is supposed to accomplish, what goal or combination of goals it must satisfy, under what conditions it must operate, and what general approach to solving the problem is to be taken. In some applications this step may be trivial; in others it may take months or years. In any case, the step obviously demands full knowledge of the problem area; there is usually little the computer can do to help us with it.

Mathematical description

In many although not all of the kinds of applications that will be considered in this book, it is necessary to formulate a mathematical description of the process under study. This can generally be done in a variety of ways; an approach must be chosen or a new one developed if no standard method applies. This step, in which the computer is not involved, requires full knowledge of the problem and of the relevant branches of mathematics.

Numerical analysis

The mathematical formulation of the problem may not be directly translatable to the language of the computer, since the computer can only do arithmetic on rational numbers and make simple decisions. Differential equations, integrals, and trigonometric functions, to name a few common examples, must be expressed in terms of arithmetic operations. Furthermore, it must be established that any errors inherent in the data, or introduced by such operations as expressing continuous functions in terms of finite approximations, do not invalidate the results.

This entire branch of modern mathematics is largely outside the scope of this book. We shall assume that the reader approaches the problem-solving process with a method of solution—in this sense—in hand. We shall, however, illustrate quite a number of elementary numerical methods in the course of demonstrating various programming concepts.

Algorithm formulation

The next step is to devise a precise and unambiguous statement of exactly what we want the computer to do, expressed as a finite sequence of the operations of which it is capable. A computer cannot follow the order "solve this equation"—at least not in any *direct* sense, in Fortran. (But see the short discussion in Chapter 10 of muMATH, where we shall see a few examples of operations on equations in symbolic form.) It *can* follow the order "square the number identified by the variable named X2REAL," or "go back to the beginning if the current value of the variable named EPS is greater than 0.0001," or "divide SUM by N." Furthermore, and this is crucial, the exact sequence of actions must be specified in complete detail in advance, especially at all points where the computer is required to make a "decision" based on relationships among values in the computation. An unambiguous definition of the actions to be carried out in solving a problem is called an *algorithm*. An algorithm might be expressed in English sentences, or in a computer language such as Fortran, or in a notation called *pseudocode*.

Computer programming

The next task, assuming the algorithm was not originally expressed in a programming language, is to express it so. Fortran is one language among many others used for this purpose. It is the language most commonly used for the types of technical problems that we address in this book. Fortran was developed in the mid-1950s by IBM and some of its customers, with a team led by John W. Backus.

A computer language imposes restrictions of its own in terms of what kinds of commands it can "understand" and carry out; different languages have different capabilities. Furthermore, the exact form in which our orders to the computer are expressed is prescribed for each language, and the rules are generally rather inflexible.

One major purpose of this book is to enable you to construct correct Fortran programs to solve problems of interest to you. Even if, as will probably be the case, you do most of your computer work using programs written by others, knowing Fortran will give you a basis for understanding what computers can (and cannot) do. It will also give you a vocabulary for talking about your needs with computing experts.

Program testing

There are so many opportunities to make mistakes in programming that most programs do not operate as intended when first tried, because of errors in any of the steps listed above. Mistakes must be located and corrected, and the program must be thoroughly tested to establish as fully as possible that it actually does what the programmer meant it to do. The computer is used heavily in this step, which can easily take longer than writing the program in the first place.

We shall place considerable emphasis in this book on writing programs in a way that minimizes programming errors and facilitates locating those that do occur. The techniques suggested will also greatly facilitate the maintenance of programs. (All programs that are used over an extended period of time have to be modified as requirements and computer equipment change.) Programs written with maintenance in mind are very much easier to maintain than those where this consideration has been ignored.

Production

Now, finally, the program can be used to process "real" data, that is, data other than the sample values used in program testing. This may be done in a variety of ways, depending upon the computer and the application. Sometimes data is entered from a computer "terminal" (a typewriterlike device that may also have been used to prepare the program) with results presented on a video display or other device that is part of the terminal. Sometimes many sets of data values are punched into cards and all of the cases run consecutively, with results printed on a paper listing. In other cases the computer is *on-line* as part of a process control system. Here, the data is obtained by direct input from sensors located in the process equipment; the output is a combination of printed logs and of signals controlling the operation of the process equipment.

Interpretation

Except in the process control situation just noted, results produced by the computer in response to our program do not always constitute an "answer" to the problem. The computer user must often now interpret the results to see what they mean in terms of the combination of goals of the proposed system. It is often necessary to repeat some or all of the preceding steps before the problem is really "solved."

Several conclusions may be drawn from this discussion. First, the computer does not, by itself, solve problems. It only follows exactly the computational procedures (i.e., the program) given to it. Second, a computer does not relieve the user of the responsibility of planning the work carefully; in fact, use of the computer demands more careful planning than noncomputerized methods ordinarily do. This is an important secondary benefit of using the computer. Third, a computer does not in any way eliminate or even reduce the need for a full and detailed understanding of the problem area, or for a thorough knowledge of the related mathematics.

The emphasis in the first part of this book is on the programming step. Problem identification, goal definition, and mathematical formulation are in the province of the technical area under consideration: electrical engineering, physics, statistics, operations research, or whatever. Numerical analysis is a branch of modern mathematics in its own right. Program testing is discussed, but not at great length and mostly in the context of how to write programs that minimize the probability of programming errors. Production is not ordinarily the programmer's responsibility, except possibly in the case of student exercises. The interpretation of results brings us back into the specialized field of the problem area.

But don't write a program at all if one already exists!

Anyone who needs to know Fortran should also know that there are often alternatives to writing Fortran programs. For a great many of the things that a user might wish to do with a computer, other people have already written programs which, possibly with slight modification, will serve the need. Programs of this type fall into two categories: software tools and applications packages.

Examples of software tools:

■ A *text editor* permits the entry, modification, and other processing of textual data, such as letters, reports, and computer programs. A related *text formatter* can be used to print documents in an attractive form or used to drive photo-composition equipment.

■ The output of a program is often most useful if displayed graphically. A *graphics package* produces pictures on a video display terminal or plotter. It is seldom in the user's best interest to write his or her own graphics programs, since they can be exceedingly complex and since such programs already exist.

■ There are many hundreds of *databases* on a vast range of subjects such as, new chemical compounds, abstracts of publications in psychology, stock and bond prices, or the AP and UPI news wires. Access to these databases is usually through a *time-sharing network,* with a telephone connection between the user's terminal and the database.

Examples of applications packages:

■ An electrical engineer wishing to study the performance of a proposed electrical network containing various kinds of components, including nonlinear devices such as diodes and transistors, can choose from among perhaps a dozen different packages that deal with such problems. Some of the packages analyze various special aspects of the network, such as the effect of electrical noise, or the identification of the critical components in terms of a worst-case analysis of component characteristic variations.

■ The designer of an optical lens can turn to any of several packages that may "know" more about the intricacies of lens design than the average designer does.

■ A chemical engineer wishing to study the probable performance of a proposed piping network can choose the package that best fits his or her precise needs: Transient or steady-state? Compressible or incompressible? With or without loops in the piping?

■ The designer of a new integrated circuit chip has available a number of packages that help lay out the components so as to minimize fabrication problems, power consumption, time delays due to capacitance, or whatever else may be of concern.

There are literally thousands of applications packages of the general sort suggested by these examples. When the problem to be solved fits an existing package, it almost always makes sense to buy or lease it rather than writing a program yourself. The package, in most cases, will have been written by top experts in the field. It will have been thoroughly checked out—in part by other users!—and therefore be much less likely to contain significant errors than whatever you might write yourself. And there will be an organization standing behind it for maintenance if problems do develop, or if modifications are needed to make the package more effective or to make it work on new computing equipment.

So why learn Fortran?

Then what is the purpose of learning Fortran at all, if you will more commonly use software tools and applications packages instead of writing Fortran programs? Four reasons:

■ You need to know what a computer is capable of, and, sometimes just as important, what it is *not* capable of. Learning a programming language and solving some problems yourself is the best way to gain an understanding of the power and limitations of a computer.

■ It will often happen that the program someone else has written will not *quite* do what you want. What then? In some cases, it will be your task to modify the package to fit your needs.

■ If the modifications are beyond your nonprofessional abilities, or if no existing program is suitable and a new one must be written, you will need to call upon the services of experts. To interact effectively with them you will need to know not only what is reasonable to ask (in terms of what a computer can do) but also enough about *their* field to be able to talk intelligently. Learning Fortran and writing some programs yourself provides that background.

■ Many students, especially those in electrical engineering, will need to take further courses in computing to be able to design systems that incorporate microcomputers. Studying Fortran is one good way to get started learning enough about computers and computing to be able to do such design work.

This book deals with both aspects of getting work done with a computer, i.e., writing Fortran programs yourself and using programs that others have written. In the first eight chapters of the book you are given enough information about Fortran to enable you to write meaningful programs. Using this knowledge, you will be able to write programs in your other courses (if you are in school), and this experience will serve your need to know what a computer can do and how to talk with computer experts. Chapter 9 emphasizes the use of preprogrammed libraries whenever possible. The last chapter is a sampling of various ways to get work done with a computer without doing any "programming" at all, at least not as "programming" has conventionally been understood.

A simple program

Let us begin the study of Fortran programming by considering a simple example of a program, one where the required processing is so short and easily stated that the algorithm is a matter of one sentence.

The task is to compute the value of a certain fourth-degree polynomial for a value of X that is to be obtained from the computer terminal or from a punched card. Let us turn to the program shown in Figure 1.1 to see how this job might be done.

NOTE! The presentation of this first program is designed only to give you an overview of what a program is and some of the concepts in writing and running a program. *All of the Fortran details will be covered again, most of them in the next chapter.* Don't worry if you don't understand every detail on this first exposure!

Fortran statements

A Fortran program consists of *statements*, and almost always, *comment lines*. The latter are identified by having the letter C or an asterisk in the first position of the line. We see that this program has five statements and one comment line. Blank lines may be used freely to improve readability; they are in effect comment lines. Comments in this book are printed in upper- and lowercase letters to make them stand out; you will probably be able to use only uppercase letters on your computer.

Variables and the REAL statement

The term *variable* is used in Fortran to denote a quantity that is referred to by name, such as X and POLY. It is our responsibility to be sure that the program gives values to variables, since Fortran does not do so automatically. A variable is able to take on many values during the execution of a program, but can have only one value at any specific time. The values of variables are changed under the control of Fortran statements that we write in the program.

Fortran provides a variety of *types* of variables, all of which we shall investigate in due course. With the REAL statement we specify that X and POLY are variables that are permitted to have fractional parts, and that they are stored in a particular manner inside the computer. The most commonly used other variable type is INTEGER, which we shall study in the next chapter. We shall also learn there that Fortran would have interpreted X and POLY to be of the type REAL even if the REAL statement had not been present. This is a flexibility that will never be used in this book, however, since careless application of it can cause serious difficulties. In this book all variables will be listed in *type statements:* REAL, INTEGER, or one of the others.

The REAL statement is called a *specification statement*, or a *declaration*. It provides information about the program without itself calling for any action when the program is executed. The READ statement that comes next, on the other hand, *does* call for action when the program is executed. It and the others are called *executable statements*.

```
* A program to evaluate a polynomial for a value of X read as input

      REAL X, POLY

      READ *, X
      POLY = 2.0*X**4 - 15.0*X**3 - 2.0*X**2 + 120.0*X - 130.0
      PRINT *, ' FOR X = ', X, ' THE VALUE OF THE POLYNOMIAL IS ', POLY

      END
```

Figure 1.1 A Fortran program to evaluate a particular polynomial for a value of X read as input, then print X and the result.

The READ statement

One way to give a value to a variable is with the READ statement, which we see is the first executable statement of this program. As written here, with an asterisk following the word READ, the READ will take its input from the standard input device, and give that value to the variable, X, named in the READ. The asterisk also implies that we may enter the value of X in almost any convenient form—with or without a decimal point, for example. We shall see later (Chapter 6) that there is much more to the subject of input than this. Luckily, we can post-pone these details, which greatly enhance the power and flexibility of input and output in Fortran, but which are rather confusing if brought in at this stage.

As noted, the READ obtains a value for X from "the standard input device." If you are running the program interactively at a terminal, the READ will tell you that it is ready to accept data and then wait for you to type in a value. If the standard input device on your computer is a card reader, the READ will expect to find a card containing a value for X. Representative ways to carry out these operations are shown at the end of the chapter.

Once X has a value, we can use it in computing the value of the polynomial.

The assignment statement

The Fortran *assignment statement* is a command to "assign" a value to a variable. The statement has the form:

```
variable = expression
```

This means "evaluate the expression on the right and assign that value to the variable named on the left." If the variable named on the left had a value before this operation, it is lost.

In our case, the expression specifies a series of operations involving X and several constants. A Fortran *constant* is a quantity whose value cannot change during the execution of the program. For example, in a statement such as:

```
XNEW = X**2 + 14.78
```

XNEW and X are variables, whereas 2 is an integer constant and 14.78 is a real constant.

Arithmetic operations

Presumably it is reasonably obvious that plus and minus signs mean what you would expect them to mean, and that an asterisk (∗) means multiplication. Two asterisks side by side mean "raise to a power," and division, although it does not occur in this program, is represented by a slash (/).

Blank spaces in Fortran are almost never of any significance; the decision to put blanks around the plus and minus signs but not around the others was purely and simply a matter of trying to make the meaning clear and yet avoid having the line exceed the space available on the page in this book. A more or less consistent style will be followed on this matter in this book, and defended in terms of (human) readability—but unless you are instructed otherwise, the question of blank spaces is up to you.

The PRINT **statement**

With the value of the polynomial now computed and assigned to the variable named POLY, we are ready to print the values of X and POLY. As with the READ, the asterisk in the PRINT statement permits us to postpone a sizable body of details about formatting that will be interesting and useful later, but which would be distracting now.

Following the asterisk there is a *list* of what we want printed. Any material enclosed in apostrophes is printed exactly as it appears, providing identification of the output. The X and POLY, which are not enclosed in apostrophes, are the same variable names used earlier. Their appearance in the PRINT statement means that we wish their values to be printed.

When the PRINT is followed by an asterisk, the output is printed in a form that is chosen by the designer of the Fortran system. If you execute this program, your output may have more or fewer trailing zeros, for example, or the spacing may be different.

The END **statement**

Implicit in the foregoing description of the program is the idea of *sequential execution*. This means that—in the absence of instructions to the contrary—statements are carried out one after the other, beginning with the first executable statement and continuing in sequence. When control reaches the END statement, our program is ordered to relinquish control of the computer to other tasks.

We shall see in Chapter 3 and following that there are several ways to alter sequential execution of statements. And when it is necessary to stop program execution somewhere other than at the end of the program (upon detection of errors in the data, for example) the STOP statement is available.

The END statement also has the function of designating the physical end of the program, whether or not that is where execution terminates. This information is needed by the compiler, whose function we must now investigate.

The compilation of a Fortran program

"Fortran," as the term is commonly used, refers both to a language for expressing algorithms and to a *compiler* that translates from the language of Fortran into the rather more elementary language of the computer itself, called its *machine language*. A computer's own internal language deals with basic operations of arithmetic, transferring data between different parts of the computer, making various tests on data at a very low level, and the details of reading and writing data. The Fortran language, as we have already seen to some extent, deals with the processing of data at a higher level. Exponentiation, for example, is not done directly by the computer; rather, depending on the size and type of the exponent, it is done by repeated multiplication or by the use of the exponential and logarithm functions.

The translation from the language of Fortran to the language of the computer is the task of the Fortran compiler. Indeed, the name Fortran itself—usually written FORTRAN at first—stood for FORmula TRANslation.

An additional bit of terminology will make it easier to clarify whether a reference to "the program" means the Fortran program as we wrote it, or the machine language program that results from compilation. We shall speak of the Fortran program as we wrote it as the *source program.* The result of compiling (translating) the source program into machine language is referred to as the *object program.*

The Fortran compiler is itself a large computer program. It in turn is one of an integrated collection of programs known as an *operating system.*

The compiled object program may or may not be executed immediately. In the popular version of Fortran called WATFIV, for example, the object program is executed immediately. Indeed, in WATFIV there is no way to save the object program; every WATFIV source program must be compiled every time it is used. Most compilers, however, do produce an object program that may be saved for use at a later time without recompilation.

The interactive compilation and execution of our program

Figure 1.2 shows the sequence of operations in compiling and running the Fortran source program of Figure 1.1 at an interactive terminal, using the time-sharing system of National CSS, Inc., on which most programs for this book were developed and run. (Naturally, if you are using some other interactive system, your commands will differ, at least in details. And if you are using cards, you will be interested in the discussion of batch operation that follows this section.)

Figure 1.2 contains a number of commands to the National CSS operating system; each such command appears on a line that begins with the time of day. The first line, therefore, says that at 44 seconds after 11:05 on the day I did this work, I issued a command called "listf," which means "list file." A *file* can be a source program, an object program, some data, or perhaps a chapter of a book. Every file has a *filename* and a *filetype.* (The "mode" and "items" columns do not concern us.)

Note that what I typed is in lowercase; the computer's responses are in uppercase. Such details might be different on other systems, although this convention is fairly common.

With the "listf" command on the first line I said that I wanted to know about all files named POLY1. (That's the name I specified when I created the Fortran program.) The response is that I have one file named POLY1 and its filetype is FORTRAN, which is the required filetype for a Fortran source program.

The next command is "vsfortran," which in the National CSS system means to compile the program named in the command. The "vs" in the command identifies one of several Fortran compilers available on this system, the one that translates Fortran 77 source programs. Most large computers have several variations of Fortran available; if this is true in your computer installation you will be instructed which one to use.

Now I ask again for a listing of the files having the filename POLY1. This time we see that along with the source program, which has a filetype of FORTRAN, there is another name with a filetype of TEXT. This new file is the object program that was created by the Fortran compiler.

Now I say to run the object program. The line EXECUTION: means that the object program has been loaded and is ready to be executed. In fact, it is waiting

```
11.05.44 >listf poly1
FILENAME FILETYPE MODE      ITEMS
POLY1     FORTRAN    P         9

11.05.50 >vsfortran poly1
VSFORTRAN-REL2-MOD0
VS FORTRAN COMPILER ENTERED.  11:06:04

*STATISTICS*    SOURCE STATEMENTS = 5, PROGRAM SIZE = 452 BYTES

*STATISTICS*    NO DIAGNOSTICS GENERATED.

****** END OF COMPILATION 1 ******
VS FORTRAN COMPILER EXITED.   11:06:06


11.06.12 >listf poly1
FILENAME FILETYPE MODE      ITEMS
POLY1     FORTRAN    P         9
POLY1     TEXT       P        20

11.06.27 >run poly1
EXECUTION:
>0.0
 FOR X =  0.000000000     THE VALUE OF THE POLYNOMIAL IS  -130.000000

11.06.52 >start
EXECUTION:
>1.
 FOR X =   1.00000000     THE VALUE OF THE POLYNOMIAL IS  -25.0000000

11.07.00 >start
EXECUTION:
>2
 FOR X =   2.00000000     THE VALUE OF THE POLYNOMIAL IS   14.0000000

11.07.13 >start
EXECUTION:
>3.0000
 FOR X =   3.00000000     THE VALUE OF THE POLYNOMIAL IS  -31.0000000
```

Figure 1.2 The output produced at an interactive terminal when the source program of Figure 1.1 was compiled and executed.

for input, because, as we remember, the first operation in the program was to ask for a value for X. (Whenever the National CSS system is waiting for input, it prints the symbol >.) I type in the value 0.0, and the object program prints the values of X and POLY. The value of the polynomial when X is zero is seen to be − 130, which is just the constant term in the polynomial.

With the object program still available in the computer, I can use the command "start," which I do several times for different values of X. Note the flexibility (in the entering of input values) that the use of the asterisk in the READ statement gives me. But the format of the output values is standard.

We see that the polynomial changes signs between 1 and 2 and again between 2 and 3, indicating the presence of roots in those intervals. We shall visit with this polynomial several times as we go along and, among other things, shall find all of its roots.

Running in a batch system

What we have just seen is an example of a specific system of interactive computing, where each command is executed immediately, and where we can decide as we go along what to do next. This way of working with a computer is becoming more and more common, but it is not available to all programmers.

The other major way of working with a computer, which was the *only* way in earlier years, is called *batch operation*. In this way of doing things, a *job* is submitted to the computer along with many other jobs, and all are run without the possibility of intervention by the programmer.

In the case of our program, we would ordinarily punch the program into a deck of cards and submit it for compilation. With such a small program we can have reasonable hope that it will run the first time, so we would probably provide a data card containing our input data (X value), as well. The program cards, the data card, and certain *control cards* (instructions to the operating system that we shall consider shortly) would be submitted as a deck. At some later point— maybe seconds, maybe hours—the deck would be read, the program compiled, and provided there were no errors the object program would read the data card and print the results. The original deck and the printed output would be returned to us.

Job control cards

When a deck is submitted for batch operation, how is the computer supposed to know what we want it to do? What takes the place of the commands in the interactive version, where we specified compilation, execution, etc.?

The answer is that along with cards containing the program and data, we supply a few *job control cards* that serve these and other functions. Since there is a great deal that an operating system can do besides what we have seen so far, a complete *job control language* exists for each operating system, although that exact language is not always used. "Job Control Language" gets abbreviated to JCL, and we speak then of "JCL cards."

Figure 1.3 is a listing of the program as it could be submitted for batch execution at my school, The City College of New York, on our IBM 4341. The first two lines, since they begin with two slashes, are recognized as JCL cards. *Yours will almost certainly be different!* The details vary from one installation to the next.

The first JCL line gives my account number and specifies that the file containing the program is named POLY1. The second JCL line says to *execute* WATFIV, which is a variation of Fortran developed at the University of Waterloo, in Ontario, Canada. WATFIV is widely used for introductory programming courses because it gives good diagnostic messages that help in locating errors and because it compiles quickly. The latest version of WATFIV is sufficiently close to Fortran 77 that readers of this book who are using WATFIV should have little difficulty in adapting. A few additional comments at the beginning of the program point out the only differences in this case. (The STOP statement is actually optional, since execution of the END statement terminates program execution. However, when this is done, some WATFIV systems give a warning message that becomes tiresome.)

```
//DANCC JOB POLY1
//      EXEC WATFIV
$JOB
C A PROGRAM TO EVALUATE A POLYNOMIAL FOR A VALUE OF X READ AS INPUT
C THIS VERSION OF THE PROGRAM RUNS UNDER WATFIV
C NOTE FOLLOWING DIFFERENCES BETWEEN WATFIV AND FORTRAN 77:
C      WATFIV USES C INSTEAD OF * TO DESIGNATE COMMENTS
C           (FORTRAN 77 PERMITS EITHER)
C      WATFIV DOES NOT PERMIT BLANK LINES; MUST MAKE INTO COMMENTS
C      WATFIV DOES NOT USE THE * IN THE READ AND PRINT STATEMENTS
C      WATFIV PREFERS A STOP BEFORE THE END
C
       REAL X, POLY
C
       READ, X
       POLY = 2.0*X**4 - 15.0*X**3 - 2.0*X**2 + 120.0*X - 130.0
       PRINT, ' FOR X = ', X, ' THE VALUE OF THE POLYNOMIAL IS ', POLY
C
       STOP
       END
$ENTRY
-1.0
/*
```

Figure 1.3 The program of Figure 1.1, with Job Control Language (JCL) lines added, and modified to use the WATFIV version of Fortran.

The third line is a WATFIV control. If you are not using WATFIV, you will not have this line, although you might have something else. Then comes the program, with the slight modifications noted. The comments are entirely in uppercase, because lowercase letters cannot be entered into the particular system I used here. After the program, there is another WATFIV control indicating that data follows. (Actually, you must have this card even if there is no data!) Following this is the one data value and finally, a JCL line that marks the end of the submission.

A program in this form can be punched on cards and submitted for compilation and execution. (Actually, the program is compiled and, *if there are no disabling errors*, it is immediately executed.) Alternatively, it can be entered into the computer using a terminal. In fact, the latter was done here, using the WYLBUR system developed at Stanford University and marketed by On-Line Business Systems, Inc. The program as printed in Figure 1.3 was in fact listed at my terminal.

Whether entered on cards or through a terminal, the result is of the form shown in Figure 1.4. We see that the JCL lines have been deleted, that WATFIV has added numbers to the statements (mostly for making error messages easier to understand), and that the last line of the listing is the output of the program. (Actually, there were a few additional lines, giving summary statistics about the program and its execution, which I deleted because they were too long to print on a page here and because they weren't very interesting in this case.)

The manner and physical form in which you get the compiled program will depend on how you are running your program. With the program on cards, you will get a printed listing produced by a printer that is part of the computer

```
        $JOB
        C A PROGRAM TO EVALUATE A POLYNOMIAL FOR A VALUE OF X READ AS INPUT
        C THIS VERSION OF THE PROGRAM RUNS UNDER WATFIV
        C NOTE FOLLOWING DIFFERENCES BETWEEN WATFIV AND FORTRAN 77:
        C      WATFIV USES C INSTEAD OF * TO DESIGNATE COMMENTS
        C          (FORTRAN 77 PERMITS EITHER)
        C      WATFIV DOES NOT PERMIT BLANK LINES; MUST MAKE INTO COMMENTS
        C      WATFIV DOES NOT USE THE * IN THE READ AND PRINT STATEMENTS
        C      WATFIV PREFERS A STOP BEFORE THE END
        C
  1            REAL X, POLY
        C
  2            READ, X
  3            POLY = 2.0*X**4 - 15.0*X**3 - 2.0*X**2 + 120.0*X - 130.0
  4            PRINT, ' FOR X = ', X, ' THE VALUE OF THE POLYNOMIAL IS ', POLY
        C
  5            STOP
  6            END
        $ENTRY
FOR X =             -1.0000000   THE VALUE OF THE POLYNOMIAL IS        -235.0000000
```

Figure 1.4 The listing produced when the program of Figure 1.3 was submitted and run. (A few nonessential lines at the end have been deleted.)

system. If you are using a typewriter-type terminal, the listing can be printed at your terminal. If you are using a CRT-type terminal, you can look at the output on your screen, and, in most cases, command the system to produce a paper listing on the computer's printer.

We observe minor variations in the form of the output, which are to be expected.

Conclusion

This chapter has contained a significant amount of new material that may have been quite unfamiliar to you. The best way to master the subject is to write and run programs! (And it's fun, too!) You are strongly urged to run the program of this chapter on your computer. It is small, of course, but getting it running will be highly educational. Naturally, you or your instructor may prefer to use some other simple task instead of this one. The point is simply that actually running a program as early as possible is essential to getting a good start in programming.

Unfortunately, a large part of what you need to know more about right now depends heavily on the details of your computer, its operating system, and the organization and operation of your computer center. There would be no possible way a book of this sort could provide these details for enough different computers to be useful without becoming an encyclopedia—and one that would rapidly become out of date, at that.

Here is a partial checklist of the things you will need to know to run the program.

■ How to obtain authorization to use the computer. You will need an account number and a password, typically. The account number may be for your entire class or project, or it may be yours alone.

■ How much computer time and other resources you are authorized to use, and how usage is charged. In a classroom situation you will probably have a limit on the amount of computer usage authorized. If you burn it up early in the course looking for big prime numbers or playing Starwars . . .

■ How to punch cards, if you are running under a batch operating system based on cards. Part of what you will need to learn here is where Fortran statements are permitted to be punched on the cards in your system. In most systems, a statement cannot begin before column 7 or extend past column 72. (We shall see later that columns 1 to 5 and column 6 have other functions.)

■ Exactly what job control language (JCL) is needed. You will probably be given a sample to imitate, but you will still probably have to insert your name and account number. (A common error is to omit the control card between the end of the program and the start of the data. If you get a weird error message, check this possibility first.)

■ The routine for submitting decks and picking up output, in a card-oriented batch system. Unless you operate the card reader and the printer yourself, which is done in some installations, you will probably need to fill out a simple form—possibly printed on a card—to accompany the deck. Output is often returned wrapped around the deck. The output may be placed in boxes for you to pick up, or, if there has been too much poaching of programs at your school or organization, you may have to identify yourself somehow to get things back.

■ How to gain access to the interactive time-sharing system, if that is how you will be working.

■ How to use the interactive text editor with which you will enter and modify programs and (sometimes) prepare data.

■ The basic commands for compiling and executing programs, together with the most basic commands for manipulating files.

■ How to use a terminal to build a disk file for submission to a batch operating system, if that is how your computer center operates.

Exercise

Do it! Punch or enter the program of Figure 1.1, compile it, and run it with a value for X of 2.5. Modify the program slightly if you feel you must exercise your creativity—but don't get carried away. You have a *lot* of things to learn in order to do even this much.

TWO

THE ASSIGNMENT STATEMENT AND RELATED MATTERS

Introduction

Now that you have seen one small but complete program in operation, and we hope have run it yourself, it is time to consolidate the basics of some matters that we passed over rather lightly before. The topics are fundamental concepts that we shall use in several examples in this chapter and in all of our subsequent work: constants, integer and real quantities, variables and their names, the type-statements to distinguish real and integer variables, operations and expressions, the assignment statement, and mathematical functions.

Integer and real quantities

Let us begin by looking at the difference between integer and real quantities.

A Fortran integer is just what it sounds like: a whole number. It may be zero or any positive or negative number of less than, typically, eight decimal digits. The limit on the size of an integer varies from computer to computer; in some cases there may be a choice on the limits within one Fortran system. These details for your system may be obtained from a suitable reference manual, which it is helpful to have available.

Because integers are restricted to whole number values, they are ordinarily used only in special situations.

Most numbers used in Fortran computations are real. Inside the computer Fortran real numbers are represented in *floating point* form. This is similar to what is commonly called *scientific notation*, in which a number is treated as a fraction times a power of 10. (Actually, in many computers a real quantity is represented as a fraction times a power of 2 or 16, which creates some problems that we shall explore in the illustrative programs below.) The *magnitude* (sign not considered) of the number so represented must be zero or lie between limits in the range of approximately 10^{-50} to 10^{+50}. Again, the exact limits vary.

A Fortran integer is always an integer in the mathematical sense, whereas a Fortran real number may be an integer or have a fractional part as well. Furthermore, Fortran carries out computations on real numbers in such a way that we do not have to be concerned with the location of decimal points. All questions of lining up decimal points before addition or subtraction, for example, are automatically taken care of by the computer.

A Fortran real number is a *rational* real number in the mathematical sense. Irrational numbers cannot be represented; they can only be approximated to some degree of precision. Indeed, not all *rational* numbers can be represented exactly in a computer. The simple fraction 1/3, for instance, cannot be represented exactly in any number base that does not contain a factor of 3. Since the only number bases of practical importance are binary (base 2), decimal, and hexadecimal (base 16), the fraction 1/3 can only be approximated as a Fortran real quantity. We shall see some of the consequences of this fact later.

(Fortran provides another type of number, called *double precision,* to improve the accuracy of approximation. We shall, however, postpone consideration of this type of number until Chapter 8, where we shall also consider three other types of Fortran quantities: complex, logical, and character.)

Constants

Any number that appears in a statement in explicit form is called a *constant,* whereas a quantity that is given a name and is allowed to vary during program execution is a *variable.* For instance, here are two simple assignment statements:

```
K = 2
XNEW = -130.0 + X * 12.7
```

In these statements 2, −130.0 and 12.7 are constants; K, XNEW, and X are variables.

Fortran distinguishes between integer and real constants by the presence or absence of a decimal point or exponent (see below). If the constant contains neither a decimal point nor an exponent, it is taken to be an integer. If it contains a decimal point or an exponent or both, it is taken to be real.

As noted in Chapter 1, there is a significant difference between integer and real quantities in the way they are stored inside the machine. *Sometimes* they can be used interchangeably, but traps await the unwary who use this flexibility carelessly. We shall return to this issue shortly.

If a constant is positive, it may or may not be preceded by a plus sign, as desired. If it is negative, it must be preceded by a minus sign. The following are acceptable integer constants:

```
0
6
+400
-1234
10000
-2000
```

The following are not acceptable integer constants:

```
12.78   (decimal point not allowed in integer constant)
-10,000    (comma not allowed)
12345678900000   (too large in most systems)
```

The decimal point in a real constant may appear at the beginning of a number, between any two digits, or at the end. In principle, a real constant may have any number of digits, but there are at least three practical restrictions:

1. Many Fortran systems place a limit on the number of digits permitted, such as eight or twelve. As usual, details vary among different implementations and different computers; you should consult a reference manual on your Fortran to get this information.

2. Some systems use the number of digits as a signal that the constant is double precision: if there are more than, say, seven digits, the constant is taken to be double precision. Sometimes it is of no consequence if this is done unintentionally, but in other cases it can cause serious errors or confusing compiler diagnostic error indications.

3. Finally, there is the practical consideration that any machine has a limit on the amount of storage space allocated to a constant. If this limit is seven digits, it makes no sense to write constants with 20 or 30 decimal digits even if the Fortran system permits it. The computer simply cannot retain that much information in the space available.

It is possible to write a real constant with an *exponent,* which consists of the letter E (for "exponent"), and a one- or two-digit positive or negative power of 10 by which the constant is to be multiplied. When this is done, the result is a real constant, whether or not it contains a decimal point. The power of 10 is then called the *exponent* of the constant. This simplifies the writing of very large or very small numbers.

The following are acceptable real constants:

```
0.0
0.1
.1
6.0
6.
-20000.0
-0.0002783
+15.083
5.0E-12
-7.E+6
6.215E13
-.1E7
11E2
```

The following are not acceptable real constants:

```
12,345.6      (comma not allowed)
+234          (no decimal point)
1.6E97        (too large in most systems)
5.862E2.5     (exponent part not an integer)
E+7           (exponent alone not allowed)
```

Exercises

Answers to starred exercises are given at the back of the book.

*1. Write the following numbers as Fortran real constants. 256 2.56 $-43{,}000$ 10^{12} 0.000000492 -10 10^{-16}

2. Write the following numbers as Fortran real constants. 16 4.59016 $-10{,}000$ 10^{19} 0.000006 -1 -10^{19}

*3. All of the following are unacceptable as Fortran real constants. Why?

87,654.3   +987   9.2E+98   7E-9.3

4. All of the following are unacceptable as Fortran real constants. Why?

−10000   1E−99   2.34−E12   2E5.1

*5. Do the following pairs of real constants represent the same number in each case?

16.9	+16.9
23000.	2.3E4
0.000007	.7E−5
1.0	1.
.906E5	906.0E+02
110.0	11E1

*6. Some of the following are unacceptable as integer constants. Identify the errors.

+234.   −234   23,400   1E12   100000000000

7. Some of the following are unacceptable as integer constatns. Identify the errors.

−16.5   16000   16,000   2.E12.5   0.01

Variables and the names of variables

The term *variable* is used in Fortran to denote any quantity that is referred to by name instead of by explicit value. A variable is able to take on many values during the execution of a program, whereas a constant is restricted to just one value.

Variables may be *integer* or *real*.

An integer variable is one that may take on any of the values permitted for an integer constant, namely, zero or any positive or negative integer in the range permitted for the particular version of Fortran being used.

The *name* of a variable is composed of one to six letters or digits, the first of which must be a letter. A variable that is not mentioned in a *type-statement* (see below) is considered to be of the integer type if its first letter is I, J, K, L, M, or N, and real otherwise. We shall not use this *default naming convention*, also called the *implied naming convention*, however, but instead shall name all program variables in type-statements—for reasons that will be developed as we go along.

The value of a real variable is represented inside the computer in the same form as a real constant, that is, as a fraction times a power of 10 (or other number base, such as 2 or 16). If a real variable is not mentioned in a REAL type-statement, its name must *not* begin with one of the letters I through N.

Examples of acceptable variable names: I, KLM, L123, I6M2K, KAPPA, MATRIX (all of these would be considered integer variables if not named in a REAL statement), AVAR, R5ITX, FRONT, G, GAMMA, SVECTR, AMATRX, PASCAL (all of these would be considered real variables if not named in an INTEGER statement) Examples of unacceptable variable names: J123456 (too many characters), 5M (does not begin with a letter), *J78 (contains a character other than a letter or a digit), J34.5 (ditto), HI−LO (ditto), PL/I (ditto).

The assignment of names to the variables appearing in a program is entirely under the control of the programmer. Both common sense and long experience argue for naming variables so their intended meaning is as clear as possible, but you can certainly name variables after old boyfriends or basketball stars if you wish. Such cuteness is to be avoided, however, since it not only complicates program testing and maintenance but also wears thin rather quickly.

It should be noted that a Fortran compiler places no significance on the name of a variable beyond inspecting its first letter to determine whether the variable is integer or real, and not even then if the variable is named in a REAL or INTEGER statement. A name such as B7, for example, does *not* mean B times 7, B to the seventh power, or the seventh element in a vector named B. Furthermore, every combination of letters and digits is a separate name. Thus ABC is not the same as BAC, and A, AB, and ABC are all different and distinct names.

Exercises

*1. Which of the following are acceptable names for integer variables (in the absence of a type-statement), which are acceptable names for real variables (ditto), and which are not acceptable names for *any* variable?

```
G    GAMMA   GAMMA421   I    IJK
J79    J79-1   LARGE    R(2)19    BT07TH
ZSQUARED    ZCUBED    12AT7    2N173
B6700   CDC6600   S/370   IBM370
DELTA   LAMBDA   EPSILON    EPSLON
ALEPH   NUN   DAYAN    CHAYA    L'CHAYIM
A1.4    A1POINT4   A1P4    AONEP4
FORTRAN    ALGOL    PL/I    ADA    PASCAL
```

2. Same as Exercise 1.

```
K    I12G    VOLTS2    FLOW6    X+2
XPLUS2    NEXT    42G    CURRENT    LASTQ
NUMBER    RHMU    INDEX    A*B
X1.4    (X61)    GAMMA81    AI    IA
X12    1X2    XFIFTH    AVER    MEAN
VARIANCE    SIGMA2    KURTOSIS    MOMENT3
COBOL    LISP1.5    PL/M    PLZ    NOMAD
```

Type-statements

We shall make a practice in this text of always *declaring* variables in *type-statements*. Doing so is required if we wish to override the initial-letter naming convention for real and integer variables or if we are using variables of types other than real or integer. Even if not strictly required, however, we recommend it as good practice, to avoid a variety of mishaps to which programs are liable without explicit type-statements.

To declare a variable to be of type real, we write the word REAL followed by the names of variables separated by commas. The integer type-statement is the same, with the word INTEGER.

Thus we might write:

```
REAL NN, IOTA, TEMP2
INTEGER YEAR, MONTH, EASTER
```

The REAL declaration makes NN and IOTA real variables; TEMP2 would have been real anyway. The INTEGER declaration makes YEAR and EASTER integer variables; MONTH would have been so anyway.

All type-statements must appear before the first executable statement.

Arithmetic operations

Fortran provides five basic arithmetic operations: addition, subtraction, multiplication, division, and exponentiation. (Exponentiation is not usually considered an "arithmetic" operation in mathematics, but that is the conventional Fortran terminology.) Each operation is represented by a symbol:

```
Addition           +
Subtraction        -
Multiplication     *
Division           /
Exponentiation     **
```

Notice that the combination of two asterisks for exponentiation is considered to be one symbol; there is no confusion between * and **, because it is never correct to write two operation symbols side by side. These are the only mathematical operations for which symbols are provided; any others must be built up from sequences of the basic five or computed using the functions that are discussed below.

Arithmetic expressions

A Fortran *arithmetic expression* is a rule for computing a numerical value. In many cases an expression consists of a single constant or a single variable. Two or more of these elements may be combined, using operation symbols and parentheses, to build up more complex expressions. Some examples of expressions and their meanings are given in Table 2.1.

The programmer must observe certain rules to convey his or her intentions correctly to Fortran.

1. Two operation symbols must never appear next to each other. Thus A*−B is not a valid expression, although A*(−B) is.

2. Parentheses are used to indicate groupings, just as in ordinary mathematical notation. Thus $(X + Y)^3$ must be written (X + Y)**3 to convey the correct meaning. X + Y**3 would be a valid expression, but of course the meaning is not the same, since it calls for only Y to be cubed, rather than the sum of X and Y. Again, A − B + C and A − (B + C) are both legitimate expressions, but the meanings are different. Parentheses force the *inner* operation to be done first, just as in conventional mathematical notation.

3. When the sequence of arithmetic operations is not completely specified by parentheses, the "strength" of the operations is given by the following *operator precedence* rules:

■ all exponentiations are performed first

■ then all multiplications and divisions

■ and finally, all additions and subtractions

Thus these two expressions are equivalent:

```
A*B + C/D - E**F
(A*B) + (C/D) - (E**F)
```

4. Within a sequence of consecutive multiplications and/or divisions, or additions and/or subtractions, in which the order of operations is not fully specified by parentheses, the meaning is that of left-to-right evaluation. Thus the expression A/B*C would be taken to mean:

$$\frac{A}{B} \cdot C$$

and not:

$$\frac{A}{B \cdot C}$$

and I − J + K means (I − J) + K, not I − (J + K).

Free use of parentheses is encouraged; there is no important penalty in doing so, and the grief saved can be substantial. *When in doubt, parenthesize.*

5. Any expression may be raised to a power that is a positive or negative integer quantity, but only a real expression may be raised to a real power. An exponent may itself be any expression, of either integer or real type. Thus X**(I + 2) is perfectly acceptable. In no case, however, is it permissible to raise a negative value to a real power or to raise zero to the zero power.

6. Any operation having one operand of integer type and the other of real type produces a result of real type. This is called a *mixed-mode* expression. Except for the operation of raising a real value to an integer exponent, the integer operand in a mixed-mode expression is automatically converted to real representation before the operation is carried out. Mixed-mode arithmetic has tricky hazards associated with it, as we shall see in an example program later. It is best avoided unless done very deliberately and for a good reason.

Table 2.1

Expression	Meaning
K	The value of the integer variable K
3.14159	The value of the real constant 3.14159
A + 2.1828	The sum of the value of A and 2.1828
RHO − SIGMA	The difference in the values of RHO and SIGMA
X*Y	The product of the values of X and Y
OMEGA / 6.2832	The quotient of the value of OMEGA and 6.2832
C**2	The value of C raised to the second power
(A + F) / (X + 2.0)	The sum of the values of A and F divided by the sum of the value of X and 2.0
1. / (X**2 + Y**3)	The reciprocal of $(X^2 + Y^3)$

7. Parentheses in an expression indicate grouping. They do *not* imply multiplication. Thus the expression:

```
(A + B)(C + D)
```

is incorrect. Fortran does not interpret it as meaning:

```
(A + B) * (C + D)
```

Table 2.2 lists some examples of mathematical expressions, correct expressions corresponding to them, and incorrect expressions containing common mistakes. The correct expressions are not unique, which is to say that there are always many ways of writing correct expressions having the same meaning. In the first example, for instance, B∗A would give exactly the same result.

Does it matter?

The rules stated above are important for a number of reasons. For one, it is necessary to convey our intentions to Fortran correctly. Just as in ordinary mathematical notation, the expression A ∗(B + C) *must* be written with parentheses since multiplication is a "stronger" operation than addition. For a second reason, some things are impossible because of the way the computer and the compiler operate. An integer cannot be raised to a real power because the result would, in general, have a fractional part that could not be expressed in integer form.

The impact of finite number representation

The third reason for adhering to these rules is less obvious, and sometimes comes as a rather nasty shock to beginners in computing: arithmetic operations with operands of finite length do not obey all of the "normal" rules of mathe-

Table 2.2

Mathematical Notation	Correct Expression	Incorrect Expression	
$a \cdot b$	A∗B	AB	(no operation)
$a \cdot (-b)$	A∗(-B) or -A∗B	A∗-B	(two operations side by side)
$-(a + b)$	-(A+B) or -A-B	-A+B or -+A+B	$(= a^i + 2)$
a^{i+2}	A∗∗(I+2)	A∗∗I + 2	
$a^{b+2} \cdot c$	A∗∗(B+2.0)∗C	A∗∗B + 2.0∗C	$(= a^b + 2 \cdot c)$
$\dfrac{a \cdot b}{c \cdot d}$	A∗B/(C∗D)	A∗B/C∗D	$\left(= \dfrac{a \cdot b}{c} \cdot d \right)$
$\left(\dfrac{a + b}{c} \right)^{2.5}$	((A+B)/C)∗∗2.5	(A+B)/C∗∗2.5	$\left(= \dfrac{a + b}{c^{2.5}} \right)$
$a[x + b(x + c)]$	A∗(X+B∗(X+C))	A(X+B(X+C))	(missing operators)
$\dfrac{a}{1 + [b/(2.7 + c)]}$	A/(1.0+B/(2.7+C))	A/(1.0+B/2.7+C)	$\left(= \dfrac{a}{1 + (b/2.7) + c} \right)$

matics, such as associativity. For a simple example of what can happen, suppose we are working with a computer in which Fortran real quantities are represented with eight decimal digits. Now consider this expression:

```
0.40000000 + 12345678. - 12345677.
```

If this is evaluated from left to right the result of the addition, to eight decimal digits, is simply 12345678—the 0.4 has been lost entirely since retaining it would require more digit positions than our hypothetical computer has available. Then, when the 12345677 is subtracted, the final result is 1.0000000.

Suppose, on the other hand, that the expression had been written:

```
0.40000000 + (12345678. - 12345677.)
```

The parentheses force the subtraction to be done first, giving 1.0000000. Now, when the 0.40000000 is added, the result is 1.4000000. In other words, in the original form the addition of a small number to a large number caused complete loss of significance of the small number.

The order of arithmetic operations can lead not only to loss of significance but also to a failure to get any answer, as the following example shows. Suppose we wanted to evaluate A*B/C, in which the values of A, B, and C are all in the range of 10^{40}. The multiplication is done first, following (in the absence of parentheses) the left-to-right rule, giving 10^{80}. This is too large a number to be stored in most computers, leading to what is called *floating-point overflow*. If we could store such an intermediate result, the final result would be within limits—but, since we cannot do so, the answer is meaningless. Most compilers would produce an object program that would stop execution in such a case; others would give utterly meaningless results with no warning.

The simple solution in this case is to use parentheses to force the division to be done first: A*(B/C). The result of the division is a number within the allowable range, and so is the final result.

Integer division

Integer division raises a special problem of its own. When two integers are divided, the quotient is usually not an integer. Integer division is arranged to discard the remainder, which can also be expressed by saying that an integer quotient is *truncated* to an integer. *It is not rounded.* Thus the result of the integer division 5/3 is 1, not 1.6666667 or 2.

As it happens, most calculations do not involve the kinds of considerations that we have been discussing—but they can arise inadvertently to trap the unwary. Consider another example.

In the integer expression 5/3*6 the left-to-right rule says that the division will be done first; this truncated intermediate result of 1 is multiplied by 6 to give the final result of 6. The result is *not* 10, which we would get from multiplying 5 by 6 and then dividing by 3, or 12, which we would get if the quotient were rounded instead of truncated. On the other hand, the result *is* 10 if the expression is written as 5*6/3, 6*5/3, 5*(6/3), or (6/3)*5.

All of this applies only to integer arithmetic. Any of the forms in the preceding paragraph, if written using only real constants, i.e., with decimal points included, would give 10.000000. (Except . . . there might be one incorrect digit in the last position; see the next paragraph.)

Another finite representation example

Even with real arithmetic things can happen that may not be expected. Suppose we were to form this sum:

```
1.0/3.0 + 1.0/3.0 + 1.0/3.0
```

The real representation of 1.0/3.0, still thinking in terms of a hypothetical computer with eight decimal digits in the fractional part of a floating point number, is 0.33333333. The result of the additions, then, is 0.99999999. If we were to write a program that compared this result with 1.0000000, which we might have expected to be the sum, the answer, of course, would be "not equal," which might come as something of a shock to the unsuspecting programmer.

These problems are not insuperable, and most of them are not so different from things that may happen when working with paper and pencil or a hand calculator. When a program is executed, however, we are not ordinarily in a position to watch the results develop step by step, looking for unexpected actions. With a computer we have to try to anticipate such difficulties, for, by the time the object program has begun execution, we have become bystanders.

Exercises

1. Write Fortran expressions corresponding to each of the following mathematical expressions.

*a. $x + y^3$

b. $(x + y)^3$

c. x^4

*d. $a + \dfrac{b}{c}$

e. $\dfrac{a + b}{c}$

f. $a + \dfrac{b}{c + d}$

g. $\dfrac{a + b}{c + d}$

h. $\left(\dfrac{a + b}{c + d}\right)^2 + x^2$

i. $\dfrac{a + b}{c + [d/(e + f)]}$

*j. $1 + x + \dfrac{x^2}{2!} + \dfrac{x^3}{3!}$

*k. $\left(\dfrac{x}{y}\right)^{g-1}$

l. $\dfrac{(a/b) - 1}{g[(g/d) - 1]}$

2. Following are a number of mathematical expressions and corresponding Fortran expressions, each of which contains at least one error. Point out the errors and write correct expressions.

   a. $(x + y)^4$ X + Y**4

  *b. $\dfrac{x + 2}{y + 4}$ X + 2.0/Y + 4.0

   c. $\dfrac{a \cdot b}{c + 2}$ AB/(C + 2.)

   d. $-\dfrac{(-x + y - 16)}{y^3}$ (-X+Y-16)/Y**3

  *e. $\left(\dfrac{x + a + \pi}{2z}\right)^2$ (X+A+3.14)/(2.*Z)**2

   f. $\left(\dfrac{x}{y}\right)^{n-1}$ (X/Y)**N - 1

  *g. -2^{r-1} (-2.0)**(R-1.0)

   h. $\dfrac{a}{b} + \dfrac{c \cdot d}{f \cdot g \cdot h}$ A/B + CD/FGH

   i. $(a + b)(c + d)$ A + B*C + D

  *j. $a + bx + cx^2 + dx^3$
     which can be rewritten $a + x[b + x(c + dx)]$ A+X*(B+X(C+DX)

   k. $\dfrac{1,600,042x + 10^5}{4,309,992x + 10^5}$ (1,600,042X+1E5)/(4,309,992X+1E5)

   l. $\dfrac{1}{a^2}\left(\dfrac{r}{10}\right)^a$ (R/(10.0*A**2))**A

The assignment statement

The basic Fortran elements we have discussed so far have many applications in writing source programs. The most important is in computing a new value of a variable, which is done with an *assignment statement*. Its general form is:

```
variable = expression
```

Here *variable* is a single variable name written without a sign, and *expression* is any expression, as was already described. An assignment statement is an order to Fortran to compute the value of the expression on the right of the equal sign and to give that value to the variable named on the left.

The equal sign in an assignment statement is not used as it is in ordinary mathematical notation. We are not allowed to write statements such as:

```
Z - RHO = ALPHA + BETA
```

for example, in which Z is the unknown, and expect Fortran to "solve" the *equation* for us. *A Fortran assignment statement is not an equation!*

The only legitimate form for an assignment statement is one in which the left side of the statement is the name of a single variable, and all the variables on the right have previously received values either from assignment statements or from input operations. The precise meaning is then: *replace the value of the variable named on the left with the value of the expression on the right.* Thus the statement:

```
A = B + C
```

is an order to form the sum of the values of the variables B and C and to replace the value of the variable A with this sum. The previous value of A is lost, but the values of B and C are unchanged. For the statement to be meaningful, values must previously have been given to B and C. Whatever statements give values to B and C may be executed more than once during the running of the program. Therefore, when we say "the value of B," for example, we mean the value *most recently assigned to the variable named* B.

Another example of an assignment statement brings out very forcefully the special meaning of the equal sign as used here. A statement such as:

```
N = N + 1
```

has the meaning: *Replace the value of the variable* N *with its old value plus 1.* This kind of statement, which is clearly not an equation, finds frequent use. For example, we might use it to count how many times something has been done, or to move through the elements of a list.

Sometimes it is useful to convert between integer and real representations of quantities. One simple way to do this is to write an expression of one type on the right of an assignment statement and a variable of the other type on the left. When this is done, all arithmetic is done in the type of the expression (on the right) and then the value is converted to the other type before assigning it to the variable (on the left).

A few examples may help to clarify the uses of assignment statements. Suppose A, B, C, D, and X have already been given values by statements executed earlier, and that we need to compute a new value of R from:

$$R = \frac{A + BX}{C + DX}$$

The following statement will do what is required:

```
R = (A + B*X) / (C + D*X)
```

None of the variables on the right will be changed by the statement; the previous value of R will be lost.

Suppose we want to convert a Celsius temperature to Fahrenheit:

```
F = 1.8 * C + 32.0
```

The operator precedence rule says that the multiplication will be performed before the addition, so parentheses are not required. On the other hand, this statement would have exactly the same meaning:

```
F = (1.8 * C) + 32.0
```

For conversion in the other direction, parentheses *are* required:

```
C = (F - 32.0) / 1.8
```

Without parentheses, the division would be performed first, which is obviously not what we want.

Table 2.3 shows a number of mathematical formulas, together with acceptable equivalent assignment statements. Variable names have been chosen to suggest the quantities represented, but naturally, other choices could have been made. We assume, of course, that previously executed statements have given values to all of the variables in the expressions on the right hand sides of the assignment statements.

The examples in Table 2.4 emphasize the importance of writing expressions and statements in the prescribed form. Fortran demands precise adherence to its rules of syntax, that is, its rules for forming correct statements from the elements of the language. Each of the statements in Table 2.4 contains at least one error.

Table 2.3

Arithmetic Assignment Statement	Original Formula
(a) BETA = -1./(2.*X) + A**2/(4.*X**2)	$\beta = \dfrac{-1}{2x} + \dfrac{a^2}{4x^2}$
(b) C = 1.112*D*R1*R2/(R1 - R2)	$C = 1.112D \dfrac{r_1 r_2}{r_1 - r_2}$
FY = X*(X**2 - Y**2)/(X**2 + Y**2)	$F_y = x \cdot \dfrac{x^2 - y^2}{x^2 + y^2}$
(c) Y = (1E-6 + A*X**3)**(2.0/3.0)	$y = (10^{-6} + ax^3)^{2/3}$
J = 4*K - 6*K1*k2	$j = 4K - 6k_1 k_2$
I = I + 1	$i_{new} = i_{old} + 1$
K = 12	$k = 12$
PI = 3.141593	$\pi = 3.141593$
M = 2*M + 10*J	$m_{new} = 2m_{old} + 10j$

Table 2.4

Incorrect Statement	Error
Y = 2.X + A	* Missing
3.14 = X - A	Left side must be a variable name
A = ((X + Y)A**2	Not the same number of right and left parentheses; * missing
X = 1,624,009.*DELTA	Commas not permitted in constants
-J = I**2.	Integer quantities may not be raised to real powers; variable on left must not be written with a sign
BX6 = 1./-2.*A**6	Two operation symbols side-by-side not permitted, even though the minus sign here is not intended to indicate subtraction
A*X + B = Q	Left side must be a single variable; should be Q = A*X + B

Exercises

1. State the value of A or I stored as the result of each of the following arithmetic assignment statements and show whether the result is in integer or real form. A is a real variable and I is an integer variable.

   *a.   A = 2*6 + 1

   *b.   A = 2/3

   c.   A = 2.*6./4.

   d.   I = 2*10/4

   e.   I = 2(10/4)

   f.   A = 2(10/4)

   g.   A = 2.*(10./4.)

   h.   A = 2.0*(1.0E1/4.0)

   i.   A = 6.0*1.0/6.0

   j.   A = 6.0*(1.0/6.0)

   *k.   A = 1./3. + 1./3. + 1./3. + 1./3.

   l.   A = (4.0)**(3/2)

   m.   A = (4.0)**3./2.

   *n.   A = (4.0)**(3./2.)

   *o.   I = 19/4 + 5/4

   p.   A = 19/4 + 5/4

   q.   I = 100*(99/100)

2. Each of the following arithmetic assignment statements contains at least one error. Identify them.

   a.   -V = A + B

   b.   4 = I

   c.   V - 3.96 = X**1.67

   d.   X = (A + 6)**-2

   e.   A*X**2 + B*X + C

   f.   K6 = I**A

   g.   Z2 = A*-B + C**4

   h.   X = Y + 2.0 = Z + 9.0

   i.   R = 16.9X + AB

3. Write arithmetic assignment statements to do the following.

   *a. Add 2 to the current value of the variable named BETA; make the sum the new value of a variable named DELTA.

   b. Subtract the value of a variable named B from the value of a variable named A, square the difference, and assign it as the new value of W.

   *c. Square A, add to the square of B, and make the new value of C the square root of the sum.

*d. A variable named R is to have its present value replaced by the square root of 2.

e. Multiply THETA by π and store half of the product as the new value of RHO.

f. Add the values of F and G, divide by the sum of the values of R and S, and square the quotient; assign this result to P.

*g. Add the square of two times X to the square root of one-half of X; set Y equal to the result.

*h. Increase the present value of G by 2 and replace the present value of G with the sum.

i. Multiply the present value of A by -1.0 and replace the present value of A with the product.

j. Assign to OMEGA the value of 2π.

k. Assign to the variable named D a value 1.1 times as great as the present value of the variable named D.

Example: Tension in a rope on a pulley

Let us take a simple example to see some of these ideas in action. Figure 2.1 shows two masses connected by a weightless rope passing over a frictionless and massless pulley. The formula for the tension in the rope is:

$$T = 2g \frac{m_1 m_2}{m_1 + m_2}$$

where m_1, m_2 = mass, slugs

$\qquad g$ = gravitational constant, 32.2 ft/sec^2

$\qquad T$ = tension, lb

The data consists simply of the two masses. We are to compute the tension and print it along with the values of the two masses.

The program in Figure 2.2 illustrates a number of the concepts discussed earlier in the chapter.

The program begins with a *prologue,* as we shall call it, which consists of a brief description of the program and its variables. Provision of something along these lines in all programs is strongly recommended.

Recall that the REAL statement with which the program begins is not "executed." Instead, it "declares" facts about the variables named in it, information which the compiler uses in setting up the object program correctly. As we proceed, we shall find other examples of *program declaratives,* which must always be placed before the first executable statement.

The first executable statement in the program assigns a value to a variable G, the gravitational constant. This value is indeed a constant in the ordinary

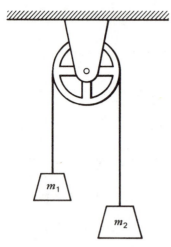

Figure 2.1 The system of
two weights and a pulley
that is the basis of the
program of Figure 2.2

mathematical sense of that term, at least to the precision used here, so why give
its value to a variable? But suppose this were a large program in which the
gravitational constant appeared dozens of times—and that we needed to change
to metric units? We would then be faced with finding and changing all the
instances of 32.2, which would be even more error-prone if some of them were
buried in combinations like 64.4. As this program is written, on the other hand,
changing units would require modifying only the one assignment statement in
which G is given a value. We shall tend to follow this practice except for constants

```
* Tension in a rope on a pulley
* Variables:
*     M1, M2: two masses, measured in slugs
*     (A mass of 1 slug weighs 32.2 pounds at the earth's surface)
*     T: tension in the rope, measured in pounds
*     G: Gravitational constant, 32.2 ft/sec/sec in English units
*

      REAL M1, M2, T, G

      G = 32.2

      READ *, M1, M2
      T = 2.0 * G * (M1 * M2) / (M1 + M2)
      PRINT *, ' WITH MASSES ', M1, M2, ' SLUGS '
      PRINT *, ' THE TENSION IN THE ROPE IS ', T, ' POUNDS '

      END
```

Figure 2.2 A program to compute the tension in the rope shown in Figure 2.1

that really could not change unless the entire problem changed, such as the 2 in this case.

The READ statement takes two values from the input device, giving the first value to M1 (since it is named first) and the second to M2.

The assignment statement is a fairly direct translation of the formula. Only the parentheses around the denominator are essential to achieve correct compilation; those around the numerator are intended only for readability.

The rest of the program contains no new concepts. Two PRINT statements are used solely for the pragmatic reason that a single line of output would be too wide to fit on a page in this book. Here are the lines of output produced when this program—after compilation, of course—was executed with input data specifying masses of 2 slugs and 1 slug:

```
19.10.55 >run c2pull
EXECUTION:
>2.0 1.0
  WITH MASSES    2.00000000        1.00000000       SLUGS
  THE TENSION IN THE ROPE IS     42.9333190        POUNDS
```

Note that the two input values were entered with one blank separating them. With the free form input we are using until Chapter 6, we may separate values either by blanks or commas.

Example: Temperature conversion

Let's look more closely at a familiar problem, the conversion of a Fahrenheit temperature to Celsius, which will provide some mild surprises. The conversion formula can be written:

$$C = \tfrac{5}{9} (F - 32)$$

Figure 2.3 is a program to do this job. In order to make a few points about the use of integer and real quantities, the conversion formula has been written in the assignment statement with the 5 and the 9 separated; algebraically the meaning is, of course, the same. But we do have to be clear on what happens when real and integer quantities are mixed in the same expression. Since F is real, the constant 32 will be converted to real form before the subtraction. Then

```
* A program to convert a temperature from Fahrenheit to Celsius
* Variables:
*     F: temperature in degrees Fahrenheit
*     C: temperature in degrees Celsius

      REAL F, C

      READ *, F
      C = 5 * (F - 32) / 9
      PRINT *, F, ' FAHRENHEIT = ', C, ' CELSIUS '

      END
```

Figure 2.3 First version of a program to convert temperatures.

the 5 will be similarly converted before the multiplication, and the 9 before the division. The result, which is clearly real, is assigned to C.

Here is the terminal session when this program was compiled and run:

```
19.15.45 >run c2tempa
EXECUTION:
>32.
   32.0000000      FAHRENHEIT =   0.000000000       CELSIUS

19.16.42 >start
EXECUTION:
>50.
   50.0000000      FAHRENHEIT =   10.0000000        CELSIUS

19.16.49 >start
EXECUTION:
>60.
   60.0000000      FAHRENHEIT =   15.5555553        CELSIUS

19.16.56 >start
EXECUTION:
>68.
   68.0000000      FAHRENHEIT =   20.0000000        CELSIUS

19.17.02 >start
EXECUTION:
>-40.
  -40.0000000      FAHRENHEIT =  -40.0000000        CELSIUS
```

Everything seems to be in order. 32° F certainly is 0° C, and the others are mostly what we would expect.

Well, let's try something else. Figure 2.4 is a program that is identical except that the constants 5 and 9 have been combined into one factor enclosed in parentheses.

Before reading on, try to predict what will happen!

Here is a printout of the terminal session:

```
19.20.06 >run c2tempb
EXECUTION:
>32.
   32.0000000      FAHRENHEIT =   0.000000000       CELSIUS
```

```
* A program to convert a temperature from Fahrenheit to Celsius
* Variables:
*     F: temperature in degrees Fahrenheit
*     C: temperature in degrees Celsius
* A modified version, combining integer constants
* THIS PROGRAM CONTAINS A TRAP!!!

      REAL F, C

      READ *, F
      C = (5 / 9) * (F - 32)
      PRINT *, F, ' FAHRENHEIT = ', C, ' CELSIUS '

      END
```

Figure 2.4 Second version of a program to convert temperatures. *This program does not operate as intended, because of an incorrect use of integer division.*

```
19.20.21 >start
EXECUTION:
>50.
  50.0000000      FAHRENHEIT =   0.000000000      CELSIUS

19.20.29 >start
EXECUTION:
>60.
  60.0000000      FAHRENHEIT =   0.000000000      CELSIUS

19.20.38 >start
EXECUTION:
>68.
  68.0000000      FAHRENHEIT =   0.000000000      CELSIUS

19.20.44 >start
EXECUTION:
>-40.
  -40.0000000     FAHRENHEIT =   0.000000000      CELSIUS
```

32° F came out as expected, but what about the others? The answer is that the division of 5 by 9, being enclosed in parentheses, was done as a separate integer operation. Remember what we said about integer division? The quotient is *truncated* to an integer. The division of 5 by 9 gives a quotient of zero with a remainder of 5. One way of viewing truncation is to say that the remainder is simply ignored. The subsequent multiplication involving this quotient—zero—always gives zero, no matter what the input temperature is.

That's not so good. Let's try another version, shown in Figure 2.5. This would seem to be the most direct way to express the conversion formula, and in a way it is—but it holds surprises, too. Here is the terminal output:

```
19.24.05 >run c2tempc
EXECUTION:
>32.
  32.0000000      FAHRENHEIT =   0.000000000      CELSIUS

19.24.35 >start
EXECUTION:
>50.
  50.0000000      FAHRENHEIT =   9.99999904       CELSIUS

19.24.44 >start
EXECUTION:
>60.
  60.0000000      FAHRENHEIT =   15.5555543       CELSIUS

19.24.52 >start
EXECUTION:
>68.
  68.0000000      FAHRENHEIT =   19.9999847       CELSIUS

19.25.00 >start
EXECUTION:
>-40.
  -40.0000000     FAHRENHEIT =   -39.9999847      CELSIUS
```

The first result is as expected. But why has 50° F not been converted to its exact equivalent of 10° C? The problem is that the quotient 5.0/9.0 cannot be expressed exactly in the computer being used to run the program, which expresses real numbers as a fraction times a power of 16. The inexactness of the constant leads to the inexact results.

```
* A program to convert a temperature from Fahrenheit to Celsius
* Variables:
*    F: temperature in degrees Fahrenheit
*    C: temperature in degrees Celsius
* A second modified version, using real constants

      REAL F, C

      READ *, F
      C = (5.0 / 9.0) * (F - 32.0)
      PRINT *, F, ' FAHRENHEIT = ', C, ' CELSIUS '

      END
```

Figure 2.5 Final version of a program to convert temperatures.

Maybe this is a problem and maybe it isn't. Nobody would expect that a temperature of 50° F was to be regarded as accurate to six decimal places, so why should we expect that of its Celsius equivalent? On the other hand, if the intention were to present the results to laymen who may not be comfortable with the concept of significant figures, the results might not really be the best we could do.

Summary of the lessons taught by the three versions of this example:

■ *Watch out for expressions that mix real and integer arithmetic! They can lead to nasty surprises!* For the most part, it is best to use only real constants in real expressions, except for integer exponents. (This requires merely writing the constants with decimal points.)

■ Number base conversion and the finite representation of fractions can lead to results that are not *quite* what you expect.

Mathematical functions

Many engineering and scientific applications of computers require common mathematical functions, such as square root, logarithm, exponential, sine, cosine, arctangent, and absolute value, among others. The developers of Fortran responded to this need by providing pieces of machine language code to do such tasks, and which could be "compiled" into our programs with no more effort than writing their names. Indeed, this is the origin of the term *compiler*, which is used in this context as it is in literary usage.

Every function has a preassigned name. Some of those we shall use, and their names, are given in Table 2.5.

To use a mathematical function in Fortran, we simply write its name and follow that with an expression enclosed in parentheses. The expression within parentheses is called the *argument* of the function. This directs the Fortran compiler to incorporate a set of computer instructions that will compute the named function of the expression value (argument). (Some functions have more than one argument, such as one that returns the largest of its arguments. In such cases, we separate the arguments by commas.)

Table 2.5

Mathematical Function	Fortran Name
Exponential (base e)	EXP
Natural logarithm (base e)	ALOG
Common logarithm (base 10)	ALOG10
Sine of an angle in radians	SIN
Cosine of an angle in radians	COS
Hyperbolic tangent	TANH
Square root	SQRT
Arctangent; angle computed in radians	ATAN
Absolute value	ABS

As an example, suppose it is necessary to compute the cosine of an angle named THETA. Writing COS(THETA) in a statement will result in the computation of the cosine of the angle. That value will then be used in whatever way the rest of the statement directs. In this example, the *argument* of the function is the single variable THETA. But the argument is not limited to a single variable; in fact, it may be *any* expression. Thus, for example, if we want the square root of $b^2 - 4ac$, we can write:

```
SQRT(B**2 - 4.0*A*C)
```

The only restriction is that, for all of the functions listed in Table 2.5, the argument must be a real quantity and the function value is returned in real form. There are also functions that expect arguments of the various other types. For example, there is a function, named DCOS, that finds the cosine of an angle expressed in double precision, and a function, CSQRT, that takes a complex square root. The Appendix lists all of the standard functions.

The argument of a function may well involve another function. Suppose, for instance, that we need to compute V from:

$$V = \frac{1}{\cos x} + \log \left| \tan \frac{x}{2} \right|$$

An inspection of the Appendix reveals the existence of a function to find the tangent, along with those for the logarithm and absolute value. So we are able to write:

```
V = 1.0/COS(X) + ALOG(ABS(TAN(X/2.0)))
```

This one statement has the same effect as the following three statements, executed in the order written:

```
T = TAN(X/2.0)
ABSVAL = ABS(T)
V = 1.0/COS(X) + ALOG(ABSVAL)
```

The latter form is perhaps a little easier to read, and therefore slightly preferred. "Easier to read" *by us*, of course! The compiler doesn't care! But anything that makes a program easier for us to read makes it more likely to be correct and makes modifying it less error-prone, goals that we choose to emphasize.

Example: Radioactive decay

The rate of emission of particles by a substance undergoing a single type of radioactive decay is given by:

$$R = R_0 e^{-\lambda t}$$

in which

$$\lambda = 0.693/T \left(= \frac{\log_e 2}{T} \right)$$

where R = rate of emission at time t, any convenient units
R_0 = rate of emission at time zero, same units as R
λ = disintegration constant, seconds^{-1}
t = time, any convenient units
T = half-life, same units as t

We wish to write a program that will read values of R_0, t, and T, then compute λ and R, and print everything except λ. Doing this job will require the Fortran exponential function.

The program shown in Figure 2.6 does the job we want. Once the idea of a function is clear, the program is actually quite simple. LAMDBA is used as an intermediate variable, computed before the statement that finds the rate of emission. Since we are not required to print LAMBDA, the entire computation could have been achieved with one statement:

```
R = RO * EXP((-ALOG(2.0)/THALF) * T)
```

The use of the intermediate variable makes the program easier to understand and wastes no time in the execution of the object program. It would have been possible to write the natural logarithm of 2, approximately 0.6931472, instead of using a Fortran function to compute it. In this case, there is little to choose between the two; using the function may make the program a trifle clearer.

```
*  A program to compute radioactive decay rate
*  Variables:
*       RO: rate of emission at start, when T = 0
*       R: rate of emission at time T
*       LAMBDA: disintegration constant
*       T: time
*       THALF: half-life of substance

    REAL R, RO, LAMBDA, T, THALF

    READ *, RO, THALF, T
    LAMBDA = ALOG(2.0) / THALF
    R = RO * EXP(-LAMBDA * T)
    PRINT *, ' INITIAL RATE OF EMISSION = ', RO, ' PER MINUTE '
    PRINT *, ' HALF-LIFE OF SUBSTANCE = ', THALF, ' MINUTES '
    PRINT *, ' AT TIME ', T, ' MINUTES LATER, '
    PRINT *, ' RATE OF EMISSION = ', R, ' PER MINUTE '

    END
```

Figure 2.6 A program to compute the radioactivity of a sample of material.

Here is the terminal output when this was run:

```
19.37.06 >run c2rad
EXECUTION:
>437.0 57.0 73.0
  INITIAL RATE OF EMISSION =    437.000000       PER MINUTE
  HALF-LIFE OF SUBSTANCE =    57.0000000        MINUTES
  AT TIME    73.0000000         MINUTES LATER,
  RATE OF EMISSION =    179.867126        PER MINUTE
```

The substance in question is Zinc-69, which disintegrates by electron emission to Gallium-69 with a half-life of 57 minutes.

This program would clearly be somewhat more useful if it could accept a designation of the time unit (second, hour, etc.) and print it appropriately. The techniques required to do this sort of thing will be investigated in Chapter 8.

Exercises

1. Write arithmetic assignment statements to compute the values of the following formulas. Use the letters and names shown for variable names.

  ✓ *a. $\text{AREA} = 2 \cdot P \cdot R \cdot \sin \dfrac{\pi}{P}$

  b. $\text{CHORD} = 2R \sin \dfrac{A}{2}$

  ✓ *c. $\text{ARC} = 2\sqrt{Y^2 + \dfrac{4X^2}{3}}$

  d. $s = \dfrac{-\cos^4 x}{x}$

  *e. $s = \dfrac{-\cos^{p+1} x}{p + 1}$

  *f. $g = \dfrac{1}{2} \log \dfrac{1 + \sin x}{1 - \sin x}$

  g. $R = \dfrac{\sin^3 x \cos^2 x}{5} + \dfrac{2}{15} \sin^3 x$

  h. $D = \log |\sec x + \tan x|$

  *i. $e = x \arctan \dfrac{x}{a} - \dfrac{a}{2} \log (a^2 + x^2)$

  j. $f = -\dfrac{\pi}{2} \log |x| + \dfrac{a}{x} - \dfrac{a^3}{9x^3}$

  k. $Z = -\dfrac{1}{\sqrt{x^2 - a^2}} - \dfrac{2a^2}{3(\sqrt{x^2 - a^2})^3}$

  *l. $Q = \left(\dfrac{2}{\pi x}\right)^{1/2} \sin x$

  m. $B = \dfrac{e^{x/\sqrt{2}} \cos (\sqrt{x/2} + \pi/8)}{\sqrt{2\pi x}}$

*n. $Y = (2\pi)^{1/2}x^{x+1}e^{-x}$

o. $t = a \cdot e^{-\sqrt{w/2p} \cdot x}$

2. A sample X, with a mean of M and a standard deviaiton of S, can be transformed into a sample Z, with mean zero and standard deviation 1 by applying to each sample value X the transformation

$$Z = \frac{X - M}{S}$$

The transformed sample can be converted to a sample with mean 500 and standard deviation 100 by applying the further transformation

$$T = 100Z + 500$$

Given the variables X, M, and S, write assignment statements to compute and store Z and T.

3. If a wheel of radius a rolls along a straight line, a particle on its circumference describes a cycloid, as sketched. The parametric equations of the cycloid are

$x = a(\theta + \sin\theta)$
$y = a(1 + \cos\theta)$

where θ is the angle that the particle makes at the center with the highest point of the circle. Given A and THETA, compute X and Y.

*4. If the lengths of the sides of a triangle are given by the values of the variables A, B, and C, then the area of the triangle can be computed from

$$\text{AREA} = \sqrt{S(S - A)(S - B)(S - C)}$$

where

$$S = \frac{A + B + C}{2}$$

Write two statements that compute S and AREA, given A, B, and C.

The Fortran coding form

The information on each line of a program is either punched into cards for running under a batch system, or entered into the computer from a terminal using a time-sharing system.

The numbers shown above the first statement line on a coding form, such as 1, 5, 6, 7, 10, 15, etc., are called *character positions*. They may be thought of either as card columns or typing positions, depending on how the program is entered into the computer.

The first *field* (group of character positions) in a statement is called the *state-*

ment number field. We shall see several uses for statement numbers later, although most statements do not require them.

The first character position in a statement has another function: it indicates a *comment line.* As you may recall, if the first character in a line is either a C or an asterisk, Fortran does not process the information on that line but simply transmits it to the *program listing,* a printed version of the program that is ordinarily a by-product of compilation. Free use of comments is strongly encouraged to make the program easily readable by people—yourself and all the others who will need to be able to read it.

Position 6 is used to indicate a *continuation line.* If a statement can be written entirely on one line, position 6 is left blank. (Technically, a zero may be written there, but doing so is pointless and confusing.) If more than one line is needed for a statement, each line after the first must have a nonblank and nonzero character in position 6. Some programmers like to number continuation lines (even though the compiler attaches no meaning to such a convention) as a possible aid to understanding. We choose to employ a character, the dollar sign, that has little use otherwise, to minimize the possibility of confusion with other parts of the statement.

The statement itself is written in positions 7 to 72. Blanks in this field are entirely ignored, except within character literals, such as the identifications that we have used in PRINT statements. Blanks may thus be used freely to improve readability, and you are strongly urged to do so. The statement need not begin in position 7. In this book, for example, we shall frequently indent statements a few positions to show the scope of influence of the control structures discussed in the next chapter, and to set off continuation lines. We shall also, somewhat less consistently, leave a space around plus and minus signs but not always around asterisks and slashes, in the hope of conveying the strength of the arithmetic operators. The use of such conventions is, of course, at the discretion of the programmer.

When a program is to be punched into cards, the punching is often done by a card punch operator, also called a keypunch operator. In this case, it is essential to be absolutely clear how many spaces are desired at each point in a line. It is for this reason that most Fortran coding forms have a box or short vertical line to indicate the character positions.

When someone else is punching your program, it also becomes absolutely essential to distinguish unambiguously between 1 and l, and between O (the letter) and 0 (the digit). Unfortunately, there are no universally accepted conventions for making such distinctions. Some installations specify that the letter O should be written with a slash through it, whereas others put a slash through the zero! Just find out how things are done where you are.

As a matter of fact, even if you punch or otherwise key in your own programs and data, it is worthwhile to double check for errors in this regard. If you punch C0S where you meant COS (see how hard it is to tell the difference?), the error messages may be on the confusing side. (Consider yourself lucky if the message is as explicit as "THE FOLLOWING NAMES ARE UNDEFINED: C0S.")

Columns 73 to 80 are not processed by Fortran and may be used for any desired card or program identification. Some time-sharing systems automatically insert the first three characters of the program name in columns 73 to 75 and a five-digit sequence number in columns 76 to 80.

Free-form programs

Users of time-sharing systems should be aware that many such systems permit *free-form* programs, in which columns 1 to 5, 6, and 7 to 72 do not have the fixed functions just described. Different conventions have to be established for identifying a statement number and a continuation line.

Some programmers find this flexibility attractive, but we shall not use it in this book. It is not part of the Fortran standard, for one thing, and it is not clear that its use promotes program readability.

Exercises

In each of the following exercises data values are to be read, the values used in a computation, and the results printed. The data values are to be printed with the results for easy reference. Include REAL and END statements.

*1. Read: a, b, c
Evaluate:

$$F = \frac{1 + a}{1 + [b/(c + 6)]}$$

Print: a, b, c, F

2. Read: s, x
Evaluate:

$$g = (12.7 - x)^{s+2}$$

Print: s, x, g

3. Read: x, y
Evaluate:

$$h = \frac{x \cos^4 x}{2y}$$

Print: x, y, h

*4. Read: a, b, c
Evaluate:

$$X1 = \frac{-b + \sqrt{b^2 - 4ac}}{2a}$$

$$X2 = \frac{-b - \sqrt{b^2 - 4ac}}{2a}$$

Print: $a, b, c, X1, X2$

5. Read: a, b, c, x
Evaluate:

$$r = \frac{b \cdot c}{12}\left[6x^2\left(1 - \frac{x}{a}\right) + b^2\left(1 - \frac{x}{a}\right)^3\right]$$

Print: a, b, c, x, r

*6. Read: a, e, h, p
Evaluate:

$$x = \frac{e \cdot h \cdot P}{(\sin a)[(h^4/16) + h^2 p^2]}$$

Print: a, e, h, p, x

*7. Read: a, x, s
Evaluate:

$$y = \sqrt{x^2 - a^2}$$

$$z = \frac{x \cdot s}{2} - \frac{a^2}{2}\log|x + s|$$

Print: a, x, s, y, z

THREE

CONTROL STRUCTURES

Introduction

All of the illustrative programs we have seen so far have required the computer only to carry out—unconditionally—a simple sequence of operations, then stop. Seldom will this capability be adequate for our needs. We also need to be able to specify that some action, or group of actions, is to be carried out *conditionally*, only *if* some condition is true. And we need to be able to carry out some action or group of actions repeatedly in what is called a *loop*.

Examples of conditional tests:

- Is this the last data value?

- Are two values equal?

- Does the data show that an employee worked over 100 hours in a week, suggesting invalid (and possibly fraudulent) data?

- Have we tried some computational process 50 times without getting a satisfactory answer?

Examples of loops:

- Read data values and evaluate some function of the data, until reading a negative data value.

- Given a list of 200 data items, replace each item except the first and last with the average of it and its two neighbors (which is called *smoothing*).

- Carry out an approximation process for finding a root of an equation, until two approximations are within 10^{-6} of each other.

Almost all examples in the rest of the book will involve conditional tests and loops of many kinds. Each time such a capability is needed, we draw upon the Fortran features provided for the purpose. In fact, Fortran provides more features of this type than we really need, and they can be combined in a bewildering variety of ways. The resulting potential for confusion is not a problem solely for beginners; it plagues experienced programmers as well. A great many programming errors—by neophytes and professionals alike—involve the *control logic* of programs, i.e., the conditional tests and the loops.

43

Our response to this problem is twofold:

■ We restrict the use of the available Fortran control features to a carefully controlled set, which we can reasonably hope to master fully.

■ We introduce a method of planning the control logic of a program, called *pseudocode*.

In this chapter we take up the study of the most important logical control elements in Fortran, the IF-THEN-ELSE structure and the WHILE structure. The latter is directly available in WATFIV and many other programming languages, and is readily implemented in Fortran 77 using standard Fortran elements. We also use Fortran 77 features to implement a less commonly used structure, the REPEAT. Another specialized but widely useful Fortran element, the DO statement for loop control, is taken up in Chapter 7.

Let us begin our study with the IF-THEN-ELSE structure and several examples of its use.

The IF-THEN-ELSE **structure in pseudocode form**

This heavily used logic control feature lets us choose between two alternative sets of operations, on the basis of some kind of *logical test*. The pseudocode representation is shown in Figure 3.1. We see that the structure begins with the word IF followed by some kind of logical expression. We shall see several examples of logical expressions shortly. The word THEN introduces the action(s) to be taken if the logical expression is true. If the logical expression is true, the action(s) specified in statement-group 1 are carried out, and those in statement-group 2 are *not* carried out. If, on the other hand, the logical expression is false, the action(s) in statement-group 1 are *not* carried out and the action(s) in statement-group 2, those preceded by the word ELSE, *are* carried out. The END IF makes it clear where statement-group 2 ends.

Indentation is conventionally used to indicate the "scope of control" of each branch of the IF-THEN-ELSE, thus helping to make the intended meaning clearer. Indentation will also be used in the Fortran representation of this structure.

The IF-THEN-ELSE **structure in flowchart form**

Some people find a graphical representation of the basic logic control structures helpful in understanding them. We accordingly present in Figure 3.2 a *flowchart* of the IF-THEN-ELSE structure.

```
IF logical test is true THEN
    carry out statement-group 1
ELSE
    carry out statement-group 2
END IF
```

Figure 3.1 Pseudocode representation of the IF-THEN-ELSE logic control structure.

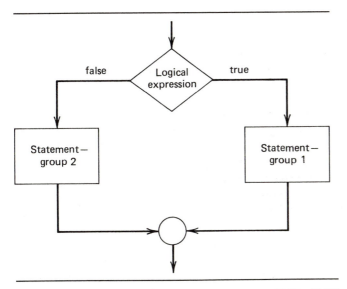

Figure 3.2 Flowchart representation of the IF-THEN-ELSE logic control structure.

Since we shall use flowcharts in this book only to illustrate the basic control structures, there is no need to become concerned about the details of how flowcharts should be drawn. (Many programmers have come to the conclusion that full-scale flowcharts do not help one to understand the logic of large programs.)

An example from mechanical engineering

To see the IF-THEN-ELSE structure in a practical application, let us turn to a simple problem in mechanical engineering.

A designer wishes to obtain data for plotting a curve of the safe load of a certain type of load-bearing column as a function of the *slimness ratio* of the column, which is defined as the ratio of its length to its width. A handbook provides two empirical formulas that give the safe load in two ranges of the slimness ratio:

$$S = \begin{cases} 17,000 - 0.485R^2 & \text{for } R < 120 \\[2ex] \dfrac{18,000}{1 + (R^2/18,000)} & \text{for } R \geq 120 \end{cases}$$

where S = safe load, pounds per square inch
R = slimness ratio, dimensionless

The safe load is to be calculated for a value of the slimness ratio obtained by a READ.

This is a natural application of the IF-THEN-ELSE, since we need to choose between the two formulas based on a two-way decision about the value of a variable. Figure 3.3 is a pseudocode of the actions required.

```
Column design 1:

READ a value of the slimness ratio

IF slimness ratio is < 120 THEN
    Compute safe loading from first formula
ELSE
    Compute safe loading from second formula
END IF

PRINT slimness ratio and safe load
```

Figure 3.3 Pseudocode for a column design
calculation. One of two formulas is to be evaluated,
depending upon the value of a data item.

The first line is simply an identification, permitting easier correlation with the program shown later. We see that the actions to be carried out have been described in a mixture of ordinary English, common mathematical notation, and borrowings from Fortran. Such notational flexibility is one of the advantages of pseudocode: we are able to focus our attention on the logic rather than on the syntax of a programming language. Furthermore, the notation can be adjusted to the needs of the application and of the individual programmer, so long as we restrict ourselves to the basic set of logic control structures.

Note that in this case there is only one action controlled by the THEN branch of the IF and only one controlled by the ELSE branch.

A Fortran program that carries out the actions specified by this pseudocode appears in Figure 3.4. After the usual prologue, we have a REAL statement

```
* Column design
* The program finds the safe loading of a structural column as a
*     function of its slimness ratio, i.e., the ratio of its
*     length to its width
* Variables:
*     RATIO:  the slimness ratio of the column
*     SAFLOD: the safe load, in pounds per square inch

      REAL RATIO, SAFLOD

      READ *, RATIO

* Compute safe load from whichever formula applies
      IF ( RATIO .LT. 120.0 ) THEN
          SAFLOD = 1.7E4 - 0.485 * RATIO**2
      ELSE
          SAFLOD = 1.8E4 / (1.0 + RATIO**2 / 1.8E4)
      END IF

      PRINT *, ' FOR A SLIMNESS RATIO OF ', RATIO
      PRINT *, ' THE SAFE LOAD IS ', SAFLOD, ' LBS PER SQ IN '

      END
```

Figure 3.4 A program for the column design calculation, corresponding to the pseudocode of Figure 3.3

naming the variables, as usual, and a READ statement to get a value for the slimness ratio. Next comes a comment describing the action of the new Fortran feature; this is the first time we have seen a comment other than as a part of the prologue. But, in fact, comments may appear anywhere in a program.

The next statement, the IF, introduces new Fortran material that we must investigate.

The Fortran IF statement

The Fortran IF appears in this program in the form that we shall find most commonly useful, which is very similar to the pseudocode form displayed earlier. The Fortran syntax requires merely that the logical expression being tested be enclosed in parentheses. The "logical expression" enclosed in parentheses in the IF statement will most commonly be a simple relation, such as we have in our program (is RATIO less than 120.0?). We shall see later that more complex— and correspondingly more powerful—forms of the logical expression are also available.

The logical expression in our program is an example of a *relation test,* which has the general form:

```
expression-1   relational-operator   expression-2
```

The "relational operator" here is any one of the following:

```
.EQ.    Equal to
.NE.    Not equal to
.LT.    Less than
.LE.    Less than or equal to
.GT.    Greater than
.GE.    Greater than or equal to
```

The periods before and after the letters that define the relation are a Fortran requirement, necessary to distinguish the operators from possible variables named GT, EQ, etc. (Fortran was defined in the mid-1950s, when most keyboards did not have symbols for greater-than, less-than-or-equal-to, etc. Some still do not, and most programming languages provide alternatives.)

The operation of the IF statement in our program can be expressed thus: *if* the value of RATIO is less than 120, *then* evaluate the safe load from the statement that follows THEN. *Otherwise* ("else"), evaluate the safe load from the statement that follows ELSE. The END IF means "that's all; there is no more to this control structure."

Now we can print the data and the results, which completes the work of the program. The operation of the program is shown in Figure 3.5. Recall that in the National CSS operating system, "run" means to bring into execution an object program that has already been compiled; the compilation step is not shown. Remember also that, after an object program has been executed once, it can be executed again with the "start" command. Details like these may be different if you are using some other system. As noted at the end of Chapter 1, we assume you have received instructions in how to use your interactive or batch system.

The results show that a thick column can bear a greater load, for its size, than a slender one, which seems reasonable: a thin column buckles more readily than a thick one.

```
19.49.12 >run c3col1
EXECUTION:
>20
 FOR A SLIMNESS RATIO OF    20.0000000
 THE SAFE LOAD IS    16806.0000      LBS PER SQ IN

19.49.40 >start
EXECUTION:
>25
 FOR A SLIMNESS RATIO OF    25.0000000
 THE SAFE LOAD IS    16696.8750      LBS PER SQ IN

19.49.47 >start
EXECUTION:
>120
 FOR A SLIMNESS RATIO OF    120.000000
 THE SAFE LOAD IS    10000.0039      LBS PER SQ IN

19.49.56 >start
EXECUTION:
>200
 FOR A SLIMNESS RATIO OF    200.000000
 THE SAFE LOAD IS    5586.20703      LBS PER SQ IN
```

Figure 3.5 The operation of the column design program of Figure 3.4 when executed interactively.

The IF-THEN structure

Fairly often it turns out that we need to make a decision whether to do something or not, with no "else." That is, if a condition is true we want to carry out some action(s) but, if it is false, do nothing. This situation can be illustrated by adding a requirement to the previous program: it must reject any slimness ratio less than 20 or greater than 200.

The IF-THEN structure is not difficult, being a simple modification of the IF-THEN-ELSE structure. The pseudocode is shown in Figure 3.6 and a flowchart in Figure 3.7. We see that there is a specification of what to do if the logic test is true, but no directions on what to do if it is false; in the latter case we are to do *nothing*.

A pseudocode for the modified program is shown in Figure 3.8, in which the required validity tests have been added to the actions depicted in Figure 3.3. Figure 3.9 is a program to carry out these calculations. Here we see two IF-THEN structures, each of which controls two statements. Thus, in this case, we have a statement "group" that in fact has more than one statement in it.

```
IF logical test is true THEN
   carry out statement-group
END IF
```

Figure 3.6 Pseudocode representation of the IF-THEN logic control structure.

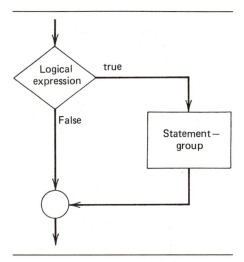

Figure 3.7 Flowchart representation of the IF-THEN logic control structure.

The second action, in each case, is to stop program execution: there is no point proceeding if the data is bad. The STOP statement terminates program execution and returns control of the computer to the operating system, which proceeds with other work that is waiting. We have not needed the STOP statement until now, because "executing" the END statement has the same effect in most Fortran systems.

Figure 3.10 demonstrates the operation of the program.

```
Column design 2:

READ a value of the slimness ratio

IF slimness ratio < 20 THEN
    PRINT error message
    STOP
END IF

IF slimness ratio > 200 THEN
    PRINT error message
    STOP
END IF

IF slimness ratio < 120 THEN
    Compute safe loading from first formula
ELSE
    Compute safe loading from second formula
END IF

PRINT slimness ratio and safe loading
```

Figure 3.8 Pseudocode for a second column design calculation, in which input data is to be checked for validity.

```
* Column design
* The program finds the safe loading of a structural column as a
*     function of its slimness ratio, i.e., the ratio of its
*     length to its width
*
* Version 2: checks input for validity
*
* Variables:
*     RATIO:  the slimness ratio of the column
*     SAFLOD: the safe load, in pounds per square inch

      REAL RATIO, SAFLOD

      READ *, RATIO

* Check input data for validity
      IF ( RATIO .LT. 20.0 ) THEN
         PRINT *, ' FORMULA NOT VALID FOR A SLIMNESS RATIO OF ', RATIO
         STOP
      END IF
      IF ( RATIO .GT. 200.0 ) THEN
         PRINT *, ' FORMULA NOT VALID FOR A SLIMNESS RATIO OF ', RATIO
         STOP
      END IF


* Compute safe load from whichever formula applies
      IF ( RATIO .LT. 120.0 ) THEN
         SAFLOD = 1.7E4 - 0.485 * RATIO**2
      ELSE
         SAFLOD = 1.8E4 / (1.0 + RATIO**2 / 1.8E4)
      END IF

      PRINT *, ' FOR A SLIMNESS RATIO OF ', RATIO
      PRINT *, ' THE SAFE LOAD IS ', SAFLOD, ' LBS PER SQ IN '

      END
```

Figure 3.9 A program for column design calculation with input validity checking, corresponding to the pseudocode of Figure 3.8.

Exercises

1. State in a few words what the following programs would do.

*a.
```
      REAL A, B

      READ *, A, B

      IF ( A .GT. B) THEN
         PRINT *, ' THE FIRST NUMBER IS LARGER'
      ELSE
         PRINT *, ' THE SECOND NUMBER IS LARGER'
      END IF

      END
```

b.
```
      REAL A, B

      READ *, A, B

      IF ( ABS(A) .EQ. ABS(B) ) THEN
         PRINT *, ' A AND B ARE EQUAL IN ABSOLUTE VALUE'
```

```
19.55.51 >run c3col2
EXECUTION:
>18.5
 FORMULA NOT VALID FOR A SLIMNESS RATIO OF    18.5000000

19.56.18 >start
EXECUTION:
>2000
 FORMULA NOT VALID FOR A SLIMNESS RATIO OF    2000.00000

19.56.26 >start
EXECUTION:
>25
 FOR A SLIMNESS RATIO OF    25.0000000
 THE SAFE LOAD IS    16696.8750      LBS PER SQ IN

19.56.34 >start
EXECUTION:
>30
 FOR A SLIMNESS RATIO OF    30.0000000
 THE SAFE LOAD IS    16563.5000      LBS PER SQ IN

19.57.21 >start
EXECUTION:
>195
 FOR A SLIMNESS RATIO OF    195.000000
 THE SAFE LOAD IS    5783.13281      LBS PER SQ IN
```

Figure 3.10 The operation of the program of Figure 3.9.

```
        ELSE
            PRINT *, ' A AND B ARE NOT EQUAL IN ABSOLUTE VALUE'
        END IF

        END

 *c.  INTEGER I, J, K

      READ *, I, J, K

      IF ( I**2 + J**2 .EQ. K**2 ) THEN
          PRINT *, ' THE NUMBERS DEFINE A RIGHT TRIANGLE'
      ELSE
          PRINT *, ' THE NUMBERS DO NOT DEFINE A RIGHT TRIANGLE'
      END IF

      END

  d.  INTEGER I, J, K

      READ *, I, J, K

      IF ( I**3 + J**3 .EQ. K**3 ) THEN
          PRINT *, ' I''M IMMORTAL!!!'
      ELSE
          PRINT *, ' FERMAT''S LAST THEOREM NOT DISPROVED'
      END IF

      IF ( K .EQ. 0 ) THEN
          PRINT *, ' (HMMM . . . DID I READ THE WHOLE THEOREM?)'
      END IF

      END
```

```
*e.  REAL U, V

     READ *, U, V

     IF ( U .LT. 0.0 ) THEN
         U = -U
     END IF
     IF ( V .LT. 0.0 ) THEN
         V = -V
     END IF

     PRINT *, ' THE SUM OF THE ABSOLUTE VALUES IS ', U + V

     END

 f.  REAL A, B, C

     READ *, A, B, C

     IF ( A .GT. B + C ) THEN
         PRINT *, ' THESE NUMBERS CANNOT BE THE SIDES OF A TRIANGL
     END IF
     IF ( B .GT. A + C ) THEN
         PRINT *, ' THESE NUMBERS CANNOT BE THE SIDES OF A TRIANGL
     END IF
     IF ( C .GT. A + B ) THEN
         PRINT *, ' THESE NUMBERS CANNOT BE THE SIDES OF A TRIANGL
     END IF

     END
```

2. Each of the following programs has at least one syntax error. Identify the errors and suggest what the compiler diagnostic error message would be.

```
*a.  REAL R, S

     READ *, R, S

     IF ( R**2 = S**2 ) THEN
         PRINT *, ' THE NUMBERS ARE EQUAL IN ABSOLUTE VALUE'
     ELSE
         PRINT *, ' THE NUMBERS ARE NOT EQUAL IN ABSOLUTE VALUE'
     END IF

     END

 b.  REAL GROSS

     READ *, GROSS

     IF ( GROSS .LT. 2000.0 ) THEN
         PRINT *, ' NOT TAXABLE
     END IF

     END

*c.  REAL A, AMAX

     READ *, A, AMAX

     IF A .LT. AMAX THEN
         PRINT *, ' OK'
     ELSE
         PRINT *, ' NO GOOD'
     END IF

     END
```

d.   `REAL SAMAGE, JOEAGE`

```
READ *, SAMAGE, JOEAGE

IF ( SAMAGE .LT. JOEAGE ) THEN
    PRINT *, JOE IS THE OLDER OF THE TWO
ELSE
    PRINT *, JOE IS YOUNGER OR THEY ARE THE SAME AGE
END IF

END
```

3. Describe how the program of Figure 3.9 would operate if the two STOP statements were removed.

4. For each of the following problem statements, write a pseudocode and a complete program to carry out the specified computations. "Complete" means that the program should be compilable, and, in particular, that it should include REAL and/or INTEGER statements, and READ, PRINT and END statements.

   *a. A point in the plane is defined by the values of x and y, and r is the radius of a circle with its center at the origin. Determine whether the point lies within the circle.

   b. A point in the plane is defined by the values of x and y. Determine whether the point lies on a line through the origin having a slope of 45°.

   *c. A certain city imposes a tax of 2% on earnings over $100 in a week. Compute the tax, which may be zero.

   d. Another city imposes a tax of 2% on the amount by which a week's earnings exceed $20 per dependent. Compute the tax, which may be zero.

   *e. A point in the plane is defined by the values of x and y, and another point is defined by r and s. Determine whether the two points lie within a distance d of each other.

   f. Two points in the plane are defined by (x_1, y_1) and (x_2, y_2). Print HORIZONTAL if $y_1 = y_2$, and ALMOST HORIZONTAL if the line joining the points makes an angle of less than 0.01 rad with the horizontal.

The WHILE logic control structure

For the frequent situations where it is necessary to carry out some sequence of operations repetitively, we shall most commonly use the WHILE logic control structure. (And the DO statement, when we reach it in Chapter 7.)

The WHILE is shown in pseudocode form in Figure 3.11 and in flowchart form in Figure 3.12. The essence of the operation is that a test is made at the *beginning* of the loop. If the test gives a "true" answer, the statement group controlled by the WHILE is executed and the test is repeated. If and when the test gives a "false" answer, the statement group is *not* executed and control passes to whatever follows the END WHILE. Two points need emphasis:

■ The logic test is made *first*, before anything in the statement group is executed. Specifically, this means that, if the logic test is false the first time it is

```
WHILE logical test is true
   statement group
END WHILE
```

Figure 3.11 Pseudocode representation of the WHILE logic control structure.

tested, *the statements controlled by the WHILE are not executed at all*. Something in the statement group controlled by the WHILE must change the value of a variable in the logic test, or else the statement group will be executed forever! (Well, at least until the operating system throws you off the machine.)

The WHILE structure is not, regrettably, part of standard Fortran 77. (It *is* available in WATFIV, which is one of WATFIV's major advantages.) To gain the advantages of the WHILE structure when using Fortran 77, we implement it using a combination of available language elements in a way that is perfectly straightforward, and which rapidly becomes second nature. The form is shown informally in Figure 3.13. (This implementation was suggested by Professor John L. Lowther, Michigan Technological University.)

The comment lines, those beginning with *****, are, of course, optional; they will be used here to highlight the logic structure. The CONTINUE statement is an interesting feature of Fortran: it doesn't do anything! Its function, as we shall use it in this book, is simply to provide something to which we can attach a *statement number*, which is needed whenever we need to refer to a statement from elsewhere in a program. The 10 on the CONTINUE statement is an example of a statement number. You will use statement numbers other than 10, of course,

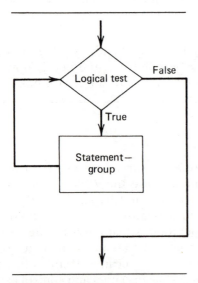

Figure 3.12 Flowchart representation of the WHILE logic control structure.

```
***** WHILE logical test is true
   10  CONTINUE
       IF logical test is true THEN

          statement group

       GO TO 10
       END IF
***** END WHILE
```

Figure 3.13 Informal representation of the implementation of the **WHILE** structure in Fortran 77. *The statement number will not always be 10.*

because you may have several statements requiring statement numbers, and no statement number may appear more than once in any one program.

We choose to put statement numbers only on **CONTINUE** statements because doing so makes modifying a program safer and easier. If statements other than **CONTINUE** (and **FORMAT**, as we shall see in Chapter 6) have statement numbers, it is all too easy to change such a statement and forget to repeat the statement number, or to move the statement and forget that doing so destroys the logic, or to make a variety of other mistakes. Hence, we put statement numbers only on **CONTINUE**.

With the **CONTINUE** statement doing nothing, the **IF** statement immediately makes the logical test. If the answer is "true," the "statement group" controlled by the **THEN** is executed. The final statement in the group is a **GO TO** statement, which returns control to the **CONTINUE**, after which the logical test is made again.

This process is repeated until a change in some variable involved in the logical test causes the answer to be "false." At that point, control proceeds to whatever follows the **WHILE** structure.

Of course, readers using WATFIV need not bother with this implementation. They will put the logical test directly in the **WHILE** statement, omit the **CONTINUE**, **IF**, and **GO TO** statements, and remember to include the **END WHILE**.

The WHILE **structure in the column design example**

All of this may seem more complex than it really is. Let's bring it down to earth with an example. The task is to evaluate the column design formula for a series of values of the slimness ratio, from 20 to 200, inclusive, in steps of 5.

Since we have a complete statement of the values of the slimness ratio, and since they fall into a simple pattern, we shall generate them with the program and not read any data. The pseudocode representation of the plan of attack is given in Figure 3.14. The only new aspect is in the notation. The line:

```
RATIO <-- 20
```

is one way of representing the action of a Fortran assignment statement. It can be read as "RATIO is replaced by 20," or, "RATIO gets the value 20." In the Fortran program it will become:

```
Column design 3:

PRINT column headings

RATIO <-- 20

WHILE RATIO <= 200
    IF RATIO < 120 THEN
        Compute safe load from first formula
    ELSE
        Compute safe load from second formula
    END IF
    PRINT slimness ratio and safe load
    RATIO <-- RATIO + 5
END WHILE
```

Figure 3.14 Pseudocode of the column design
calculation for a range of slimness ratios generated by
the program itself; no data is read.

```
RATIO = 20.0
```

with its usual potential for confusion with an equation. And since the terminal used to prepare the manuscript for this book does not have a symbol for "less than or equal to," the "less than" and "equal to" symbols have been written side-by-side (which is exactly how some programming languages would represent the relation, rather than with Fortran's .LE. notation).

Observe that a value is given to RATIO before entering the WHILE structure, that the WHILE test involves RATIO, and that the value of RATIO is changed within the WHILE structure. This is a typical pattern, reflecting the idea that no variable can usefully be referenced until it has been given a value, and the idea that *some* variable involved in the WHILE test must change if we are ever to get out of the loop.

This version of the program will not involve any validity testing, since there is no input. Any errors in the values of the slimness ratio would have been created by the program itself, which we are writing. Needless to say, errors are possible in this process, too. But finding them is part of the program checkout process, which we take up in Chapter 5.

Finally, before we look at the program, observe that the WHILE structure "contains" an IF-THEN-ELSE structure. This combination of logic control structures, which is sometimes called *nesting*, will be common in almost everything that follows. Not only can a WHILE contain an IF, but an IF can contain a WHILE, one WHILE can contain another WHILE, etc.

Figure 3.15 implements this pseudocode. Most of it is a reasonably direct parallel of the pseudocode, and closely related to previous versions of this calculation. The test at the beginning of the WHILE structure has been implemented with the Fortran statement:

```
IF ( RATIO .LE. 200.0 ) THEN
```

The program output is shown in Figure 3.16. Since the interactive aspect of running the program is not instructive in this case, the commands to produce it have not been reproduced. This, then, is how the output would appear in

```
* Column design
* The program finds the safe loading of a structural column as a
*    function of its slimness ratio, i.e., the ratio of its
*    length to its width
*
* Version 3: generates the values of RATIO with a WHILE structure
* Does not include a READ or validity checking
*
* Variables:
*    RATIO:  the slimness ratio of the column
*    SAFLOD: the safe load, in pounds per square inch

      REAL RATIO, SAFLOD

* Print column headings
      PRINT *, ' SLIMNESS RATIO      SAFE LOAD '

* Give RATIO its starting value
      RATIO = 20.0

***** WHILE RATIO is not greater than 200
   10 CONTINUE
      IF ( RATIO .LE. 200.0 ) THEN

*         Compute safe load from whichever formula applies
          IF ( RATIO .LT. 120.0 ) THEN
              SAFLOD = 1.7E4 - 0.485 * RATIO**2
          ELSE
              SAFLOD = 1.8E4 / (1.0 + RATIO**2 / 1.8E4)
          END IF

          PRINT *, RATIO, SAFLOD

*         Increment RATIO
          RATIO = RATIO + 5.0

      GO TO 10
      END IF
***** END WHILE

      END
```

Figure 3.15 A program for a column design calculation, based on the pseudocode of Figure 3.14.

either interactive or batch mode—always remembering that the details of spacing and number format in your system may be slightly different.

We see that the range of slimness ratios is as specified, which is always worth checking. One of the commonest mistakes in programming is to execute a loop one too few or one too many times.

There is one issue, actually, that is even more urgent than making sure that a program terminates with the right value: making sure that it stops at all! For an example of what can go wrong, suppose we were working on a hexadecimal computer, and that the step size were some fraction that does not have an exact hexadecimal equivalent—such as the simple decimal number 0.1—and that we had tested for exact equality by writing:

```
RATIO .EQ. 200.0
```

SLIMNESS RATIO	SAFE LOAD
20.0000000	16806.0000
25.0000000	16696.8750
30.0000000	16563.5000
35.0000000	16405.8750
40.0000000	16224.0000
45.0000000	16017.8750
50.0000000	15787.5000
55.0000000	15532.8750
60.0000000	15254.0000
65.0000000	14950.8750
70.0000000	14623.5000
75.0000000	14271.8750
80.0000000	13896.0000
85.0000000	13495.8750
90.0000000	13071.5000
95.0000000	12622.8750
100.000000	12150.0000
105.000000	11652.8750
110.000000	11131.5000
115.000000	10585.8750
120.000000	10000.0039
125.000000	9635.68750
130.000000	9283.66796
135.000000	8944.09765
140.000000	8617.02343
145.000000	8302.36718
150.000000	8000.00000
155.000000	7709.69531
160.000000	7431.19140
165.000000	7164.17578
170.000000	6908.31250
175.000000	6663.23828
180.000000	6428.57031
185.000000	6203.92578
190.000000	5988.91015
195.000000	5783.13281
200.000000	5586.20703

Figure 3.16 The output of the program of Figure 3.15.

This program would never terminate, at least not of its own accord; the operating system would eventually interrupt it. The problem is that since the hexadecimal approximation of decimal 0.1 is slightly different from the true value, adding that approximation repeatedly would never give a value exactly equal to 200. It is almost never a good idea to test for the completion of a loop by demanding that two real quantities be equal. (This applies whether the test is made with .EQ. or .NE., of course.)

Let's explore a related aspect of this problem by rewriting the program to use an increment of 0.1 (maintaining the .LE. test, of course). We shall also modify the program so that it prints only after RATIO reaches a value greater than 198; there would otherwise be about 25 pages of output, which would be a bit much for our purposes. The modified program is shown in Figure 3.17. The only changes from the previous version are to make the step size 0.1 instead of 5.0 and to embed the printing of the output in an IF statement that permits printing only after RATIO exceeds 198. We note here, incidentally, that this time the WHILE structure has two IF structures imbedded within it.

```
*  Column design
*  The program finds the safe loading of a structural column as a
*     function of its slimness ratio, i.e., the ratio of its
*     length to its width
*
*  Version 4: uses an increment of 0.1, printing only when RATIO > 198
*
*  Variables:
*     RATIO:  the slimness ratio of the column
*     SAFLOD: the safe load, in pounds per square inch

       REAL RATIO, SAFLOD

*  Print column headings
       PRINT *, ' SLIMNESS RATIO       SAFE LOAD '


*  Give RATIO its starting value
       RATIO = 20.0

***** WHILE RATIO is less than or equal to 200
   10  CONTINUE
       IF ( RATIO .LE. 200.0 ) THEN

*          Compute safe load from whichever formula applies
           IF ( RATIO .LT. 120.0 ) THEN
               SAFLOD = 1.7E4 - 0.485 * RATIO**2
           ELSE
               SAFLOD = 1.8E4 / (1.0 + RATIO**2 / 1.8E4)
           END IF

           IF ( RATIO .GT. 198.0 ) THEN
               PRINT *, RATIO, SAFLOD
           END IF

*          Increment RATIO
           RATIO = RATIO + 0.1

       GO TO 10
       END IF
***** END WHILE

       END
```

Figure 3.17 The program of Figure 3.15, modified to use an increment of 0.1 and print only when RATIO exceeds 198.

The output is shown in Figure 3.18.

This is really not very satisfactory: the error in the representation of decimal 0.1 has accumulated, as the starting value of 20.0 was incremented, giving values of the slimness ratio that are rather different from those intended.

Is there a way out, or must we simply accept this state of affairs as an unavoidable quirk of working with computers? Well, things are not quite as bad as all that. A simple solution in this case is to set up an auxiliary integer variable that runs from 200 to 2000, and divide it by 10.0 to get the value of RATIO. Most of the values of RATIO thus generated will not have exact representations, either, *but now the errors do not accumulate.*

The modified program in Figure 3.19 introduces a new variable named TEN-RAT ("ten times the ratio"), that is declared to be of type integer and initialized

SLIMNESS RATIO	SAFE LOAD
198.083694	5660.66015
198.183685	5656.74218
198.283676	5652.82812
198.383667	5648.92187
198.483657	5645.01562
198.583648	5641.11328
198.683639	5637.21484
198.783630	5633.32031
198.883621	5629.42578
198.983612	5625.53906
199.083602	5621.65234
199.183593	5617.77343
199.283584	5613.89453
199.383575	5610.01953
199.483566	5606.14843
199.583557	5602.27734
199.683548	5598.41406
199.783538	5594.55468
199.883529	5590.69531
199.983520	5586.83984

Figure 3.18 The output of the
program of Figure 3.17.

to 200. The first action within the WHILE loop is then to divide TENRAT by 10.0 to get the value of RATIO. Recall that, when integer and real quantities are combined in an arithmetic operation, the integer quantity is converted to real beforehand. So we have none of the problems with integer division that we explored in the previous chapter. The test for completion has been changed so that TENRAT is compared against 2000, since this test needs to be made at the beginning of the WHILE structure, before RATIO has been derived from TENRAT. The only other difference, and it is a crucial one, is that here TENRAT is incremented by 1, whereas before RATIO was incremented by 0.1. The integer constant 1 can be represented exactly, so there is no error to accumulate.

```
* Column design
* The program finds the safe loading of a structural column as a
*     function of its slimness ratio, i.e., the ratio of its
*     length to its width
*
* Version 5: solves the accumulation of roundoff error problem
*
* Variables:
*     RATIO:  the slimness ratio of the column
*     TENRAT: 10 times RATIO
*     SAFLOD: the safe load, in pounds per square inch

        REAL RATIO, SAFLOD
        INTEGER TENRAT

* Print column headings
        PRINT *, ' SLIMNESS RATIO      SAFE LOAD '

* Give TENRAT its starting value
        TENRAT = 200
```

Figure 3.19 A modified version of the program 3.17; this one does not accumulate roundoff errors.

```
***** WHILE TENRAT is less than or equal to 2000
   10  CONTINUE
       IF ( TENRAT .LE. 2000 ) THEN

*          Convert from TENRAT to RATIO
           RATIO = TENRAT / 10.0

*          Compute safe load from whichever formula applies
           IF ( RATIO .LT. 120.0 ) THEN
               SAFLOD = 1.7E4 - 0.485 * RATIO**2
           ELSE
               SAFLOD = 1.8E4 / (1.0 + RATIO**2 / 1.8E4)
           END IF

           IF ( RATIO .GT. 198.0 ) THEN
               PRINT *, RATIO, SAFLOD
           END IF

*          Increment TENRAT
           TENRAT = TENRAT + 1

       GO TO 10
       END IF
***** END WHILE

       END
```

Figure 3.19 *(cont'd)*

The output is shown in Figure 3.20. We see that the problem with the re-peated addition of an approximation to decimal 0.1 has been solved. All integer values of RATIO are now exact. Interestingly, so are values ending in .5. This is because small negative powers of 2 *do* have exact representations in hexadecimal.

SLIMNESS RATIO	SAFE LOAD
198.099990	5660.01953
198.199996	5656.10546
198.299987	5652.19140
198.399993	5648.28125
198.500000	5644.37500
198.599990	5640.47656
198.699996	5636.57812
198.799987	5632.68359
198.899993	5628.78906
199.000000	5624.90234
199.099990	5621.01562
199.199996	5617.13671
199.299987	5613.25781
199.399993	5609.38281
199.500000	5605.51171
199.599990	5601.64453
199.699996	5597.77734
199.799987	5593.91796
199.899993	5590.06250
200.000000	5586.20703

Figure 3.20 The output of the program of Figure 3.19.

(If you feel that printing 198.099990 is a poor "solution," and that you would prefer to see 198.1, please be patient until Chapter 6. We shall learn there how simple it is to specify the number of decimal places desired in results. These same computations, with output rounded to one decimal place, would be as you expect.)

Exercises

1. What would the following programs do?

*a.
```
      INTEGER I, SUM

      I = 1
      SUM = 0

10    CONTINUE
      IF ( I .LE. 10 ) THEN

          SUM = SUM + I
          I = I + 1

      GO TO 10
      END IF

      PRINT *, SUM

      END
```

b.
```
      INTEGER I, SUM, N

      READ *, N

      I = 1
      SUM = 0

10    CONTINUE
      IF ( I .LE. N ) THEN

          SUM = SUM + I
          I = I + 1

      GO TO 10
      END IF

      PRINT *, N, SUM

      END
```

*c.
```
      INTEGER I, N

      READ *, N

      I = 1

10    CONTINUE
      IF ( I .LE. N ) THEN

          PRINT *, I, I**2
          I = I + 2
```

```
        GO TO 10
        END IF

        END
```

d.
```
        INTEGER I, N

        READ *, N

        I = N

   10   CONTINUE
        IF ( I .GE. 1 ) THEN

            PRINT *, I, I**3
            I = I - 1

        GO TO 10
        END IF

        END
```

*e.
```
        REAL X

        X = 0.0
        PRINT *, '  X (DEGREES)            SIN(X)'

   10   CONTINUE
        IF ( X .LE. 180.0 ) THEN

            PRINT *, X, SIN(X/(180.0/3.141596))
            X = X + 10.0

        GO TO 10
        END IF

        END
```

f.
```
        INTEGER I, N

        READ *, N

        I = 1

   10   CONTINUE
        IF ( I**2 .LE. N ) THEN

            PRINT *, I, I**2
            I = I + 2

        GO TO 10
        END IF

        END
```

2. Some of the following programs contain syntactic errors, i.e., violations of the rules for writing correct programs; such programs will not compile. Others of the programs are syntactically correct and would compile, but would operate in a way that the programmer could not have intended. Identify the errors.

*a.
```
        INTEGER I, N

        READ *, N
```

```
      20   CONTINUE
           IF ( I .LT. N ) THEN

               PRINT *, I, I**4
               I = I + 3

           GO TO 10
           END IF

           END

b.         INTEGER I, N, M

           READ *, N, M

           I = N

      30   CONTINUE
           IF ( I .LE. M ) THEN

               PRINT *, I, I/2

           GO TO 30
           END IF

           END

*c.        REAL A, B, X, DELTAX

           READ *, A, B

           DELTAX = (B - A) / 10.0

      40   CONTINUE
           IF ( X .LE. B ) THEN

               PRINT *, X, TAN(X)
               X = X + DELTAX

           GO TO 40
           END IF

           END

d.         REAL A, B
           INTEGER SINCOS

           READ *, SINCOS, A, B

           IF ( SINCOS .EQ. 1 ) THEN
               PRINT *, ' X (RADIANS)        SIN(X)
           ELSE
               PRINT *, ' X (RADIANS)        COS(X)
           END IF

           DELTAX = (B - A) / 10.0
           X = A

      40   CONTINUE
           IF ( A .LE. B ) THEN

               IF ( SINCOS .EQ. 1 ) THEN
                   PRINT *, X, SIN(X)
               ELSE
                   PRINT *, X, COS(X)
```

```
          END IF
          X = X + DELTAX

      GO TO 40
      END IF

      END
```

*3. Modify the program of Figure 3.19 so that it produces values of RATIO from 20 to 200 in steps of 0.5.

4. Modify the program of Figure 3.19 so that the step size is 0.2, and so that it prints only when the value of RATIO is in the range of 20 to 21.5 inclusive.

*5. Write a program to read a value of a and then print the value of the following integral.

$$\int_0^\pi \frac{x \sin x \, dx}{1 - 2a \cos x + a^2} = \begin{cases} \dfrac{\pi}{a} \ln (1 + a) & |a| < 1 \\ \pi \ln (1 + 1/a) & |a| > 1 \end{cases}$$

The program should stop without printing the value of the integral if a is equal to zero, or if a is equal to 1.

6. Write a program to read values of a and b, and then print the value of the following integral:

$$\int_0^\pi \ln (a^2 - 2ab \cos x + b^2) dx = \begin{cases} 2\pi \ln a & a \geq b > 0 \\ 2\pi \ln b & b \geq a > 0 \end{cases}$$

The program should stop with an error message if either a or b is not greater than zero.

*7. If λ is the mean arrival rate of customers requiring service at a window and μ is the mean rate at which customers can be serviced, then the average length of the queue at the window is given by

$$\bar{n} = \frac{\lambda/\mu}{1 - \lambda/\mu} \qquad \text{if } \frac{\lambda}{\mu} < 1$$

Write a pseudocode and a program to set $\mu = 15$; then compute and print the average queue length for $\lambda = 0, 1, 2, \ldots, 14$.

8. With the definitions of Exercise 7, the probability that the queue is of length n is

$$P_n = \left(\frac{\lambda}{\mu}\right)^n \left(1 - \frac{\lambda}{\mu}\right) \qquad \text{if } \frac{\lambda}{\mu} < 1$$

Write a pseudocode and a program to do the following. Let $\mu = 10/\text{hr}$. Read a value of λ; then compute P_n for $n = 0, 1, 2, \ldots, 15$. Run the program with data values of $\lambda = 4, 5, \ldots, 9$. Verify the integrity of the data by checking that $\lambda/\mu < 1$; stop with an error message if not.

*9. Write a pseudocode and a program to read a value of n; then compute and print

$$\sum_{i=1}^{n} i$$

along with the value of $n(n + 1)/2$.

10. Write a pseudocode and a program to compute and print the value of the following series:

$$\frac{4}{\pi}\left(\frac{\sin 1}{1} + \frac{\sin 3}{3} + \cdots + \frac{\sin 51}{51}\right)$$

(The correct result is approximately 1.002. The value of this *Fourier series* may be made arbitrarily close to 1 by taking more terms of the series.)

*11. Write a pseudocode and a program to compute and print the value of the following series:

$$\frac{4}{\pi}\left(\frac{\sin x}{1} + \frac{\sin 3x}{3} + \cdots\right)$$

for angles from $-300°$ to $300°$ in steps of $20°$. (These angles must be converted to radians for use with the Fortran SIN function, of course.) Take terms of the series through $\sin 99x/99$. This is the Fourier series for a *step function* of the following form:

$$f(x) = \begin{cases} 1 & 0 < x < \pi \\ -1 & -\pi < x < 0 \end{cases}$$

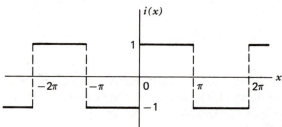

12. The Fourier series for $|\sin x|$ is given by

$$|\sin x| = \frac{2}{\pi} - \frac{4}{\pi}\left(\frac{\cos 2x}{1 \cdot 3} + \frac{\cos 4x}{3 \cdot 5} + \frac{\cos 6x}{5 \cdot 7} + \cdots\right)$$

Write a pseudocode and a program to compute the value of this series, through terms in $\dfrac{\cos 100x}{99 \cdot 101}$, for angles in degrees from $-300°$ to $300°$ in steps of $20°$.

*13. State what output would be produced by the following program.

```
INTEGER J, K

J = 3
K = 10
```

```
10   CONTINUE
     IF ( J .LE. 8 ) THEN

         K = K + J
         PRINT *, J, K
         J = J + 1

     GO TO 10
     END IF

     END
```

This is a "nonsense" program, in that it has no relation to any interesting physical or mathematical problem. Nevertheless, taking it as an exercise in the Fortran expression of a problem in logic, you should be able to determine its output in no more than five to ten minutes. Simply imitate the program, giving values to each of the variables as the program would do, and write down a line of output every time the PRINT is encountered.

If you cannot do that, and if the reason is not uncertainty about how Fortran works, then you are probably behind the majority of your classmates. If, two weeks from now, you still can't do it in a very few minutes, you should probably have a meeting with your instructor. Programming may not be for you.

14. State what output would be produced by the following program.

```
     INTEGER I, N1, N2, TEMP

     I = 1
     N1 = 1
     PRINT *, I, N1
     I = 2
     N2 = 1
     PRINT *, I, N2
     I = 3

10   CONTINUE
     IF ( I .LE. 10 ) THEN

         TEMP = N1 + N2
         PRINT *, I, TEMP
         N1 = N2
         N2 = TEMP
         I = I + 1

     GO TO 10
     END IF

     END
```

This *is* related to an interesting mathematical problem. Despite the presence of more variables, it is not much harder than the previous exercise. Just "play computer" with the values of the variables.

Evaluating a polynomial for a range of X values

For a little more practice in the use of the WHILE construct, let us write a program to evaluate the polynomial of Chapter 1 for a range of X values read as input. The input will consist of the left and right ends of the interval, together with the increment on X. The program will check the input for validity, assuring that the left end value is actually less than the right end and that the increment is positive and nonzero. Assuming the values pass these tests, the polynomial will be evaluated for the left end value, then for that value plus the increment, that value plus the increment, etc., until it has been evaluated for the largest value that does not exceed the right end value. This kind of table preparation, with the independent variable at equally spaced intervals, is a common computer application. And, since the range and step size could change, we prefer to read them as input rather than generating them within the program as was done with the column design calculation.

The pseudocode of Figure 3.21 and the program of Figure 3.22 contain no new concepts, but do provide an opportunity for you to review the materials presented earlier in the chapter.

The terminal printout reproduced in Figure 3.23 displays the program in operation for three different sets of data. The first shows that there are four changes of sign in the interval from -10 to $+10$, indicating the presence of four real roots. The next two sets narrow down the interval within which one of those roots lies.

An example from electricity

One final example of the use of the WHILE structure will provide further reinforcement of the concepts, and introduce a useful technique.

Suppose that we are required to compute the current flowing in an alternat-

```
Polynomial evaluation for a range of X values:

READ left (LEFT) and right (RIGHT) ends of interval,
     and increment (DELTAX)

IF LEFT >= RIGHT THEN
    PRINT error message
    STOP
END IF
IF DELTAX <= zero THEN
    PRINT error message
    STOP
END IF

PRINT column headings

X <-- LEFT

WHILE X <= RIGHT
    Compute the value of the polynomial
    PRINT X and the value of the polynomial
    X <-- X + DELTAX
END WHILE
```

Figure 3.21 Pseudocode of a method for evaluating a polynomial at equally spaced values of the independent variable.

```
* A program to evaluate a polynomial for a range of values
* The input is checked for validity
*
* Variables:
*     LEFT:   Value of X at left end of range of values
*     RIGHT:  Value of X at right end of range
*     DELTAX: Increment on X
*     X:      Independent variable
*     POLY:   Value of given polynomial for current value of X

      REAL LEFT, RIGHT, DELTAX, X, POLY

* Read range ends and step size
      READ *, LEFT, RIGHT, DELTAX

      IF ( LEFT .GE. RIGHT ) THEN
          PRINT *, ' LEFT END MUST BE LESS THAN RIGHT; EXECUTION HALTED '
          STOP
      END IF

      IF ( DELTAX .LE. 0.0 ) THEN
          PRINT *, ' DELTAX MUST BE GREATER THAN ZERO; EXECUTION HALTED '
          STOP
      END IF

* Print column headings
      PRINT *, '         X                POLYNOMIAL '

* Start X at left end of range, then work to right
      X = LEFT

***** WHILE X <= RIGHT
  10  CONTINUE
      IF ( X .LE. RIGHT ) THEN

          POLY = 2.0*X**4 - 15.0*X**3 - 2.0*X**2 + 120.0*X - 130.0
          PRINT *, X, POLY
          X = X + DELTAX

      GO TO 10
      END IF
***** END WHILE

      END
```

Figure 3.22 A program for polynomial evaluation, corresponding to the pseudocode of Figure 3.21.

ing current (ac) circuit that contains resistance, capacitance, and inductance in series with an ac voltage source. The current in the circuit is given by:

$$I = \frac{E}{\sqrt{R^2 + [2\pi fL - 1/(2\pi fC)]^2}}$$

where I = current, amperes
E = voltage, volts
R = resistance, ohms
L = inductance, henrys
C = capacitance, farads
f = frequency, hertz

```
10.22.41 >run c3poly
EXECUTION:
>-10 10 1
         X                POLYNOMIAL
  -10.0000000             33470.0000
   -9.00000000            22685.0000
   -8.00000000            14654.0000
   -7.00000000             8879.00000
   -6.00000000             4910.00000
   -5.00000000             2345.00000
   -4.00000000              830.000000
   -3.00000000               59.0000000
   -2.00000000             -226.000000
   -1.00000000             -235.000000
    0.000000000            -130.000000
    1.00000000             -25.0000000
    2.00000000               14.0000000
    3.00000000             -31.0000000
    4.00000000             -130.000000
    5.00000000             -205.000000
    6.00000000             -130.000000
    7.00000000              269.000000
    8.00000000             1214.00000
    9.00000000             2975.00000
   10.0000000              5870.00000

10.23.25 >start
EXECUTION:
>1  2 0.1
         X                POLYNOMIAL
    1.00000000             -25.0000000
    1.09999942             -17.4568023
    1.19999885             -10.6528472
    1.29999828              -4.62287902
    1.39999771              0.603103637
    1.49999713              4.99990844
    1.59999656              8.54711914
    1.69999599             11.2291107
    1.79999542             13.0351562
    1.89999485             13.9591980
    1.99999427             14.0000305

10.23.39 >start
EXECUTION:
>1.3 1.4 0.02
         X                POLYNOMIAL
    1.30000019              -4.62275695
    1.31999969              -3.51237487
    1.33999919              -2.43440246
    1.35999870              -1.38903808
    1.37999820             -0.376464843
    1.39999771              0.603103637
```

Figure 3.23 The output of the program of Figure 3.22, run interactively for three sets of data.

(An aside—the examples used in this book are taken from various areas of the application of computers. As outlined in Chapter 1, the complete process of problem solution requires a full understanding of the subject matter of a problem and its formulation. We are assuming here, however, that these preliminary steps have already been completed, and that we begin our work with a complete statement of what is to be done. You are not, therefore, required to know how formulas like the one for this example are derived or, for that matter, what they mean. The emphasis is on programming; it should not matter if a particular example is from an application area that is unfamiliar to you.)

We shall assume that the purpose of the computation is to provide the data for drawing a graph of the relation between current and the frequency of the voltage source. Therefore we shall arrange to read the fixed values of voltage, resistance, inductance, and capacitance, and then a series of values of the frequency. For each frequency value, we will print a line giving the frequency and the corresponding current. It is not necessary that the frequency values be equally spaced. If equal spacing were a program requirement, the job could be handled more simply by the techniques used with the polynomial in the previous section.

The end of the data will be signaled this time by entering a negative frequency, a physically meaningless value that would never need to be read as normal data. A data value that signals the end of data is commonly called a *sentinel*. We shall see as we proceed that there are other ways to deal with this question of telling the program that the end of the data has been reached. Looking for an otherwise meaningless data value is one common way, and one that is suitable for our purposes at this stage.

The pseudocode in Figure 3.24 uses a WHILE structure to detect a negative frequency, which requires that the first frequency be read before entering the WHILE structure. The last action controlled by the WHILE is the reading of another data value. When a negative frequency is read and control returns to the WHILE test, the body of the WHILE loop will not be executed and the program will stop.

The program in Figure 3.25 implements this program design in a straightforward manner. The only variation from previous practice is to give a name to 2π, to clarify the meaning of what would otherwise be a constant.

```
RLC circuit:

READ parameters (E, R, L, C)
PRINT parameters

READ first frequency

WHILE frequency is not negative
    Compute current
    PRINT frequency and current
    READ a frequency
END WHILE
```

Figure 3.24 Pseudocode of a method for calculating current in an ac circuit.

```
* Current in an AC circuit
* Series circuit contains a voltage source, a resistance,
*    a capacitance, and an inductance; frequency of voltage
*    source is read as data, with a negative value as sentinel
*
* Variables:
*    I: Current, amperes
*    E: Voltage, volts
*    R: Resistance, ohms
*    F: Frequency, hertz
*    L: Inductance, henrys
*    C: Capacitance, farads
*    TWOPI: 2 times pi

      REAL I, E, R, F, L, C, TWOPI
      TWOPI = 2.0 * 3.1415963

* Read the unchanging parameters

      READ *, E, R, L, C

* Print parameters and column headings

      PRINT *, ' E = ', E, ' VOLT '
      PRINT *, ' R = ', R, ' OHM '
      PRINT *, ' L = ', L, ' HENRY '
      PRINT *, ' C = ', C, ' FARAD '
      PRINT *, '    FREQUENCY              CURRENT '

* Get the first value of the frequency
      READ *, F

***** WHILE frequency is greater than zero
   10 CONTINUE
      IF ( F .GE. 0.0 ) THEN

         I = E / SQRT( R**2 + (TWOPI*F*L - 1.0/(TWOPI*F*C))**2 )
         PRINT *, F, I
         READ *, F

      GO TO 10
      END IF
***** END WHILE

      END
```

Figure 3.25 A program for an ac circuit calculation.

The output in Figure 3.26 is for a series of unequally spaced frequency values surrounding the resonant frequency of about 2250 hertz. The frequency values were set up as a separate data file for running the program, so that the printed results do not include the entry of the data. This is the same appearance that the output would have if the program were run with a batch system and with data on cards.

Logical operators

Often we wish to combine two logical tests in an IF statement, to inquire whether either condition is true (or whether both of them are). Sometimes we wish to

```
E =      10.0000000        VOLT
R =      1000.00000        OHM
L =      0.100000023       HENRY
C =      0.500000005E-07   FARAD
  FREQUENCY                CURRENT
  500.000000       0.163023360E-02
  1000.00000       0.364495650E-02
  1200.00000       0.466015562E-02
  1400.00000       0.582892447E-02
  1600.00000       0.712741166E-02
  1800.00000       0.843260437E-02
  2000.00000       0.948233902E-02
  2100.00000       0.981274247E-02
  2150.00000       0.991703197E-02
  2200.00000       0.997922197E-02
  2250.00000       0.999999418E-02
  2300.00000       0.998133793E-02
  2350.00000       0.992634892E-02
  2400.00000       0.983894243E-02
  2500.00000       0.958467647E-02
  2700.00000       0.888110324E-02
  2900.00000       0.809800252E-02
  3100.00000       0.735570117E-02
  4000.00000       0.503166019E-02
  5000.00000       0.370754255E-02
```

Figure 3.26 The output of the
program of Figure 3.25, when run with
data specifying a range of unequally
spaced frequency values.

ask whether a condition is *not* true. For these situations, Fortran provides the
logical operators .OR., .AND., and .NOT..

The .OR. is used to connect two relational expressions, asking whether either
(or both) is true. Thus we might have an IF statement like this:

```
IF ( A .GT. B .OR. N .NE. 9 ) GO TO 10
```

This says that, if A is greater than B *or* N is not equal to 9 *or both,* transfer control
to statement 10.

The relational expressions connected by logical operators are often more
complex than in this example, in which case it may clarify meaning to enclose
the relational expressions in parentheses and use a continuation line, this way:

```
   IF (      (A .GT. B)
  $      .OR. (N .NE. 9) ) GO TO 10
```

Here is an example of the use of the .AND. operator:

```
   IF (       ((A - B) .GT. 1E-6)
  $      .AND. (ITER    .LE. 15  ) ) GO TO 10
```

This says that, if A minus B is greater than 10^{-6} *and* ITER is less than or equal
to 15, go to statement 10. Both conditions joined by .AND. must be true for the
combination to be true.

The .NOT. operator reverses the "truth value" of whatever it precedes. These
two IF statements, therefore, have exactly the same effect:

```
IF ( .NOT. (X .LT. 10.0) ) GO TO 10

IF ( X .GE. 10.0 ) GO TO 10
```

"Not-logic" can be rather confusing; we shall find few occasions to use it.

Logical operators may be combined in the same expression, as in this example:

```
IF ( (A .EQ. B) .AND. (C .EQ. D) .OR. (E .EQ. F) ) GO TO 20
```

The transfer of control will take place if both A equals B *and* C equals D, *or* if E equals F. *All of the inner parentheses here could be removed without changing the meaning.* That is to say, relational expressions need not be enclosed in parentheses, because relational operators are "stronger" than logical operators, and the .AND. operator is "stronger" than the .OR.. This latter is a direct analog of the fact that multiplication is a "stronger" operation than addition, as regards the interpretation of expressions written without parentheses. Considering the high potential for confusion in complex logical expressions written without parentheses, however, and considering that there is no penalty for inserting "unnecessary" parentheses, we shall parenthesize freely.

Having said this, we should also note that this kind of *compound logic* is potentially confusing and is to be avoided whenever possible. Fortunately, nothing we shall do in this book will involve logic more complicated than what you have just seen, and most will in fact be simpler.

The Newton-Raphson method

We can illustrate some of these ideas in a program to find a square root using the Newton-Raphson method, a powerful, elegant, and often-used numerical method.

Given a function of x, $F(x) = 0$, the Newton-Raphson method says that, subject to certain conditions, if x_i is an approximation to a root, then a better approximation is given by:

$$x_{i+1} = x_i - \frac{F(x)}{F'(x)}$$

where the prime denotes the first derivative.

For instance, suppose we have the equation $F(x) = x^2 - 25$. As a first approximation to the root, let us arbitrarily take $x_0 = 2$. Since $F'(x) = 2x$, a better approximation can be found from:

$$x_1 = x_0 - \frac{x_0^2 - 25}{2x_0} = 2 - \frac{4 - 25}{4} = 7.25$$

This is called an *iteration formula*. Continuing in the same way, now substituting 7.25 into the same formula, and so on, we get a succession of approximations:

$x_0 = 2$
$x_1 = 7.25$
$x_2 = 5.35$
$x_3 = 5.0114$
$x_4 = 5.00001$
$x_5 = 5.0000000$

The approximation can be made as accurate as we please by continuing the process. ("As accurate as we please" *within the limits of precision of representation of numbers in the computer,* that is. See the discussion later, in connection with running the program.) One root of the equation is indeed 5, the positive square root of the constant term.

The Newton-Raphson method is readily adapted to computer use. Programming the iteration formula is simple, as is starting it and determining when the process has converged. A pseudocode of the process is shown in Figure 3.27.

After reading the value of A, the number whose square root we wish to find, we give X, the "old" approximation, the starting value, A/2. Any positive value will do, actually, and convergence is so fast that it doesn't matter much what value is used. Then we initialize the iteration counter. This will be used to count the number of iterations of the process as a guard against infinite repetition if the process somehow fails.

You will now observe that we have not used the WHILE structure here. Although a program can assuredly be designed to do this job using only the WHILE structure, it turns out to be somewhat artificial and nonintuitive to make the test at the beginning of the loop. The problem is that convergence is to be determined by comparing the old and new approximations to the root, and, during the first time through the loop, the new approximation doesn't exist. Initializing it to some value that will ensure at least one pass through the loop is a possibility, but, as noted, this leads to a program that hardly seems obvious. Accordingly, we have used an alternate form that we shall call a REPEAT structure, although that term is not universally used. The idea is that the test for completion of the loop may be made anywhere in the loop—except as the first action, in which case it would be the WHILE under a different guise. When the completion test establishes that the work of the loop is finished, there is an

---

```
Newton-Raphson method for square root:

READ A, the number whose square root is desired

X <-- A/2 (first approximation)
Iteration counter <-- 1

REPEAT until converged or too many iterations
    Compute new approximation from Newton-Raphson formula
    PRINT new approximation
    IF absolute value of difference between old and new
        approximations is less than 1E-6,
        OR iteration counter exceeds 15, ESCAPE
    Old approximation <-- new approximation
    Add 1 to iteration counter
END REPEAT

IF iteration counter > 15 THEN
    PRINT non-convergence message
ELSE
    PRINT approximation to square root
END IF
```

Figure 3.27 Pseudocode for the Newton-Raphson method for finding an approximation to the square root of a number, *A*. Note the escape from the middle of the loop. See text discussion.

"escape" to a CONTINUE following the REPEAT structure. Putting all this together, we see that the REPEAT loop begins with the iteration formula. So that we may observe convergence, each new approximation is printed.

A flowchart of the REPEAT logic control structure is shown in Figure 3.28.

Now we are ready to determine whether to stop program execution. This involves two tests, connected by a logical .OR.. The first question is whether the old and new approximations are within 10^{-6} of each other. Since we have no way of ensuring that convergence will be from either above or below, we cannot know the sign of the difference. We therefore use the absolute value function to remove the sign of a negative difference. The second test establishes whether the iteration process has gone on beyond a reasonable number of iterations, which would indicate that something strange had happened such that the process is probably never going to converge. If neither of these conditions is true, that is, if we have not converged but there have not been too many iterations, we replace the value of the old approximation with the value of the new approximation, increment the iteration counter, and go around again.

A matching program is shown in Figure 3.29.

One aspect of the program may bear a word of comment. The basics of the Newton-Raphson method were explained above in terms of a succession of approximations to which we gave different subscripts. We shall see later, in Chapter 7, that Fortran variables may certainly be subscripted—but that is not needed here. All we require, actually, is one variable to hold the old approxi-

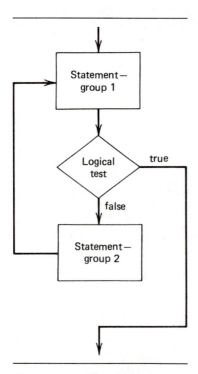

Figure 3.28 Flowchart representation of the REPEAT logic control structure.

```
*  Finding a square root using the Newton-Raphson method
*  Program uses a REPEAT structure with an escape from middle of loop
*
*  Variables:
*      A:    the number whose square root is to be found
*      X:    previous value of X--initialized to A/2
*      XNEW: new value of X, computed from Newton-Raphson formula:
*               Xnew = X - f(X) / f'(X)
*      ITER: iteration counter; limit of 15 iterations

       REAL A, X, XNEW
       INTEGER ITER

       READ *, A
       X = A/2.0
       ITER = 1

***** REPEAT until converged or too many iterations
***** NOTE escape from middle of loop
  10   CONTINUE

          XNEW = (X**2 + A) / (2.0*X)
          PRINT *, XNEW
          IF (        (ABS(X - XNEW) .LT. 1E-6)
      $            .OR. (ITER .GT. 15)            ) GO TO 20
          X = XNEW
          ITER = ITER + 1

       GO TO 10
***** END REPEAT

  20   CONTINUE

       IF ( ITER .GT. 15 ) THEN
          PRINT *, ' DID NOT CONVERGE IN 15 ITERATIONS FOR ', A
       ELSE
          PRINT *, ' THE SQUARE ROOT OF ', A, ' IS (APPROX) ', XNEW
       END IF

       END
```

Figure 3.29 A Newton-Raphson program for finding a square root.

mation to X and another to hold the new one being computed. In this program the two are called, respectively, X and XNEW. The test for convergence can be made by comparing the absolute value of the difference between the two against the convergence criterion. If convergence has not been achieved the value of X can be replaced by that of XNEW.

Observe that the iteration formula has been revised to involve a little less arithmetic and, thus, speed up the program execution a bit. The formula given before was:

$$x_{i+1} = x_i - \frac{x_i^2 - A}{2x_i}$$

Preparatory to putting everything on the right over a common denominator, we write:

$$x_{i+1} = \frac{2x_i^2}{2x_i} - \frac{x_i^2 - A}{2x_i}$$

Combining and simplifying, we get:

$$x_{i+1} = \frac{x_i^2 + A}{2x_i}$$

This is the form used in the program.

Figure 3.30 is the terminal printout when this program was run for several values of A. We see that the method converged fairly rapidly for 25 and 2, but rather more slowly for a small number. About all the latter proves is that a good square root routine for general use needs to be a little smarter about the starting approximation. For a negative number the process did not converge at all. This is no fault of the method, which cannot find a root unless there *is* a (real) root! Naturally, a better solution to this problem would be a test after reading A, stopping with a message if the number is negative.

The final case is instructive. At the end the approximations are alternating between two values that are quite close to the root. The difference between them is greater than 10^{-6} so the convergence test is not satisfied, and the iteration procedure would go on indefinitely if not stopped by the iteration counter. A proper solution in this case is to test the *relative* size of the two successive approximations, which is a matter of the numerical analysis in the original specifications. Still, the example is illuminating in showing how things that are taken for granted in "ordinary" mathematics ("you can get as close to the root as you please by continuing this process") turn out to be untrue in the world of computers.

```
10.47.50 >run c3newt
EXECUTION:
>25
   7.25000000
   5.34913730
   5.01139354
   5.00001144
   4.99999904
   4.99999904
 THE SQUARE ROOT OF    25.0000000      IS (APPROX)   4.99999904

10.48.30 >start
EXECUTION:
>2
   1.50000000
   1.41666603
   1.41421508
   1.41421318
   1.41421318
 THE SQUARE ROOT OF    2.00000000      IS (APPROX)   1.41421318

10.48.43 >start
EXECUTION:
>0.01
   1.00249958
   0.506236672
   0.262995123
   0.150509238
   0.108475148
   0.100331068
   0.100000500
 0.999999642E-01
 THE SQUARE ROOT OF   0.100000016E-01  IS (APPROX)   0.999999642E-01
```

```
10.48.54 >start
EXECUTION:
>-2
 0.500000000
-1.75000000
-0.303571403
 3.14233017
 1.25292968
-0.171664536
 5.73948001
 2.69550705
 0.976765632
-0.535404205
 1.60004425
 0.175039351
-5.62547874
-2.63497448
-0.937976896
 0.597135841
DID NOT CONVERGE IN 15 ITERATIONS FOR   -2.00000000

10.49.17 >start
EXECUTION:
>1000000
 250000.937
 125002.437
 62505.2148
 31260.6171
 15646.2968
 7855.10156
 3991.20288
 2120.87719
 1296.19018
 1033.84033
 1000.55322
 999.999755
 999.999511
 999.999755
 999.999511
 999.999755
DID NOT CONVERGE IN 15 ITERATIONS FOR    1000000.00
```

Figure 3.30 The output of the program of Figure 3.29, when run for various values of *A*. Note the nonconvergence for the last two cases; see the text discussion.

Exercises

1. In each of the following exercises, you are asked whether the two logical expressions have the same *truth value*, that is, whether they would both be true or both be false if written in a Fortran IF statement.

*a. `(A .GT. B) .AND. (C .GT. D) .OR. (E .GT. F)`
   `((A .GT. B) .AND. (C .GT. D)) .OR. (E .GT. F)`

b. `(A .GT. B) .AND. (C .GT. D) .OR. (E .GT. F)`
   `(A .GT. B) .AND. ((C .GT. D) .OR. (E .GT. F))`

*c. `(A .GT. B) .AND. (C .GT. D) .OR. (E .GT. F) .AND. (G. GT. H)`
   `((A .GT. B) .AND. (C .GT. D)) .OR. ((E .GT. F) .AND. (G .GT. H))`

d. `.NOT. (A .GT. B) .AND. (C .GT. D)`
   `.NOT. ((A .GT. B) .AND. (C .GT. D))`

```
*e.  .NOT. ((A .GT. B) .AND. (C .GT. D))
     .NOT. (A .GT. B) .OR. .NOT. (C .GT. D)

 f.  .NOT. ((A .GT. B) .AND. (C .GT. D))
     .NOT. (A .GT. B) .AND. .NOT. (C .GT. D)

*g.  (A. GT. B) .AND. (C .GT. D) .AND. (1.0 .EQ. 2.0)
     (A .GT. B) .AND. .NOT. (A .GT. B)

 h.  (A .GT. B) .OR. (C .LE. D) .OR. (C .EQ. D) .OR. (C .GT. D)
     (A .GT. B) .OR. (1.0 .EQ. 1.0)
```

2. Each of the following IF statement fragments contains a syntactic error, i.e., it could not be correct no matter what the writer might have meant. Identify the errors.

```
*a  IF ( A .GT. B AND C .EQ. 2.0 ) THEN . . .

 b. IF ( A .LT. 2.0 .AND. I = 6 ) THEN . . .

*c. IF A .EQ. 6.0 .AND. .NOT. K .NE. 7 THEN . . .

 d. IF ( A .NE. 22.5 .AND. .OR. A .EQ. 22.5 ) THEN . . .
```

3. In each of the following, you are given an English statement of a logical test, together with a fragmentary Fortran IF statement. Although each IF statement is syntactically correct, none of them correctly carries out the intention stated in words. Identify the errors.

*a. "If N is greater than 6 but less than 18, then . . ."

```
    IF ( N .GT. 6 .AND. .NOT. N .LT. 18 ) . . .
```

b. "If J is equal to 2 and either K is equal to 2 or L is greater than 9, then . . ."

```
    IF ( J .EQ. 2 .AND. K .EQ. 2 .OR. L .GT. 9 ) THEN . . .
```

*c. "If J is greater than 9 but not equal to 18, then . . ."

```
    IF ( J .GT. 9 .AND. .NOT. J .NE. 18 ) THEN . . .
```

d. "If J and K are both greater than L, then . . ."

```
    IF ( J .GT. K .AND. K .GT. L ) THEN . . .
```

*4. Write a program to read three real values and print YES or NO, depending on whether the three values could or could not be the sides of a triangle. Print the values. Use only one IF statement.

5. Modify the program of Exercise 4 so that it stops with an error message if any of the three data values is zero or negative.

*6. Modify the program of Figure 3.22 so that the first value of X is RIGHT, the second value is RIGHT − DELTAX, etc., with the last value of X being not less than LEFT.

7. Modify the program of Figure 3.22 so that instead of reading a value of the interval width, it reads the number of intervals. It should then compute the interval width from the values of LEFT, RIGHT, and the number of intervals.

8. Modify the program of Figure 3.22 so that it computes and prints values of the sine, rather than the fourth degree polynomial. Then use the modified program to print the values of the sine from 3 to 4 in steps of 0.1, and again from 3.1 to 3.2 in steps of 0.01.

9. In each of the following you are given an infinite series that converges to a value involving π. For each, if you compute the sum of the series through the term shown in parentheses, you will obtain an approximation to the value of the series that is accurate to about six decimal places. For each, write a pseudocode and a program to carry out the summation; then print that value along with the value given by the formula.

   *a.

$$\frac{1}{1^6} + \frac{1}{2^6} + \frac{1}{3^6} + \frac{1}{4^6} + \cdots = \frac{\pi^6}{945} \left(\frac{1}{11^6}\right)$$

   b.

$$\frac{1}{1^2} - \frac{1}{2^2} + \frac{1}{3^2} - \frac{1}{4^2} + \cdots = \frac{\pi^2}{12} \left(\frac{1}{1000^2}\right)$$

   Hint: This can be written in the form:

$$\sum_{i=1}^{\infty} \frac{(-1)^{i+1}}{i^2}$$

   *c.

$$\frac{1}{1^2 \cdot 3^2} + \frac{1}{3^2 \cdot 5^2} + \frac{1}{5^2 \cdot 7^2} + \frac{1}{7^2 \cdot 9^2} + \cdots = \frac{\pi^2 - 8}{16} \left(\frac{1}{33^2 \cdot 35^2}\right)$$

   d.

$$\frac{1}{1^2 \cdot 2^2 \cdot 3^2} + \frac{1}{2^2 \cdot 3^2 \cdot 4^2} + \frac{1}{3^2 \cdot 4^2 \cdot 5^2} + \cdots = \frac{4\pi^2 - 39}{16}$$
$$\left(\frac{1}{10^2 \cdot 11^2 \cdot 12^2}\right)$$

10. The program of Figure 3.25 found the steady-state root-mean-square current in the RLC circuit. This exercise asks you to compute the instantaneous value of the current at a range of times.

   Assuming that a voltage E is applied at time $t = 0$, and assuming that $R^2 - 4L/C < 0$, the current at a later time t is given by

$$i = \frac{E}{\omega L} e^{\alpha t} \sin \omega t$$

   where

$$\omega = \sqrt{\frac{1}{LC} - \frac{R^2}{4L^2}}$$

   and

$$\alpha = -\frac{R}{2L}$$

   Write a pseudocode and a program to evaluate this expression, as follows:

First read values of R, L, and C, then check that $R^2 - 4L/C < 0$ and stop with an error message if not. If the values are acceptable, print them (with identifications); then print column headings for the results that will be produced by the loop that follows.

Before going into the loop, compute the (unchanging) values of ω and α. Then, using the logic of the program of Figure 3.25, enter a loop that reads values of t, computes the corresponding value of i, and prints the values of the time and the current.

Test the program with $E = 10$ volts, $R = 100$ ohms, $L = 0.25$ henry, and $C = 0.000001$ farad. Time values in the range of zero to 0.01 second will demonstrate that the circuit oscillates, but is damped.

11. Modify the program of Figure 3.29 so that it finds a root of $\sin x = 0$. Use 3 as a starting approximation, and then try 6 and -3.

12. My U.S. social security number is of the form *aaa-bb-cccc*. Call the number formed from the first three digits A, that formed from the next two digits B, and that formed from the last four digits C. With these definitions, the roots of the quation

$$Ax^2 + Bx + C = 0.$$

are $-0.0205608 \pm 1.72002i$. Determine my social security number. (There is nothing you can legally do with this knowledge. Luckily, there isn't much you do with it *illegally*, either.)

Perhaps it is obvious that you will have to think about this one for a while, and that it is not simply a programming problem. Are you even sure that a computer will be much help? Does the problem have a unique answer? If not, can you use the fact that A, B, and C must be integers, to deduce the correct answer?

13. The infinite series

$$1 - \frac{1}{2} + \frac{1}{3} - \frac{1}{4} + \cdots$$

converges to the natural logarithm of 2. Write a pseudocode and a program to find an approximation to ln 2 that is accurate within 0.001. To show convergence, print the value of the partial sum after computing and adding each term.

(Think! The problem is completely straightforward and the program is semitrivial at this stage—but do you really want to submit it?)

14. An approximation to the factorial is given by Stirling's approximation:

$$n! \cong \sqrt{2\pi n}\; n^n e^{-n}$$

Write a pseudocode and a program to illustrate the accuracy of the approximation for large n. This will require computing $n!$ exactly (using a loop, no doubt), then computing the value of $n!$ from Stirling's formula, then computing the relative error in the approximation, and finally printing a line showing n and the three values. Do this for n from 1 up to the largest value of $n!$ that you can compute exactly in your system.

That word "exactly" is possibly ambiguous, and in any event will probably require running a preliminary program to determine maximum number sizes in your computer.

15. Write a pseudocode and a program to deal with the following problem.

You are the shipping clerk for a small company that has one product and three sizes of boxes in which to ship the product. Since it is generally cheaper—both in cost of boxes and in shipping charges—to ship a given number of items in one large carton than in two smaller ones, you would like to minimize the number of boxes used in shipping an order. The data to be processed by your program is as follows:

a. Stock number, call it NSTOCK.

b. Number of items to be shipped, NOITEM.

c. Capacity of the largest box, MAXSIZ.

d. Capacity of the middle-sized box, MIDSIZ.

e. Capacity of the smallest box, MINSIZ.

Your program should accept data records consisting of these five values, until a negative value of NSTOCK signals the end of the input. The general idea is to place as many items as possible in boxes of the largest size, then place the remaining items in the smallest box in which they will fit. But consider special cases. What if the order is too small to require the largest box? What if the number remaining after using the largest box will fit in the smallest box?

16. In this exercise you are asked to try to devise a method of solution, expressed however may be suitable. If possible, you may write a program as well. The emphasis is on finding an approach, which will be of varying difficulty, from reasonably simple to very hard. It is not guaranteed that the problem statement gives you all the information you need in every case.

In other words, the emphasis is on the problem-solving approach, whether or not a computer can help you with it.

a. Given two times, both expressed in hours and minutes since midnight, such as 0145, 1130, or 2350, you are guaranteed that h_1m_1, the first time, is earlier than h_2m_2, the second time, and that they are less than 24 hours apart. You are *not* guaranteed that they are in the same day: For example, 2350 before a midnight is earlier than 0300 after that midnight. Compute the difference between the two times, in minutes.

b. Suppose that the squares of a tic-tac-toe game are numbered as in this sketch:

1	2	3
4	5	6
7	8	9

You are given N1, N2, and N2, the numbers of three squares. If it is any help, you may assume that N1 < N2 < N3. If the three squares so designated lie in a line, print YES, and print NO otherwise.

Can you think of a way to renumber the squares so as to greatly simplify the program?

c. A certain game requires the determination of the number of identical letters in two five-letter words, neither of which has any duplicated letters. For instance, there are no common letters in BLACK and WHITE, there are two common letters in KAREN and JAMES, and there are five common letters in SNAIL and NAILS. Devise a method to determine the number of duplicated letters in the two words.

d. A checkerboard is placed on a coordinate system, as in this sketch:

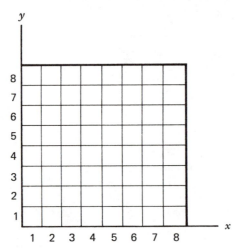

The equations of two lines are determined by the four numbers a, b, c, and d in

$$y = ax + b$$
$$y = cx + d$$

If the two lines intersect anywhere on the checkerboard, produce the square number. If they do not intersect on the checkerboard, determine whether at any point they pass through two adjacent squares and produce the square number if so.

e. You are given ten pairs of values of x and y, representing coordinate values in a bubble-chamber photograph. Devise a way to determine whether the track represented by the points is circular. If so, compute its radius of curvature. Naturally, there must be an allowance for measurement error; it is not meaningful to ask whether the 10 points lie on a circle *exactly*.

f. For each of two aircraft you are given the speed, altitude, direction, and position. If the altitudes vary by less than 1000 ft, determine whether the flight paths come within 10 minutes flying time of each other.

g. Suppose you have discovered that your Fortran system contains a function named F. You know nothing about it, except that when you invoke it with a real argument, it returns a real value. Devise a way to determine whether F represents a continuous function over the range (0, 5).

Naturally, you will have to decide what constitutes "continuity" in this situation. By the usual definitions of real analysis, no function in a digital computer is ever continuous.

*17. Write a pseudocode and a program using a loop to read 10 data values and print their average.

If you have been doing any significant amount of programming so far, this problem should take you at most five minutes

18. Same as Exercise 17, except the number of data values is variable; the end of data is signaled by a zero value.

This is a little harder, but not much.

FOUR
SUBPROGRAMS, I

Introduction

One of the most powerful features of any modern programming language is the provision of *subprograms*, also called *procedures*, which are program segments that in some sense have an independent existence, and which can be called into operation by other program segments. Subprograms are an important programming tool that we shall use heavily from now on. Their importance is based on three factors.

1. If someone else has already written a program segment that is suitable for your purposes, it obviously saves you time and effort to use it. This saving applies not only to the basic task of planning and writing the program segment, but also to the work of testing it and correcting the probable programming errors. Besides, with any luck, the programmer of the segment you borrow has done a thorough job of researching the mathematical and computational methods needed, and has, in effect, done a better job than you could hope to do in a reasonable amount of time.

This can be a major factor, as was noted in Chapter 1. In fact, when faced with any major computer application, your first impulse should be to make a serious inquiry to see if a program to do the job already exists. Or perhaps a program can be found that is close to what you need; modifying it may be a far smaller job than writing your own from scratch.

This approach will apply even in a student situation, once you have progressed beyond writing exercise programs to learn the basics. Virtually every large computing installation has a library of subroutines for computing various special mathematical functions that are not provided as Fortran intrinsic functions (cosine, square root, etc.) and for performing such tasks as solving systems of simultaneous equations. You should be planning to look into this resource for your later work. This subject will be addressed again in Chapter 9.

2. If a certain programming operation of significant size has to be carried out at several points in a program, it saves storage space in the object program to write one copy of the operation and then call it into action from all the places it is needed.

This advantage is of somewhat less importance with today's large computers than it once was, especially on jobs of "student size."

3. In many cases the most important advantage of subprograms is a dual one:

they provide an organizing tool for developing programs, and they make it possible to write programs that are much easier to understand. And, as we have emphasized before and will do again, a "clean" understandable program is much more likely to be correct and is much easier to modify. This advantage can easily outweigh the other two. We may, accordingly, make a subprogram out of some simple operation that occurs only once in a program—just to promote clarity.

We turn now to a study of the two types of subprograms in Fortran, FUNCTION and SUBROUTINE. Each will be illustrated in action. The basic plan is to provide only as much information as is needed to permit early use of these features and then pick up the rest of the details in Chapter 9.

The FUNCTION subprogram

A FUNCTION subprogram is a program segment that performs some operation and returns a value. It has a kind of "stand-alone" independence. It can be compiled by itself, if that is useful, and combined with a main program and/or other subprograms at a later time. It may be used with a variety of main programs, if that is beneficial.

FUNCTION subprograms are very similiar to the (built-in) functions of Chapter 2, except that you must define them yourself.

A FUNCTION subprogram is defined by writing the word FUNCTION and the function name at the beginning of the statements that do the work, and the word END after them. The name of the FUNCTION is formed according to the same rules as for variable names: one to six characters, the first of which must be a letter. We shall always precede the word FUNCTION by a type specification, e.g., the word REAL or INTEGER. The name of the FUNCTION is followed by parentheses enclosing arguments, which are separated by commas if there is more than one.

The name of the FUNCTION subprogram must be given a value in the subprogram, which may be done either with an assignment statement or a READ statement.

We shall always place a RETURN statement in a subprogram. When this statement is encountered, it directs that control go back to whatever program called the subprogram into action. Control is returned to the next operation after the invocation of the FUNCTION. In a subprogram that has no RETURN statement, encountering the END has the same effect—but this promotes confusion between the *definition* of a subprogram and its *execution,* a topic that is already potentially confusing enough. Let us turn to that topic.

A FUNCTION is *defined,* as just noted, by writing a group of Fortran statements between the words FUNCTION and END. *Dummy argument(s)* written in the definition are used to specify how to give a value to the subprogram name. This is the definition.

To call a FUNCTION into action and thus cause computation, its name is written in a main program, or in another subprogram, at a point where its value is desired. This is called *invoking* the FUNCTION. *Actual arguments* are written in parentheses following the name; these supply specific values to the FUNCTION subprogram, which uses them, according to the actions specified by the statements in the FUNCTION subprogram, to compute a value for the FUNCTION. This value is returned to the point of invocation of the FUNCTION.

Any main program or subprogram that invokes a FUNCTION should include a type specification statement containing the name of the FUNCTION. This is because, without such a type specification, the compiler uses the first letter of the FUNCTION name to determine whether it is INTEGER or REAL, according to the I–N rule. Mismatches of type caused by unintentional default typing can lead to extremely strange results that are difficult to diagnose.

A "main program," by the way, is simply one that does *not* begin with the word FUNCTION or the word SUBROUTINE. When a group of object programs is executed, there must always be exactly one main program, no matter how many subprograms there may be. Execution always begins with the first executable statement of the main program. As we shall use the terminology, then, a *program* consists of a main program and its associated FUNCTION and SUBROUTINE subprograms, if any.

An example using the polynomial evaluation

Let us consolidate this new material by seeing the concepts in action, once again working with the fourth-degree polynomial familiar from prior chapters. The task this time is to evaluate and print the value of the polynomial at zero, then for eleven equally spaced points from the left end of an interval to the right end, inclusive.

Figure 4.1 is a pseudocode of the program design for doing this task. The style used here is that each program segment, main program or subprogram, begins with an identifying line ending in a colon. The descriptions in some cases are more condensed than in the previous chapter. This reflects the fact that pseudocode is intended to be helpfully flexible, so that we can concentrate on logic rather than on the requirements for writing the pseudocode.

```
Polynomial evaluation with a FUNCTION:

Evaluate polynomial at X = 0, PRINT values of X and polynomial

READ LEFT and RIGHT ends of an interval

PRINT column headings
DELTAX <-- (RIGHT - LEFT) / 10
STEPER <-- 0.0  (an auxiliary variable)

WHILE STEPER <= 10
   Argument value = LEFT + STEPER * DELTAX
   PRINT argument value, polynomial function at that value
   STEPER <-- STEPER + 1
END WHILE

STOP

FUNCTION POLY4F:

Compute fourth degree polynomial function of argument
```

Figure 4.1 Pseudocode of the design for a program to evaluate a polynomial at a number of points, using a FUNCTION subprogram.

The technique for generating the 11 values of X at which the polynomial is to be evaluated is to divide the interval between the left and right end by 10, calling that DELTAX, then multiply that by a number that runs from 0 to 10, inclusive. Adding each of these 11 products to the value of the left end of the interval produces the 11 values that are supplied successively to the FUNCTION.

Figure 4.2 is a program embodying this design. It consists of a main program and a FUNCTION subprogram. Let us study the subprogram first.

After an identifying comment line, the first thing we see, as must always be the case, is a line containing the word FUNCTION, the name of the function (which is POLY4F, in this case), and parentheses surrounding the one dummy argument. The type declaration, REAL, means that the value returned by the execution of this subprogram will be of type REAL. Now the usual type declaration informs Fortran that X is to be treated as a real quantity. The assignment statement says what to do with the argument, and gives a value to the name of the function. The RETURN says to go back to wherever we came from, and the END says that the definition of the subprogram is complete.

Consider now the main program. After a REAL statement to declare the variables, we have another REAL statement naming the FUNCTION that will be invoked. This is not strictly required, since the default naming convention would regard POLY4F as type real anyway, but it is an excellent habit to develop. (Using two REAL statements here, instead of one or several, is a matter of program clarity, and is optional.) The first executable statement immediately invokes POLY4F, providing an actual argument that is a constant. The actual argument is permitted to be any real-valued expression, and a single constant satisfies that requirement. This statement causes control in the object program to be given to POLY4F, and causes the value 0.0 to be passed to it. When POLY4F gives control back to the main program, a value will have been assigned to the name of the function, which the assignment statement in the main program then assigns to ZERVAL. Then it is printed, along with identification.

Next comes a READ, to get values for the ends of the interval, followed by the printing of column headings and the computation of DELTAX and STEPER. A WHILE loop now generates the eleven equally spaced values of X—but these values are not, in fact, assigned to any variable. This demonstrates that the actual argument used in invoking a subprogram can be any expression of the proper type. By "proper type" we mean that if a dummy argument is declared to be REAL in the FUNCTION definition, the corresponding real argument must also be REAL, etc. So long as this requirement is met, the actual argument written when invoking a FUNCTION can be any expression.

(A note on terminology. Fortran usage is to speak of a *dummy argument* in the definition of a subprogram and of an *actual argument* in its invocation. In other languages these terms might be replaced by *parameter* and *argument*, respectively, or by *formal parameter* and *parameter*.)

The output when the program was run is shown in Figure 4.3.

```
* A program to illustrate the invocation of a FUNCTION
* This main program invokes POLY4F in two places,
*    one of which is in a WHILE loop, to evaluate the
*    polynomial at zero and at 11 equally spaced points
*    including the ends of an interval read as data
*
* Program does not validate input
*
* Variables:
*     LEFT:   Left end of interval
*     RIGHT:  Right end of interval
*     DELTAX: (RIGHT - LEFT)/10
*     ZERVAL: Value of the polynomial when X = 0
*     STEPER: Auxiliary variable used in WHILE loop

      REAL LEFT, RIGHT, DELTAX, ZERVAL, STEPER
      REAL POLY4F

      ZERVAL = POLY4F(0.0)
      PRINT *, ' AT X = 0, THE POLYNOMIAL = ', ZERVAL

* Read LEFT and RIGHT, print column headings, then compute
*    and print the value of the polynomial for 11 equally-spaced
*    intervals from LEFT to RIGHT, inclusive

      READ *, LEFT, RIGHT

      PRINT *, '       X                POLYNOMIAL'
      DELTAX = (RIGHT - LEFT) / 10.0
      STEPER = 0.0

***** WHILE STEPER <= 10.0
  10  CONTINUE
      IF ( STEPER .LE. 10.0 ) THEN

          PRINT *, LEFT + STEPER*DELTAX, POLY4F(LEFT + STEPER*DELTAX)
          STEPER = STEPER + 1.0

      GO TO 10
      END IF
***** END WHILE

      END

* A FUNCTION subprogram to evaluate a certain fourth-degree polynomial

      REAL FUNCTION POLY4F(X)

      REAL X

      POLY4F = 2.0*X**4 - 15.0*X**3 - 2.0*X**2 + 120.0*X - 130.0

      RETURN

      END
```

Figure 4.2 A main program and a FUNCTION to evaluate a polynomial at a number of points.

```
18.03.30 >run c4poly
EXECUTION:
 AT X = 0, THE POLYNOMIAL =   -130.000000
>1 2
         X                 POLYNOMIAL
    1.00000000            -25.0000000
    1.09999942            -17.4568023
    1.19999980            -10.6527710
    1.29999923            -4.62280273
    1.39999961            0.603225708
    1.49999904             4.99998474
    1.59999942             8.54721069
    1.69999885            11.2292022
    1.79999923            13.0352172
    1.89999961            13.9592285
    1.99999904            14.0000152
```

Figure 4.3 The output of the program of 4.2
when run in the time-sharing mode.

Exercises

*1. Describe the effect of making the following changes in the program of
Figure 4.2. (The changes are independent of each other, that is, the change
in the second part is not in addition to that in the first, etc.)

a. With X having been declared to be of type real, the PRINT statement
within the WHILE loop is replaced by these two statements:

```
X = LEFT + STEPER * DELTAX
PRINT *, X, POLY4F(X)
```

b. The variable STEPER is declared to be of type integer rather than type
real.

c. Within the WHILE loop, the incrementing of STEPER is done with the
statement

```
STEPER = STEPER + 0.01
```

d. After the WHILE loop, these two statements are inserted:

```
X = 7.5
PRINT *, POLY4F(X)
```

e. Within the FUNCTION definition, this statement is inserted:

```
STEPER = 99.6
```

f. Within the FUNCTION definition, this statement is inserted:

```
PRINT *, 2.0, POLY4F(2.0)
```

g. The variable ZERVAL is removed from the program, and the first PRINT
statement is changed to:

```
PRINT *, ' AT X = 0, THE POLYNOMIAL = ', POLY4F(0.0)
```

*2. State what the output of the following program would be.

```
REAL X, Y, A
REAL MID

PRINT *, MID(2.0, 3.0)
A = 2.0
PRINT *, MID (A, 5.0)
Y = 16.0
PRINT *, MID(A, Y)
PRINT *, MID(MID(2.0, 4.0), 7.0)
PRINT *, MID(X, Y)

END

REAL FUNCTION MID(X, Y)
REAL X, Y
MID = (X + Y)/2.0
RETURN
END
```

3. State what the output of the following program would be.

```
INTEGER P, Q, R
INTEGER AVERAG

PRINT *, AVERAG(1, 4, 10)
P = 2
Q = 6
PRINT *, AVERAG(P, 7, Q)
R = 7
PRINT *, AVERAG(R, Q, P)
PRINT *, AVERAG(P+Q, Q-P, Q/P)
PRINT *, AVERAG(2, P+Q, 2+AVERAG(P-4, 2*P, R))

END

INTEGER FUNCTION AVERAG(A, B, C)
INTEGER A, B, C
AVERAG = (A + B + C)/3
RETURN
END
```

*4. Write a FUNCTION subprogram to compute:

$$y(x) = \begin{cases} 1 + \sqrt{1 + x^2} & x < 0 \\ 0 & x = 0 \\ 1 - \sqrt{1 + x^2} & x > 0 \end{cases}$$

Then write a main program to compute and print the following:

$$F = 2 + y(A + z)$$

$$G = \frac{y(x1) + y(x2)}{2}$$

$$H = y(\cos(2\pi x)) + \sqrt{1 + y(2\pi x)}$$

the y(x) for x values from −1.0 to 1.0 in steps of 0.1

5. Write a FUNCTION subprogram to compute the value of the safe load, as given in Chapter 3:

$$S = \begin{cases} 17{,}000 - 0.485R^2 & R < 120 \\ \dfrac{18{,}000}{1 + R^2/18{,}000} & R \geq 120 \end{cases}$$

Then write a main program that prints the safe load for a slimness ratio of 22.5 and for values from 80 to 180 in steps of 10. Use a loop for the latter.

*6. For values of x less than 0.1 in absolute value, the following approximation is accurate to six decimal places:

$$e^{\sin x} \cong 1 + x + \frac{x^2}{2} - \frac{x^4}{8} - \frac{x^5}{15}$$

Write a FUNCTION that computes $e^{\sin x}$ from this approximation. Then write a main program that uses a WHILE loop to generate x values from zero to 0.1 in steps of 0.01, and for each prints a line giving x, the value of $e^{\sin x}$ from this approximation, and the value of $e^{\sin x}$ found by using the Fortran sine and exponential functions.

7. The following series is valid for $0 < x \le 2$:

$$\ln x = \frac{(x - 1)}{1} - \frac{(x - 1)^2}{2} + \frac{(x - 1)^3}{3} - \frac{(x - 1)^4}{4} + \cdots$$

It converges quite slowly for values near zero and 2, but for $0.5 \le x \le 1.5$ it gives six-decimal accuracy if terms through $(x - 1)^{16}/16$ are used. Write a FUNCTION subprogram that uses a WHILE loop to compute an approximation to $\ln x$ from this series. Then write a main program, along the lines of Exercise 6, to compare the values given by this approximation with those found using the Fortran ALOG function, for arguments from 0.5 to 1.5 in steps of 0.1.

8. Make a FUNCTION of the Newton-Raphson program for finding the square root, as shown in Chapter 3. Improve the convergence testing by either providing a second argument specifying how closely two approximations must match to stop iteration, or set up the test as:

$$\left| \frac{\text{XNEW} - \text{X}}{\text{XNEW}} \right| < 10^{-6}$$

The SUBROUTINE **subprogram**

The FUNCTION subprogram must always return a value associated with the name of the function. Often that is not what we want, and for these other situations, Fortran provides the SUBROUTINE subprogram. The two are quite similar in most ways, but with these three differences:

1. As just noted, no value is associated with the name of the SUBROUTINE. Instead, the SUBROUTINE subprogram returns its output values by modifying the values of some or all of its arguments. In a moment we shall see an example of how this works.

2. A SUBROUTINE is not invoked simply by writing its name where the value is desired—since no value is associated with its name. Instead, we write a CALL statement to bring it into action. With this statement we specify the SUBROUTINE name and its actual arguments.

3. Since the output from a SUBROUTINE may be any combination of the various types of values, none of which is associated with the name of the SUBROUTINE, there is no naming convention regarding the name of a SUBROUTINE. The naming is otherwise the same: one to six characters, the first of which is a letter. Accordingly, it is pointless, and illegal, to use a type statement in conjunction with a SUBROUTINE name.

In all other respects, the FUNCTION and SUBROUTINE subprograms are analogous.

A SUBROUTINE **to sort three numbers**

A common need in many computer applications is to arrange a set of data values into ascending or descending sequence, a process called *sorting*. In this example we are to read three numbers, rearrange them into ascending sequence, and then print them. This process is to be done for an indefinite number of sets of three data items. The sentinel signaling the end of data will be zero values for all three numbers.

One way to approach this problem is to compare the three numbers in pairs, exchanging the two numbers of a pair whenever it is discovered that they are not in ascending sequence. We shall use a SUBROUTINE subprogram named SWAP to do the exchanging. It will be called by a main program that is doing the sorting and which contains the logic to determine when the swapping is needed.

Figure 4.4 shows the logic of the approach that will be used. After reading three numbers, a test using the .AND. operator determines if they are all zero

```
Sort three values using a SUBROUTINE:

REPEAT until three values read are all zero
    READ A, B, C
    IF all three are zero ESCAPE
    IF  A > B  THEN
        SWAP A and B
    END IF
    IF  A > C  THEN
        SWAP A and C
    END IF
    IF  B > C  THEN
        SWAP B and C
    END IF
END REPEAT

SWAP SUBROUTINE:

Exchange the values of the two arguments
```

Figure 4.4 A program design for sorting groups of three numbers into ascending sequence.

and escapes from the REPEAT loop if so. If not, the first IF asks whether A is greater than B and calls the subroutine to swap (exchange) them if so. After the completion of this IF, whether or not the SUBROUTINE was invoked, we are assured that A is less than B—either because it was to begin with or because the two have been exchanged. Now we carry out the same process with A and C, exchanging them if the current value of A is larger. Now we are assured that the current value of A is the smallest of the three numbers. (Try it on examples to satisfy yourself that this is true.) Finally we do the same thing with B and C, which, if there have been exchanges, will not be the original contents of those variables, of course. The three numbers are now in ascending sequence, regardless of what sequence they were in to begin with.

A main program and SUBROUTINE to do all this are shown in Figure 4.5. The main program follows the pseudocode closely, so we turn to the SUBROUTINE.

```
* A main program to sort three numbers into ascending sequence
* Utilizes a SUBROUTINE named SWAP

* Main program reads sets of three data values until all are zero
* For each nonzero set, it prints values before and after
*     sorting into ascending sequence

      REAL A, B, C

***** REPEAT until all of values are zero
   10 CONTINUE

         READ *, A, B, C
         IF ( A .EQ. 0.0 .AND. B .EQ. 0.0 .AND. C .EQ. 0.0 ) GO TO 20
         PRINT *, ' BEFORE: ', A, B, C
         IF ( A .GT. B ) CALL SWAP (A, B)
         IF ( A .GT. C ) CALL SWAP (A, C)
         IF ( B .GT. C ) CALL SWAP (B, C)
         PRINT *, ' AFTER:  ', A, B, C

      GO TO 10
***** END REPEAT

   20 CONTINUE

      END

* A SUBROUTINE that exchanges two values sent to it by calling program

      SUBROUTINE SWAP (X, Y)

      REAL X, Y, TEMP

      TEMP = X
      X = Y
      Y = TEMP

      RETURN

      END
```

Figure 4.5 A main program and a SUBROUTINE for sorting groups of three numbers into ascending sequence.

The subprogram SWAP is not difficult. The REAL statement lists both of the dummy arguments and also a temporary variable named TEMP. The exchange process is done in three steps. First, move the first argument to TEMP, then move the second argument to the first argument, and finally move the first argument from TEMP to the second argument.

As an aside, many beginning students are puzzled by the need for the three-stage process and the temporary variable. Why not simply write this:

```
X = Y
Y = X
```

The answer, of course, is that after the first statement has been executed the previous value of X has been lost. The net effect of the two statements would be to make both variables equal to the initial value of Y.

This SUBROUTINE demonstrates that the arguments of a subprogram can be used both for transmitting values to the subprogram and for transmitting values back to the calling program. In this instance the arguments are used both for input to the subprogram and for output from it, which is permitted. Sometimes the arguments will be a combination of all three, that is, some input only, some output only, and some both.

Testing a combination of main program and subprogram of this type can be exhaustive, that is, we can try all six orderings of the input values. Exhaustive testing will seldom be feasible in real life, and even if we do it, we are not thereby guaranteed that there are not other types of errors than those in the sorting logic. Nevertheless, correct operation with various values that try all of the possible orderings of the input improves our confidence that the program is correct. Figure 4.6 shows the terminal listing when this was done. The fact that the program operates as expected with negative numbers doesn't really tell us much we didn't know before; Fortran can be depended on to handle the algebra of signed numbers properly.

```
18.43.12 >run c4sort
EXECUTION:
>3 2 1
  BEFORE:     3.00000000        2.00000000        1.00000000
  AFTER:      1.00000000        2.00000000        3.00000000
>3 1 2
  BEFORE:     3.00000000        1.00000000        2.00000000
  AFTER:      1.00000000        2.00000000        3.00000000
>2 1 3
  BEFORE:     2.00000000        1.00000000        3.00000000
  AFTER:      1.00000000        2.00000000        3.00000000
>2 3 1
  BEFORE:     2.00000000        3.00000000        1.00000000
  AFTER:      1.00000000        2.00000000        3.00000000
>-3 2 1
  BEFORE:    -3.00000000        2.00000000        1.00000000
  AFTER:     -3.00000000        1.00000000        2.00000000
>-3 1 2
  BEFORE:    -3.00000000        1.00000000        2.00000000
  AFTER:     -3.00000000        1.00000000        2.00000000
>0 0 0
```

Figure 4.6 The output when the program of Figure 4.5 was run for six sets of values that exhaust the possibilities of the initial orderings.

A FUNCTION **and a** SUBROUTINE **in root finding**

Our final example for this chapter uses the FUNCTION POLY4F that evaluates our fourth-degree polynomial, together with a SUBROUTINE subprogram to handle some minor housekeeping functions in a way that "unclutters" the main program logic.

The task is to find a root of the polynomial, given two values of X that supposedly "bracket" a root, together with a value for a convergence criterion that determines when the root has been located precisely enough. The program must establish that the first value is less than the second, that the convergence criterion is greater than zero, and that the polynomial does have different signs for the two values.

The root finding method to be used is called *interval-halving*, or *bisection:* we continually divide in half the interval in which the root lies, until the interval is less than the value of the convergence criterion, called EPS in the program.

The geometry of the method is displayed in Figure 4.7 for an arbitrary function. Given the left and right ends of the interval shown, we can establish that the root lies in the left half-interval, since the function has different signs at the left end of the interval and at the midpoint. We accordingly cut the original interval in half by "moving" the right end of the interval to the midpoint of the original interval. Now we repeat the process, this time discovering that the root lies in the (new) right half-interval. The midpoint of this second interval now becomes the left end of a new interval, which is one-fourth the width of the original interval. This process is continued either until we hit upon the root, meaning that we find a function value of zero, or until the width of the interval is less than the convergence criterion.

The program design pseudocode is shown in Figure 4.8. The form of this pseudocode is a bit more condensed than some of the previous examples. The

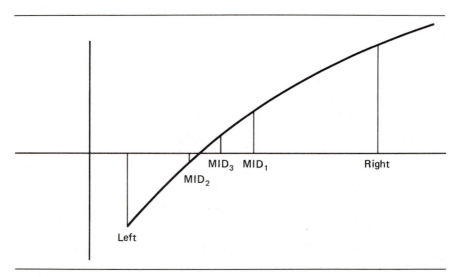

Figure 4.7 Geometrical interpretation of the method of finding the root of an equation by the method of interval-halving, also called the method of bisection.

```
Find a root by interval-halving:

READ LEFT and RIGHT values for interval, convergence criterion EPS

Check validity: RIGHT > LEFT, EPS > 0; error messages and stop if not

Evaluate function at ends of interval, check for presence of root
Error message and stop if not

Compute function values at LEFT and RIGHT

REPEAT
   CALL INTSET SUBROUTINE
   IF function value at midpoint = 0, OR interval < EPS ESCAPE
   IF root is in left half-interval THEN
      RIGHT <-- MID
      function value at RIGHT <-- function value at MID
   ELSE
      LEFT <-- MID
      function value at left <-- function value at MID
   END IF
END RPEAT

PRINT MID, = (approximate) root

STOP

SUBROUTINE INTSET:

Compute midpoint, function at midpoint, interval size

FUNCTION POLY4F:

Compute function of argument (fourth degree polynomial)
```

Figure 4.8 A program design for finding the root of an equation by the method of interval-halving.

line that begins "Check validity," for example, will become six or eight statements in the program, but since, having done it before, we have a pretty clear idea of what is involved, we don't bother to fill in the details this time. You should feel free to adapt the level of detail of a pseudocode to the needs of the application and to your own experience in a particular area.

The validity testing this time includes a check that there actually is a root in the interval supplied as input, which is tested by asking if the function has opposite signs at the ends of the interval. If the signs are opposite, the number of roots in the interval is odd; we ignore the possibility that there are more than one. We also assume that if the function values have the same signs, there is no root in the interval. In fact, there could be two, four, or any even number. This is not intended to be a complete root-finding program.

Figure 4.9 is a program that carries out the actions designed with the pseudocode.

A simple way to test whether two numbers have the same sign is to inspect the sign of their product. If their product is negative, one of the numbers must be positive and the other negative. This technique is used in the validity checking and in the REPEAT loop.

```
* A main program that finds a root of a function defined by a FUNCTION
* The method of interval-halving (bisection) is used
* *
* Variables:
*     LEFT:    Left end of interval in which root is sought
*     RIGHT:   Right end of interval
*     EPS:     Convergence criterion; when interval within which
*                 root must lie is less than EPS, the midpoint of
*                 that interval is taken as the root
*     MID:     Midpoint of interval between LEFT and RIGHT
*     FLEFT:   Function value at LEFT
*     FRIGHT:  Function value at RIGHT
*     FMID:    Function value at midpoint (if FMID happens to be zero,
*                 MID is a root and process is stopped)
*     INTRVL:  Interval, = RIGHT - LEFT
*
* Input data is LEFT, RIGHT, and EPS; values of LEFT and RIGHT
*     change as interval is narrowed
*
* Validity checks:
*     RIGHT > LEFT
*     EPS > 0
*     Existence of a root between LEFT and RIGHT, tested by requiring
*         different signs for FLEFT and FRIGHT

      REAL LEFT, RIGHT, EPS, MID, FLEFT, FRIGHT, FMID, INTRVL
      REAL POLY4F

      READ *, LEFT, RIGHT, EPS

      IF ( LEFT .GE. RIGHT ) THEN
         PRINT *, ' LEFT END MUST BE LESS THAN RIGHT; EXECUTION HALTED '
         STOP
      END IF

      IF ( EPS .LE. 0.0 ) THEN
         PRINT *, ' EPS MUST BE > ZERO; EXECUTION HALTED '
         STOP
      END IF

      FLEFT = POLY4F(LEFT)
      FRIGHT = POLY4F(RIGHT)
      IF ( FLEFT * FRIGHT .GT. 0.0 ) THEN
         PRINT *, ' INTERVAL DOES NOT BRACKET A ROOT; EXECUTION HALTED '
         STOP
      END IF
```

Figure 4.9 A program for finding the root of an equation by the method of interval-halving, corresponding to the pseudocode of Figure 4.8. The program contains a main program, a SUBROUTINE subprogram, and a FUNCTION.

The rest of the program closely follows the pseudocode. Observe that both the main program and the subprogram INTSET invoke POLY4F, which is perfectly legal and commonly done. This brings up a Fortran matter of considerable importance, mentioned briefly earlier in the chapter: Every program that invokes a FUNCTION, either a main program or a subprogram, should contain a type-statement naming the FUNCTION. Vast confusion can result from failure to follow this advice, and the Fortran compiler won't flag it. For example, a

```
* Print input data
      PRINT *, ' IN THE INTERVAL ', LEFT, RIGHT
      PRINT *, ' WITH EPSILON = ', EPS

***** REPEAT until root found or interval < EPS
  10  CONTINUE

        CALL INTSET (LEFT, RIGHT, MID, FMID, INTRVL)
        IF ( (FMID .EQ. 0.0) .OR. (INTRVL .LT. EPS) ) GO TO 20
*       does root lie in left or right half of interval?
        IF ( FLEFT * FMID .LT. 0.0 ) THEN
            RIGHT = MID
            FRIGHT = FMID
        ELSE
            LEFT = MID
            FLEFT = FMID
        END IF

      GO TO 10
***** END REPEAT

  20  CONTINUE

      PRINT *, ' THERE IS A ROOT AT   ', MID

      END         .

* A SUBROUTINE to handle some housekeeping chores:
*    Find midpoint and function value there, compute width of interval
      SUBROUTINE INTSET(LEFT, RIGHT, MID, FMID, INTRVL)

      REAL LEFT, RIGHT, MID, FMID, INTRVL
      REAL POLY4F

      MID = (LEFT + RIGHT) / 2.0
      FMID = POLY4F(MID)
      INTRVL = RIGHT - LEFT

      RETURN

      END

* A FUNCTION subprogram to evaluate a certain fourth-degree polynomial

      REAL FUNCTION POLY4F(X)

      REAL X

      POLY4F = 2.0*X**4 - 15.0*X**3 - 2.0*X**2 + 120.0*X - 130.0

      RETURN

      END
```

Figure 4.9 *(cont'd)*

FUNCTION to find the greatest common divisor of its two arguments might be named, reasonably enough, GCD. Since the arguments and the function value are all integers, the definition could begin:

```
INTEGER FUNCTION GCD (N1, N2)
```

which properly establishes the value of the result returned by the FUNCTION as being of type integer. But if this FUNCTION is invoked in a calling program that does not contain the type-statement:

```
INTEGER GCD
```

the calling program will treat the integer result transmitted to it as a real number, because of the default naming rule. Since real and integer quantities are stored inside the computer in entirely different forms, the result will be a plain mess. And a difficult mess to track down, too, because there will be no diagnostic messages from the compiler and, even if the execution of the object program gives an error message, it is likely to be misleading.

When this combination of main program and two subprograms was compiled and run, it produced this output:

```
18.55.41 >run c4root
EXECUTION:
>1 2 0.0001
 IN THE INTERVAL    1.00000000        2.00000000
 WITH EPSILON =  0.100000004E-03
 THERE IS A ROOT AT       1.38760376
```

Conclusion

Subprograms will be used in most programs in the rest of the book, because of one or more of the following reasons:

1. Someone else has already written a subprogram that we can borrow, saving time and effort, and gaining the programming and application area expertise that the subprogram contains.

2. When the same function is needed at several points in a program, a subprogram can be written once and then invoked whenever it is needed.

3. Subprograms provide a tool for organizing our work in designing a clear and correct program. Such a program is easier for others to use when that is appropriate, and easier for the original programmer or anyone else to modify during program maintenance. (More on this point in the following chapter.)

Exercises

*1. State what the output of the following program would be, assuming it is possible to know.

```
REAL X, G

CALL DOUBLE (1.0, 2.0, X)
PRINT *, X
CALL DOUBLE (X, X, G)
PRINT *, G
CALL DOUBLE (G, G, G)
PRINT *, G
X = 2.0
CALL DOUBLE (X**2, X-2.0, X)
```

```
PRINT *, X
CALL DOUBLE (X, X, 2.0)
PRINT *, X
CALL DOUBLE (2.0, 3.0, X)
PRINT *, X

END

SUBROUTINE DOUBLE (A, B, C)
REAL A, B, C
C = 2.0*(A + B)
RETURN
END
```

2. State what the output of the following program would be.

```
INTEGER J, K

CALL HALVE (5.0, J)
PRINT *, J
CALL HALVE (J+2.0, J)
PRINT *, J
J = 3
CALL HALVE (J**2+1.0, K)
PRINT *, J, K
CALL HALVE (K-1.0, J)
PRINT *, J, K
J = 4
CALL HALVE (J**2-1.0, J**2-1.0)
PRINT *, J

END

SUBROUTINE HALVE (H, I)
REAL H
INTEGER I
I = H/2
RETURN
END
```

*3. State what the output of the following program would be.

```
REAL A, B, C

READ *, A, B, C

CALL TWOF (A, B, C)

END

SUBROUTINE TWOF (A, B, C)
REAL A, B, C
PRINT *, A, B, C, SUM(A, B, C)
PRINT *, A, B, C, PROD(A, B, C)
RETURN
END

REAL FUNCTION SUM(A, B, C)
REAL A, B, C
SUM = A + B + C
RETURN
END
```

```
REAL FUNCTION PROD(A, B, C)
REAL A, B, C
PROD = A * B * C
RETURN
END
```

*4. State what the output of the following program would be, to the extent that it is possible to know.

```
REAL A, B, C, TOTAL, SUM

READ *, A, B, C
CALL ADDEM (A, B, C, TOTAL)
PRINT *, A, B, C, TOTAL

END

SUBROUTINE ADDEM (A, B, C, TOTAL)
REAL A, B, C, TOTAL, SUM
SUM = A + B + C
RETURN
END
```

5. State what the output of the following program would be.

```
REAL A, B, C, TOTAL
A = 5.0      B = 6.0,   C = 7.0
READ *, A, B, C
TOTAL = 6.0
CALL ADDEM (A, B, C, TOTAL)
PRINT *, A, B, C, TOTAL

END

SUBROUTINE ADDEM (A, B, C, TOTAL)
REAL A, B, C, TOTAL
TOTAL = A + B + C
RETURN
END
```

*6. If a main program and a subprogram are compiled together, is it permissible to omit the END for the main program?

*7. Criticize the following program.

```
INTEGER FUNCTION NFACT(N)
INTEGER N
IF ( N .EQ. 1 ) THEN
   NFACT = 1
ELSE
   NFACT = N * NFACT(N-1)
END IF
RETURN
END
```

*8. A subprogram can call another subprogram, except that a Fortran program must not call itself. How about two subprograms that call each other?

*9. How would the program of Figure 4.5 operate if the following changes were made? (The changes are independent of each other.)

    a. The statement numbers on the two CONTINUE statements are reversed.

    b. In the subprogram, X and Y are still declared to be of type real, but TEMP is declared to be of type integer.

c. In the main program, all occurrences of the variables A, B, and C are changed to D, E, and F, respectively; nothing about the subprogram is changed.

d. In the three IF statements that compare A, B, and C, change the relational operator from .GT. to .LT..

e. The first IF statement is replaced with:

```
IF ( A*B*C .EQ. 0.0 ) GO TO 20
```

f. The first IF statement is replaced with:

```
IF ( A .EQ. 0.0 ) GO TO 20
IF ( B .EQ. 0.0 ) GO TO 20
IF ( C .EQ. 0.0 ) GO TO 20
```

10. How would the program of Figure 4.9 operate if the following changes were made? (The changes are independent of each other.)

a. Within the REPEAT loop, these two statements are reversed:

```
RIGHT = MID
FRIGHT = FMID
```

b. The second IF in the REPEAT loop is replaced with:

```
IF ( FLEFT * FMID .GT. 0.0 ) THEN
    LEFT = MID
    FLEFT = FMID
ELSE
    RIGHT = MID
    FRIGHT= FMID
END IF
```

c. The assignment statement that gives a value to INTRVL is removed from the subroutine INTSET.

d. In the REPEAT loop, the CALL INTSET statement is moved to following the END IF.

e. The CALL INTSET statement is removed from its present location and placed both after the END IF statement and before the beginning of the REPEAT loop.

*11. Modify the FUNCTION in Figure 4.9 so that the polynomial is:

$2x^4 - 5.8x^3 - 0.62x^2 + 13.02x - 8.78$

Now run the program with data specifying LEFT = 1.0, RIGHT = 2.0, and EPS = 0.0001. In fact, the polynomial has roots near -1.5, 1.1, 1.4, and 1.9. What happens?

12. Modify the FUNCTION in Figure 4.9 so that the polynomial is:

$2x^4 - 11x^3 + 7x^2 + 26x - 24$

Now run the program with data specifying LEFT = 2.0, RIGHT = 3.0, and EPS = 0.0001. In fact, there is a root at $x = 2$ and the value of the polynomial at $x = 3$ is -18. What happens?

*13. Make a SUBROUTINE of the sorting operation in Figure 4.5. That is, if you write

```
CALL SORT3 (A, B, C)
```

in a main program, the result should be to reorder A, B, and C as necessary to that A ≤ B ≤ C. Your subroutine should call the subroutine SWAP as shown in Figure 4.5.

14. Modify the program of Exercise 13 so that the sorting subroutine has a fourth argument; you might call it UPDOWN. If the value of UPDOWN is 1, the first three arguments should be returned in ascending order as before; otherwise they should be returned in descending order.

15. Modify the program of Exercise 14 so that the sorting subroutine sets UPDOWN to zero on return from the subroutine if all of the first three arguments were zero.

16. Build a program consisting of a main program, the sorting subroutine of Exercise 15, the subroutine SWAP, and the polynomial evaluation subprogram POLY4F. Set up your main program to read values for three variables A, B, and C, print them, then call the sorting subroutine twice to print them in ascending and descending order. Finally, call the sorting subroutine twice more to print the value of the polynomial at these three points in ascending and descending order.

FIVE

PROGRAM DEVELOPMENT AND TESTING

Introduction

A nearly universal complaint at about this stage of a programming course runs, "I understand what you do in class, but when I go to do the homework programs I don't know where to start!" Another goes, "I've compiled this !#%**(!!# program 19 times now, and I think I'm further from getting it to work than I was at first!"

Your instructors understand, having heard the comments in every previous course they have taught. With due allowance for individual differences—some people catch on to the subject quicker than others—it seems that comments like these reflect the majority opinion.

It is not clear that we know how to teach programming so as to eliminate this problem completely. It is not even clear that experienced programmers can explain to others how they go about their work. However, in the past decade or so, approaches have been developed that seem to help many people organize their thinking about program development, so that the job can been done with less effort and frustration.

This chapter, then, is a case study in the development of a program, which demonstrates some of the things you can do to organize the work of program development.

The overall strategy

The general approach can be stated rather briefly, after which we shall apply the ideas to the development of a program to find the roots of an equation.

1. Divide and conquer: break the work of the program into subprograms that interact only in closely controlled ways. To the extent possible, each subprogram should perform one cleanly defined function. Each should, to the extent possible, be able to be tested independently of the others, so that you can concentrate your attention on one thing at a time. This is called *modularization.*

2. Start at the top; work out the logic of the main program first. A main program that consists mostly of invocations of subprograms is perfectly acceptable, and probably even desirable. This approach is usually called *top-down development.*

3. Build on small successes: compile and check out your main program first, using only rudimentary versions of the subprograms. (Examples of what is meant by "rudimentary versions" follow.) Once you are satisfied that the main program logic is correct and that it communicates correctly with the subprograms it calls, start expanding and refining the subprograms, one at a time. Now, as you work on the subprograms, you can have some level of confidence that any problems that develop are in the parts you have most recently modified. Naturally, there may still be problems in the previous versions too, but there will be a tendency for the troubles to be in new sections, helping you to localize mistakes more quickly. At the same time, the constant exercising of the top-level logic improves the chances that you will "boil out the bugs" in that critical portion of the program. This approach is usually called *stepwise refinement.*

These three points constitute the strategy; the following are more in the nature of useful tactics.

4. During development, make free use of extra PRINT statements, so that you can see everything that is happening. Once you are satisfied the program is correct, the extra statements can be removed or changed to comments by inserting an asterisk in the first position of the line.

5. Don't be embarrassed if you find yourself writing the initialization statements for loops or variables last. The instructor who confidently writes a program on the blackboard from start to finish, without ever backing up and saying, "Oops! Forgot to initialize that variable," is misleading you. He or she probably has the program memorized. Very few programmers can think through the logic of a program so thoroughly that they can anticipate all the initialization steps that will be required. And, since failure to get all the initialization steps right is highly likely and very damaging, make a habit of a last-minute check that all the initialization has been done.

6. Take full advantage of everything the compiler can do to help you find errors—but always remember that the compiler basically can find only syntactic errors. The fact that you got a "clean compile," i.e., there were no diagnostic error messages, doesn't prove that the program really does what you want it to. There may be fatal logic errors, or the program may simply not implement the specifications correctly.

7. Once you think your program is right, torture it! Be unfair! Put in bad data to see if it gets rejected. Try cases that fall right on the boundary of the program's supposed capabilities. Think up test cases that will help detect loops that are executed one too many or one too few times. One useful trick here: set up data in a way that the loop should be executed exactly once; the program's behavior here will tell you a lot.

Let's now see these ideas in operation.

The root-finding problem specification

The task that the program to be developed must accomplish is this. We are given an equation, which in the program will be expressed as a FUNCTION named F. We wish to find all the real roots of the equation $F(x) = 0$. The input to the program will consist of four numbers, which in the program will be named LEFT, RIGHT, DELTAX, and EPS, the latter to be thought of as ϵ, the convergence criterion. The program is to attempt to isolate the roots by testing all the intervals of width DELTAX, starting at LEFT and working over to RIGHT. Every time it finds that the function has different signs at the ends of an interval, it is to call a subprogram that will find the root that lies in that interval. The final version of the program will use the Newton-Raphson method to find the root.

A complete root-finder is a very large and complicated program. If you really want to find roots of anything except very simple functions, you should borrow a subroutine! As we go along, we shall observe a few of the many things that can complicate the job. Note at the outset that a sign change proves only that there is an odd number of roots in the interval, not that there is exactly one.

The top-level design

The basic program design is displayed in the pseudocode of Figure 5.1. We see that the main program has a quite simple structure, in which most actions are

```
ROOTER Version 1:
Steps along X axis from LEFT to RIGHT at intervals of DELTAX;
    calls SOLVE when it detects a sign change.
Function for which roots are desired is defined by a FUNCTION F

READ LEFT, RIGHT, DELTAX, EPS
CALL CHECK(LEFT, RIGHT, DELTAX, EPS)

X <-- LEFT

WHILE X <= RIGHT
    IF ( F(X) * F(X + DELTAX) < 0 ) THEN
        CALL SOLVE(X, X + DELTAX, EPS, ROOT)
        PRINT ROOT
    END IF
    X <-- X + DELTAX
END WHILE

STOP

CHECK (LEFT, RIGHT, DELTAX, EPS)
Version 1:
    PRINT LEFT, RIGHT, DELTAX, EPS

SOLVE (X1, X2, EPS, ROOT)
Version 1:
  ROOT <-- (X1 + X2) / 2.0
    PRINT X1, X2, ROOT

F (X)
    Compute function value
```

Figure 5.1 Pseudocode of the first version of the design for a root-finding program.

accomplished by invoking subprograms. After reading the data, it calls a SUB-ROUTINE to validate the data, then has a WHILE loop to step along the axis looking for sign changes. Whenever it finds a sign change, it calls another SUB-ROUTINE that pins down the location of the root.

The subprograms are strictly rudimentary in this first version of the design. About all they do is demonstrate that they were reached, and display the argument values that they received. Subprograms that do this little, and which are used only to test the integration of the subprograms, are sometimes called *program stubs*.

It is important to realize that, according to the overall strategy described at the beginning of the chapter, we actually program and test this version of the design. It's not just that the *design* is developed in a top-down fashion with stepwise refinement, although that is indeed the case, but that the *program* is developed in that manner, too. Large programs are inherently complex. The strategy of breaking them into pieces that are built in a stepwise fashion isn't much help unless we have some confidence that each version of the program is working correctly before going on to add refinements.

Figure 5.2 is the program corresponding to the first level of the design. No new Fortran concepts are involved, but a few details may be pointed out. The stylistic device of leaving spaces around the arguments in a SUBROUTINE call or definition is, of course, optional. Within the WHILE loop, the test for opposite signs is made by asking if the product of the two functions values is less than or equal to zero. (Naturally, if the product is *equal* to zero, at least one of the values is also zero and we have located a root. But this turns out to be sufficiently unlikely that it isn't worth testing for. Such a root will be found anyway.) Note that every time something is printed, there is first printed an identification of where in the program structure the PRINT was located. You will better appreciate this practice after you have written a few programs with lots of subprograms and found yourself puzzling over where the program was when it died!

As noted above, both CHECK and SOLVE are essentially stubs: they "identify themselves," i.e., announce that they were reached, and display their inputs. SOLVE goes one small step further and returns, as the "root," the midpoint of its two arguments. This tests communication between the subprogram and the main program that called it. Note one new feature: where we want an apostrophe mark printed, we represent it within the character literal of the PRINT by writing two successive apostrophes (single quotes).

The output, when this version of the program was run, is shown in Figure 5.3. Everything seems to be in order.

The second version of the program

Now that we have a skeletal version of the program working, it is time to start refining it, so that it does the actual task that is required of it. With something of this limited size, and in consideration of space restrictions in this book, we shall make all of the modifications in one step.

The new requirement on CHECK is that it return an error code for analysis by the calling program, which is a common approach. An error code of zero will indicate that the data is acceptable; an error code of 8 that the value of DELTAX or EPS is suspiciously small, and an error code of 16 that the values of LEFT, RIGHT, and/or EPS are such that a solution is impossible. (These values

```
* A set of programs to find the roots of an equation
* The equation is F(X) = 0, where F is a Fortran FUNCTION
*
* Version 1: checks main program logic and subprogram communication
*
* The main program reads data, then invokes CHECK to determine validity
* Main program then steps through values of X from LEFT to RIGHT,
*     in steps of DELTAX, looking for function value sign changes;
*     at each sign change it calls SOLVE to localize the root
*
* Variables:
*     X:       Independent variable
*     LEFT:    Left end of interval to be searched for roots
*     RIGHT:   Right end of interval
*     DELTAX:  Step size in root searching
*     EPS:     Convergence criterion for root finder
*
* Subprograms:
*     CHECK:   Checks input for validity
*              Version 1: prints data sent it, for inspection
*     SOLVE:   Finds root between two X values, with tolerance of EPS
*              Version 1: prints data, returns midpoint as root
*     F:       Returns function value; roots wanted for equation F(X) = 0
*
      REAL X, LEFT, RIGHT, DELTAX, EPS
      REAL F

      READ *, LEFT, RIGHT, DELTAX, EPS

      CALL CHECK ( LEFT, RIGHT, DELTAX, EPS )

      X = LEFT

***** WHILE X <= RIGHT
  10  CONTINUE
      IF ( X .LE. RIGHT ) THEN

          IF ( F(X) * F(X + DELTAX) .LE. 0.0 ) THEN
              CALL SOLVE ( X, X + DELTAX, EPS, ROOT )
              PRINT *, ' FROM MAIN PROGRAM: '
              PRINT *, ' VERY APPROX ROOT = ', ROOT
          END IF
          X = X + DELTAX

      GO TO 10
      END IF
***** END WHILE

      END

* SUBROUTINE to check input data for validity
* Version 1: merely prints data for inspection
*
      SUBROUTINE CHECK ( LEFT, RIGHT, DELTAX, EPS )

      REAL LEFT, RIGHT, DELTAX, EPS

      PRINT *, ' FROM VERSION 1 OF CHECK: '
      PRINT *, ' LEFT = ', LEFT, ' RIGHT = ', RIGHT
      PRINT *, ' DELTAX = ', DELTAX, ' EPS = ', EPS

      RETURN

      END
```

Figure 5.2 First version of a root-finding program, corresponding to the design of Figure 5.1.

```
* SUBROUTINE to find root lying between two X values, within EPS
* Version 1: prints its input arguments, returns midpoint of input
*     X values as the approximate root
*
      SUBROUTINE SOLVE ( X1, X2, EPS, ROOT )

      REAL X1, X2, EPS, ROOT

      PRINT *, ' FROM VERSION 1 OF SOLVE: '
      PRINT *, ' ROOT IS BRACKETED BY ', X1, X2
      ROOT = (X1 + X2) / 2.0
      PRINT *, ' ''ROOT'' (MIDPOINT OF ABOVE) = ', ROOT

      RETURN

      END

* FUNCTION to evaluate a function F, where roots are sought for
*     the equation F(X) = 0
*
      REAL FUNCTION F ( X )

      REAL X

      F =   2.0*X**4 - 15.0*X**3 - 2.0*X**2 + 120.0*X - 130.0

      RETURN

      END
```

Figure 5.2 *(cont'd)* First version of a root-finding program, corresponding to the design of Figure 5.1.

```
10.56.10 >run c5main1
EXECUTION:
>-10 10 1 0.001
 FROM VERSION 1 OF CHECK:
 LEFT =  -10.0000000        RIGHT =    10.0000000
 DELTAX =    1.00000000       EPS =   0.999999931E-03
 FROM VERSION 1 OF SOLVE:
 ROOT IS BRACKETED BY  -3.00000000       -2.00000000
 'ROOT' (MIDPOINT OF ABOVE) =  -2.50000000
 FROM MAIN PROGRAM:
 VERY APPROX ROOT =  -2.50000000
 FROM VERSION 1 OF SOLVE:
 ROOT IS BRACKETED BY   1.00000000        2.00000000
 'ROOT' (MIDPOINT OF ABOVE) =   1.50000000
 FROM MAIN PROGRAM:
 VERY APPROX ROOT =   1.50000000
 FROM VERSION 1 OF SOLVE:
 ROOT IS BRACKETED BY   2.00000000        3.00000000
 'ROOT' (MIDPOINT OF ABOVE) =   2.50000000
 FROM MAIN PROGRAM:
 VERY APPROX ROOT =   2.50000000
 FROM VERSION 1 OF SOLVE:
 ROOT IS BRACKETED BY   6.00000000        7.00000000
 'ROOT' (MIDPOINT OF ABOVE) =   6.50000000
 FROM MAIN PROGRAM:
 VERY APPROX ROOT =   6.50000000
```

Figure 5.3 The output of the program of Figure 5.2.

were chosen arbitrarily. However, it is common practice to set up error codes so that the more serious the error, the larger the error code.)

A new requirement on **SOLVE**, besides the fact that it should find the root using the Newton-Raphson method that we studied in Chapter 3, is that it should also return an error code: zero meaning that a root was found; 32 that convergence was not reached in ten iterations. (If convergence is not possible, the **SUBROUTINE** returns an arbitrary large value for the root.) And, of course, in order to use the Newton-Raphson method, we need another **FUNCTION** to compute the derivative of the function that defines the equation.

The only new Fortran feature is the **ELSE IF** statement, which is not difficult. We shall accordingly take this occasion to demonstrate some aspects of "checking out" a program, that is, removing all the mistakes we can find and trying to gain some confidence that the program correctly does what we want it to do. The program shown in Figure 5.4 accordingly contains a fair number of deliberate errors so that we can see how to ferret them out.

The **ELSE IF** structure is used in the subroutine **SOLVE**, to give a value to the error return code. The task here is to give the variable named **IERROR** one of the values 16, 8, or zero. We want to test for the most serious error first; if this condition is discovered, we want to give **IERROR** the value 16 and not do any further testing. But if this error is not present we need to continue testing for the less serious error. If it is found, we want to give **IERROR** the value 8 and get out of the logic structure. Only if neither error is present do we wish to give **IERROR** the value zero.

A flowchart of this control structure appears in Figure 5.5. The key thing is that if the answer to the first test is "true," the corresponding actions are taken *and the rest of the structure is ignored.* If the program were written with a simple **IF** where the **ELSE IF** appears, the syntax would be legal but the meaning different. In that case the second test would be made even when the answer to the first is "true." Thus, if both errors were present, **IERROR** would first be set to 16 and then immediately set to 8. The net effect would be to report the less serious error, which is not what is desired.

A "chain" can be built up with as many **ELSE IF** combinations as we wish. If there are n **ELSE IF**s and a final **ELSE**, exactly one out of $n+1$ actions will be taken. (Actually, the **ELSE IF** can be combined with the "ordinary" **IF** in exceedingly complex ways. Since we strive strenuously to avoid complexity in programs, and since such structures are very seldom needed, we shall not explore the subject further.)

Before proceeding, why not try to see how many errors you can find? Some will probably be fairly obvious, but it would be a pretty safe bet that you won't find them all! This is part of the problem: a mistake can in some sense "look right," and be difficult to see.

When this program was compiled it produced six error messages. Since compilers differ greatly in how they identify errors and in the format of error messages, the detailed printout will not be reproduced here. Yours would probably be different.

There were four error messages for the main program, three of them identifying the second **IF** statement as the culprit and the fourth referring to the program as a whole. The first message complained that the statement after a **STOP** statement should have a statement number, since such a statement can

```
*  A set of programs to find the roots of an equation
*  The equation is F(X) = 0, where F is a Fortran FUNCTION
*
*  Version 2: extend CHECK to do full job, use Newton-Raphson in SOLVE
*
*****************************************************************************
*  THIS VERSION CONTAINS MANY DELIBERATE ERRORS, FOR CHECKOUT PRACTICE
*****************************************************************************
*
*  The main program reads data, then invokes CHECK to determine validity
*  Main program then steps through values of X from LEFT to RIGHT,
*     in steps of DELTAX, looking for function value sign changes;
*     at each sign change it calls SOLVE to localize the root
*
*  Variables:
*     X:      Independent variable
*     LEFT:   Left end of interval to be searched for roots
*     RIGHT:  Right end of interval
*     DELTAX: Step size in root searching
*     EPS:    Convergence criterion for root finder
*     IERROR: Error code returned by CHECK and SOLVE
*
*  Subprograms:
*     CHECK:  Checks input for validity
*             Version 2: Validity checks as described with SUBROUTINE
*                Returns an error code
*     SOLVE:  Finds root between two X values, with tolerance of EPS
*             Version 2: Uses Newton-Raphson method; returns error code
*                if no convergence
*     F:      Returns function value; roots wanted for equation F(X) = 0
*     FPRIME: Returns value of function derivative, for Newton-Raphson
*

      REAL X, LEFT, RIGHT, DELTAX, EPS
      INTEGER IERROR
      REAL F

      READ *, LEFT, RIGHT, DELTAX, EPS
      PRINT *, ' LEFT = ', LEFT, ' RIGHT = ', RIGHT
      PRINT *, ' DELTAX = ', DELTAX, ' EPS = ', EPS

      CALL CHECK ( LEFT, RIGHT, DELTAX, EPS, IERROR )

      IF ( IERROR .EQ. 16 ) THEN
          PRINT *, ' IMPOSSIBLE VALUES OF LEFT, RIGHT, AND/OR DELTAX '
          PRINT *, ' PROGRAM EXECUTION HALTED '
          STOP

      IF ( IERROR = 8 ) THEN
          PRINT *, ' POSSIBLE PROBLEM: EPS AND/OR DELTAX VERY SMALL '
          PRINT *, ' EXECUTION CONTINUING '
      END IF

      X = LEFT

***** WHILE X <= RIGHT
   10 CONTINUE
      IF ( X .LE. RIGHT ) THEN

          IF ( F(X) * F(X + DELTAX) .LE. 0.0 ) THEN
              CALL SOLVE ( X, X + DELTAX, EPS, ROOT, IERROR )
              IF ( IERROR .GT. 0 ) THEN
                  PRINT *, ' NO CONVERGENCE FOR POSSIBLE ROOT NEAR ', X
              ELSE
                  PRINT *, ' THERE IS A ROOT AT ', ROOT
              END IF
          END IF
          X = X + DELTAX

      GO TO 10
      END IF
***** END WHILE

      END
```

114

```
*  SUBROUTINE to check input data for validity
*  Version 2: returns error code as follows:
*      IERROR = 0: no errors
*             = 16: DELTAX <= 0 or LEFT >= RIGHT
*             = 3:  EPS or DELTAX small; execution may be slow
*                   or convergence impossible
*
*      Action to be taken is left to calling routine
*
       SUBROUTINE CHECK ( LEFT, RIGHT, DELTAX, EPS, IERROR )

       REAL LEFT, RIGHT, DELTAX, EPS
       INTEGER IERROR

       IF ( LEFT .GE. RIGHT .OR. DELTAX .LE. 0.0 ) THEN
          IERROR = 16
       ELSE IF ( DELTAX .LT. 1E-3 .OR. EPS .LT. 1E-6 ) THEN
          IERROR = 3
       ELSE
          IERROR = 0
       END IF

       RETURN

       END

*  SUBROUTINE to find root lying between two X values, within EPS
*  Version 2: Uses Newton-Raphson method, which requires a new
*      FUNCTION, FPRIME, to compute derivative
*  Midpoint of X1 and X2 is used as starting approximation
*
*  Returns an error code:
*      IERROR = 0: no errors
*      IERROR = 32: did not converge in 10 iterations;
*                   ROOT has been set to -9E+35
*
*  Variables local to SOLVE:
*      X:     Old approximation to root
*      XNEW:  New approximation to root
*      ITER:  Iteration counter; limit 10
*
*  Subprograms invoked:
*      F:     Value of function
*      FPRIME: Value of derivative of function
*
       SUBROUTINE SOLVE ( X1, X2, EPS, ROOT, IERROR )

       REAL X1, X2, EPS, ROOT
       INTEGER IERROR
       REAL X, XNEW
       INTEGER ITER
       REAL F, FPRIME

*  Initialize
       X = (X1 + X2) / 2.0
       ITER = 1

***** REPEAT until converged or too many iterations
   10  CONTINUE

          XNEW = X - FPRIME(X) / F(X)
          IF (      (ABS(X - XNEW) .LT. 1E-6)
     $           .OR. (ITER .GT. 10)               GO TO 20
          X = XNEW
          ITER = ITER + 1

       GO TO 10
***** END REPEAT

   20  CONTINUE

       IF ( ITER .GT. 10 ) THEN
          IERROR = 32
          ROOT = -9E+35
       ELSE
          IERROR = 0
          ROOT = XNEW
       END IF

       RETURN

       END
```

```
* FUNCTION to evaluate a function F, where roots are sought for
*     the equation F(X) = 0
*
      REAL FUNCTION F ( X )

      REAL X

      F =   2.0*X**4 - 15.0*X**3 - 2.0*X**2 + 120.0*X - 130.0

      RETURN

      END

* FUNCTION to evaluate the derivative of function F
*
      REAL FUNCTION FPRIME ( X )

      REAL X

      FPRIME = 8.0*X**3 - 45.0*X**2 - 4.0X + 120.0

      RETURN

      END
```

Figure 5.4 (cont'd) An extended version of the program of Figure 5.2, intended to do the complete root-finding task as described in the text. THIS PROGRAM CONTAINS MANY DELIBERATE ERRORS, since it is used to illustrate program checkout.

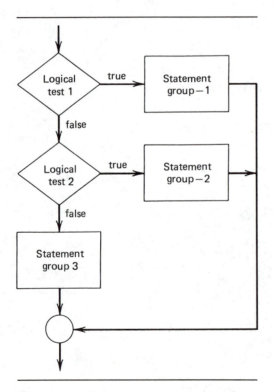

Figure 5.5 Flowchart representation of the ELSE IF structure used in the SOLVE subroutine.

never be reached unless it does have a statement number. (This is true of the GO TO and several other statements as well.) True, but not the cause of the trouble! Compilers often cannot correctly diagnose what the actual error was; they can only diagnose symptoms. Such is the case here. The second error message objected to the misplaced equal sign in the IF statement, which is indeed an error: an equal sign was written where .EQ. was intended. This diagnosis is precise. The third message said simply "error encountered in processing IF expression," which is a little hard to interpret. One assumes that correcting the equal sign matter will eliminate this one. The final message, not associated with any statement, said, "the IF blocks are incorrectly nested." And of course, the trouble is the absence of an END IF after the STOP.

There were no error messages for CHECK, but for the IF statement in SOLVE the report was "text not recognizable as a Fortran statement." This could mean a lot of things. Misplacement of the dollar sign for the continuation of the IF statement could easily do it, but in fact the placement is correct. If you don't see the problem, try counting parentheses to determine if there are the same number of left parentheses as right, which—neglecting character literals—must be true of every Fortran statement. And indeed the problem is a missing right parenthesis before the GO TO 20. The unbalanced parentheses are probably what the compiler caught; some compilers would have reported that fact with a somewhat more helpful error message.

Finally, in the new FUNCTION subprogram FPRIME, the report was "two operands follow one another with no intervening operator." That's as specific as we could wish, and points to the missing asterisk in 4.0X.

A clean compile doesn't mean there aren't any errors!

With these errors in syntax corrected, the program was compiled again, producing no error messages. Success, right? Well, maybe; perhaps we should see if the roots look right. Here is the output:

```
12.23.20 >run c5main3
EXECUTION:
>-10 10 1 0.001
 LEFT =  -10.0000000      RIGHT =    10.0000000
 DELTAX =    1.00000000      EPS =  0.999999931E-03
 NO CONVERGENCE FOR POSSIBLE ROOT NEAR  -3.00000000
 NO CONVERGENCE FOR POSSIBLE ROOT NEAR   1.00000000
 NO CONVERGENCE FOR POSSIBLE ROOT NEAR   2.00000000
 NO CONVERGENCE FOR POSSIBLE ROOT NEAR   6.00000000
```

What do we know at this point? The logic in the main program is evidently stepping along the x axis properly, but we knew that already. CHECK seems not to have located any problems, although we haven't really tested it yet. Presumably the problem is in SOLVE, which is reporting that it can't converge to any of the roots in ten iterations.

Perhaps you have long since noted the error that is causing this problem, but if not, what can you do? Basically, when totally baffled, you have to insert more PRINT statements so that you can watch the details of the computation proceeding. Alternatively, some systems permit you to *trace* program execution, which means that you get an identification of every statement as it is executed, along with any results produced.

In this case, with the finger of suspicion pointing directly at SOLVE, it makes sense to insert a PRINT after the computation of XNEW, in hopes that the sequence of values will tell us something.

We shall not reproduce the page of output when the modified program was run, but basically the results were nonsense. For the root between 2 and 3, for example, the approximations were all in the range of 38 to 40. At this point there is not really much else the computer can do to help find the error. The syntax of the program is legal Fortran, so the compiler can't help. The computer is basically doing what it was told to do, but somehow that just isn't what we really meant. If you still can't see the problem, try explaining to someone else what the program is supposed to do. You will often "see" something that you had overlooked entirely before, even before your colleague has begun to understand the program. Or your colleague will quickly spot a glaring error that had simply been invisible to you.

This "two heads are better than one" tactic has proved so useful that many organizations encourage or even require that it be done routinely. There is even a name for it: a *walk-through*. The name is meant to suggest that the program designer "walks through" the design and/or the program with others who are competent in the area but are not familiar with the particular program. The process can be as simple as two friends agreeing to read over each other's programs, or as formal as the organization wishes to make it.

Eventually, somehow, you will see that the fraction in the Newton-Raphson formula for computing XNEW is inverted: the derivative should be the denominator.

With this error corrected, the program was recompiled and, when run, produced the output in Figure 5.6. Everything is looking good, except . . . good grief! What is the meaning of the nonconvergence message for the last root? The process *did* find the root, to much better accuracy than the EPS value of 0.001.

Had you found this one? The trouble is that the Newton-Raphson loop was copied over from Chapter 3, and the test is against $1E-6$ instead of against EPS. The process did indeed converge to within 0.001, but convergence to within 10^{-6} is impossible. The approximations are alternating between two values that differ by one bit in the last place of the representation, and the difference between the two values is greater than 10^{-6}.

The "final," "correct" program

With this error corrected, we arrive at the program in Figure 5.7.

When this was run, it produced this output:

```
12.56.20 >run c5main6
EXECUTION:
>-10 10 1 0.0001
 LEFT =   -10.0000000       RIGHT =    10.0000000
 DELTAX =    1.00000000        EPS =  0.100000004E-03
 THERE IS A ROOT AT  -2.87135887
 THERE IS A ROOT AT   1.38760376
 THERE IS A ROOT AT   2.52642822
 THERE IS A ROOT AT   6.45731544
```

```
12.46.22 >run c5main5
EXECUTION:
>-10 10 1 0.001
 LEFT =  -10.0000000       RIGHT =   10.0000000
 DELTAX =   1.00000000       EPS =  0.999999931E-03
 -2.97058773
 -2.87620449
 -2.87137126
 -2.87135887
 -2.87135887
 THERE IS A ROOT AT  -2.87135887
  1.37421321
  1.38745689
  1.38760376
  1.38760376
 THERE IS A ROOT AT    1.38760376
  2.52702617
  2.52643203
  2.52642822
  2.52643394
  2.52643108
  2.52643013
 THERE IS A ROOT AT    2.52643013
  6.45830631
  6.45731830
  6.45731544
  6.45731735
  6.45731544
  6.45731735
  6.45731544
  6.45731735
  6.45731544
  6.45731735
  6.45731544
 NO CONVERGENCE FOR POSSIBLE ROOT NEAR   6.00000000
```

Figure 5.6 The output of a modified version of the program of Figure 5.2. The listing for the final root shows that there are still problems.

There are no problems here, so we must have a correct program, right? Well, maybe—let's work with it a bit.

The same program with a different equation

Now that we have a working root-finder, we ought to be able to find roots of any equation simply by changing the FUNCTION subprograms F and FPRIME. Let's do that, taking a function for which we are very sure of the roots, namely $\sin x = 0$, which clearly has roots for all multiples of π. Doing this requires only that the two assignment statements in those functions be changed to:

```
F = SIN(X)
```

and:

```
FPRIME = COS(X)
```

respectively. (From calculus we know that the derivative of the sine is the cosine.)

```
*  A set of programs to find the roots of an equation
*  The equation is F(X) = 0, where F is a Fortran FUNCTION
*
*  Final version: all known errors corrected
*  Results appear to be correct
*
*  The main program reads data, then invokes CHECK to determine validity
*  Main program then steps through values of X from LEFT to RIGHT,
*     in steps of DELTAX, looking for function value sign changes;
*     at each sign change it calls SOLVE to localize the root
*
*  Variables:
*     X:      Independent variable
*     LEFT:   Left end of interval to be searched for roots
*     RIGHT:  Right end of interval
*     DELTAX: Step size in root searching
*     EPS:    Convergence criterion for root finder
*     IERROR: Error code returned by CHECK and SOLVE
*
*  Subprograms:
*     CHECK:  Checks input for validity
*             Version 2: Validity checks as described with SUBROUTINE
*                 Returns an error code
*     SOLVE:  Finds root between two X values, with tolerance of EPS
*             Version 2: Uses Newton-Raphson method; returns error code
*                 if no convergence
*     F:      Returns function value; roots wanted for equation F(X) = 0
*     FPRIME: Returns value of function derivative, for Newton-Raphson
*

      REAL X, LEFT, RIGHT, DELTAX, EPS
      INTEGER IERROR
      REAL F

      READ *, LEFT, RIGHT, DELTAX, EPS
      PRINT *, ' LEFT = ', LEFT, ' RIGHT = ', RIGHT
      PRINT *, ' DELTAX = ', DELTAX, ' EPS = ', EPS

      CALL CHECK ( LEFT, RIGHT, DELTAX, EPS, IERROR )

      IF ( IERROR .EQ. 16 ) THEN
          PRINT *, ' IMPOSSIBLE VALUES OF LEFT, RIGHT, AND/OR DELTAX '
          PRINT *, ' PROGRAM EXECUTION HALTED '
          STOP
      END IF

      IF ( IERROR .EQ. 8 ) THEN
          PRINT *, ' POSSIBLE PROBLEM: EPS AND/OR DELTAX VERY SMALL '
          PRINT *, ' EXECUTION CONTINUING '
      END IF

      X = LEFT

***** DO WHILE X <= RIGHT
   10 CONTINUE
      IF ( X .GT. RIGHT ) GO TO 20

          IF ( F(X) * F(X + DELTAX) .LE. 0.0 ) THEN
              CALL SOLVE ( X, X + DELTAX, EPS, ROOT, IERROR )
              IF ( IERROR .GT. 0 ) THEN
                  PRINT *, ' NO CONVERGENCE FOR POSSIBLE ROOT NEAR ', X
              ELSE
                  PRINT *, ' THERE IS A ROOT AT ', ROOT
              END IF
          END IF
          X = X + DELTAX

      GO TO 10
***** END WHILE

   20 CONTINUE

      STOP

      END
```

```
*  SUBROUTINE to check input data for validity
*  Version 2: returns error code as follows:
*       IERROR = 0: no errors
*              = 16: DELTAX <= 0 or LEFT >= RIGHT
*              = 8:  EPS or DELTAX small; execution may be slow
*                    or convergence impossible
*
*       Action to be taken is left to calling routine
*
        SUBROUTINE CHECK ( LEFT, RIGHT, DELTAX, EPS, IERROR )

        REAL LEFT, RIGHT, DELTAX, EPS
        INTEGER IERROR

        IF ( LEFT .GE. RIGHT .OR. DELTAX .LE. 0.0 ) THEN
           IERROR = 16
        ELSE IF ( DELTAX .LT. 1E-3 .OR. EPS .LT. 1E-6 ) THEN
           IERROR = 8
        ELSE
           IERROR = 0
        END IF

        RETURN

        END

*  SUBROUTINE to find root lying between two X values, within EPS
*  Version 2: Uses Newton-Raphson method, which requires a new
*     FUNCTION, PRIME, to compute derivative
*  Midpoint of X1 and X2 is used as starting approximation
*
*  Returns an error code:
*       IERROR = 0: no errors
*       IERROR = 32: did not converge in 10 iterations;
*                    ROOT has been set to -9E+35
*
*  Variables local to SOLVE:
*       X:      Old approximation to root
*       XNEW:   New approximation to root
*       ITER:   Iteration counter; limit 10
*
*  Subprograms invoked:
*       F:      Value of function
*       FPRIME: Value of derivative of function
*
        SUBROUTINE SOLVE ( X1, X2, EPS, ROOT, IERROR )

        REAL X1, X2, EPS, ROOT
        INTEGER IERROR
        REAL X, XNEW
        INTEGER ITER
        REAL F, FPRIME

*  Initialize
        X = (X1 + X2) / 2.0
        ITER = 1

***** REPEAT until converged or too many iterations
   10   CONTINUE

           XNEW = X - F(X) / FPRIME(X)
           IF (     (ABS(X - XNEW) .LT. EPS)
      $         .OR. (ITER .GT. 10)                 ) GO TO 20
           X = XNEW
           ITER = ITER + 1

        GO TO 10
***** END REPEAT

   20   CONTINUE

        IF ( ITER .GT. 10 ) THEN
           IERROR = 32
           ROOT = -9E+35
        ELSE
           IERROR = 0
           ROOT = XNEW
        END IF

        RETURN

        END
```

121

```
* FUNCTION to evaluate a function F, where roots are sought for
*     the equation F(X) = 0
*
      REAL FUNCTION F ( X )

      REAL X

      F =  2.0*X**4 - 15.0*X**3 - 2.0*X**2 + 120.0*X - 130.0

      RETURN

      END

* FUNCTION to evaluate the derivative of function F
*
      REAL FUNCTION FPRIME ( X )

      REAL X

      FPRIME = 8.0*X**3 - 45.0*X**2 - 4.0*X + 120.0

      RETURN

      END
```

Figure 5.7 A final version of the root-finding program. The program contains no known errors, but treats one case of anomalous data questionably.

The program can then be recompiled and rerun. Here is one set of output when this was done:

```
13.01.00 >run c5main7
EXECUTION:
>0 10 1 0.0001
 LEFT =  0.000000000E+00  RIGHT =    10.0000000
 DELTAX =   1.00000000       EPS =  0.100000004E-03
 THERE IS A ROOT AT   0.145519152E-10
 THERE IS A ROOT AT   3.14159202
 THERE IS A ROOT AT   6.28318500
 THERE IS A ROOT AT   9.42477703
```

No problems here. It might be nice if the program had found exactly zero for the first root, but that's life with computers. It did what we told it to do.

Let's try it again, this time with values that bracket the roots at π, 2π, and 3π, and see what happens:

```
13.02.17 >run c5main7
EXECUTION:
>0 10 3 0.0001
 LEFT =  0.000000000E+00  RIGHT =    10.0000000
 DELTAX =   3.00000000       EPS =  0.100000004E-03
 THERE IS A ROOT AT  -12.5663700
 THERE IS A ROOT AT   0.000000000E+00
 THERE IS A ROOT AT   12.5663700
 THERE IS A ROOT AT   9.42477703
```

Good grief! Now what? These answers are, in fact, all multiples of π, but how did -4π and 4π get in there?

To answer this, we need a little work with a hand calculator. The program discovered that the sine function had different signs at zero and at 3, so it sent those numbers to SOLVE, which took their midpoint at the starting approximation for the Newton-Raphson process. The sine of 1.5 radians is 0.9997495 and the cosine is 0.0707372. Their quotient is 14.10142 which, subtracted from the starting guess of 1.5, gives -12.60142. It just happens that there is a root of sin x = 0 very near that point, and the Newton-Raphson process proceeded to find it. In other words, the program did exactly what we told it to do!

Is this a mistake or not? The answer depends on one's point of view. Most people would probably feel that, although the program did what we *told* it to do, it didn't do what we *wanted* it to do. But a computer can rarely, if ever, diagnose that kind of problem, unless our instructions involve a syntactic error or unless we have given the computer a great deal more information about our intentions.

There is always one more bug!

Let's try one more case. Let's take a function that definitely has two roots and definitely also has a singularity (a point where it becomes infinite). We will carefully avoid the singularity in providing data, however, so that the program should have no problems. The function is:

$$F(x) = \frac{(x - 2)(x - 3)}{x - 4}$$

This clearly has roots at 2 and 3 and a singularity at 4. We rewrite the assignment statement in F to be:

```
F = (X - 2.0) * (X - 3.0) / (X - 4.0)
```

and the assignment statement in FPRIME to be:

```
FPRIME = (X**2 - 8.0*X + 14.0)) / (X**2 - 8.0*X + 16.0)
```

Upon recompiling and rerunning, we get the news shown in Figure 5.8. What on earth! Is it time to quit for the night? The program has found the root at 2 twice, then bombed looking in a region where it wasn't even supposed to be!

You might find it instructive to trace the crime here yourself, but this is the sequence. Since the function value is zero at 2, SOLVE was invoked for the interval from 1 to 2. The main program logic asked whether the product of the two function values was less than *or equal to* zero, remember? With a starting guess of 1.5, the process found the root at 2, within the tolerance given it. The main program logic next tried the interval from 2 to 3, and discovered that the product of the function values there was also less than or equal to zero. With a starting guess of 2.5, the Newton-Raphson process again found the root at 2 which, in a certain sense, is clearly not wrong: there certainly is a root at 2, and the program was given no instructions not to find the same root twice.

But what is the rest of this gibberish? Well, it turns out that the key phrase is "FLOATING-POINT DIVIDE EXCEPTION." We can ignore the rest, which is useful for programmers who know how this machine operates and want to dig into the internals of object program. What matters is that "divide exception" is the terminology used in this part of the computing world for what happens when you attempt to divide by zero. The problem is that the function was evaluated for x = 4, which does indeed involve a division by zero.

```
13.18.03 >run c5main8
EXECUTION:
>0 3 1 0.0001
 LEFT =  0.000000000E+00   RIGHT =    3.00000000
 DELTAX =    1.00000000       EPS =  0.100000004E-03
   1.94117641
   1.99845409
   1.99999809
   1.99999904
 THERE IS A ROOT AT    1.99999904
   1.00000095
   1.85714244
   1.99212551
   1.99996852
   1.99999904
 THERE IS A ROOT AT    1.99999904
IFY240I VSTAE - ABEND CODE IS: SYSTEM 00CF, USER 0000. SCB/SDWA=00025648.
IFY240I VSTAE - IO HALTED.    PSW=FFF5000F820129EC. ENTRY POINT=00012000.
IFY240I VSTAE - REGS   0-3   00000000 000121E4 000124DE 00012030
IFY240I VSTAE - REGS   4-7   00012030 000124C0 00012486 80012214
IFY240I VSTAE - REGS   8-11  00000000 00005330 000053A0 00D47C68
IFY240I VSTAE - REGS  12-15  00012A0E 00012948 620123FE 000129B8
IFY240I VSTAE - FRGS   0-3   41200000 00000000 00000000 00000000
IFY240I VSTAE - FRGS   4-7   00000000 00000000 D9E4D540 40404040
FLOATING-POINT DIVIDE EXCEPTION AT 0129E8
```

Figure 5.8 The output of the program of Figure 5.7, modified to seek the roots of a function that has known roots at $x = 2$ and $x = 3$, and a singularity at $x = 4$. See the text discussion of error messages.

But why was the point that caused the divide exception ever evaluated? The main program logic, in the WHILE loop, is very clear that processing should stop when X exceeds RIGHT. And in fact it does. The problem, if you haven't already spotted it, is that within the WHILE loop the test for the existence of a root takes us a distance DELTAX *past* the final value of X.

It is not entirely clear what a proper correction might be; changing the test in the WHILE to read:

```
WHILE X + DELTAX <= RIGHT
```

could, depending on the values of LEFT, RIGHT, and DELTAX, result in not checking the entire interval from LEFT to RIGHT. Let us not try to resolve this one. Writing a complete root-finding program is a matter of several person-years of effort, at least. The examples here have barely scratched the surface of the problems that can arise. If you really want to find the roots of equations, you should find a good subroutine! See Chapter 9 for more on this.

Just to prove that the program can find the roots without incident if we are lucky, let's rerun it with a different value of DELTAX, one that doesn't lead to the divide exception problem. Here is the output:

```
13.38.33 >run c5main8
EXECUTION:
>0 3 0.1 0.0001
 LEFT =  0.000000000      RIGHT =    3.00000000
 DELTAX =  0.100000023      EPS =  0.100000004E-03
   1.99861335
   1.99999809
   1.99999904
```

```
THERE IS A ROOT AT    1.99999904
   3.00455284
   3.00004100
   3.00000000
THERE IS A ROOT AT    3.00000000
```

Conclusion

Program development is an imperfectly understood art, which some people do much better than others, often without being able to explain why. The strategy and tactics listed at the beginning of the chapter are worth reviewing. Although there is no magic way to write correct programs easily, you may as well start with the experience of those who have struggled with the problem before you, as summarized above.

Much research is in progress. Compiler writers provide us with better diagnostic services. People like those who developed WATFIV devise ways to diagnose many errors at execution time, errors that the compiler could not even in principle hope to detect. Language designers invent languages in which errors are less likely to occur in the first place. Developers of software tools provide us with better debugging aids, with higher level languages, and with program development techniques in which syntactic errors are essentially impossible.

But at this point, what you mostly need is practice, so that you can learn the capabilites of whatever system you are using and so that you can polish your skills at finding errors—or, much better yet, writing programs in such a way that they are less likely to have errors in them in the first place.

This last comment is one more plea for writing programs that are easy *(for people!)* to understand. If you aren't sure what your program does, you are at a *decided* disadvantage in trying to find errors in it!

Write understandable programs!

■ Break the program into bite-sized pieces, using subprograms heavily, so that you have a reasonable chance of understanding one thing at a time.

■ Use meaningful data names. Down with names like P, Q, HOTDOG, LOVER9, and Q7DG1M.

■ Keep the logic simple. Isolate complex logic in subprograms; make use of the IF-THEN-ELSE structure whenever it fits, utilize the WHILE structure freely (especially if you are using WATFIV or some other Fortran variation that has it), use the GO TO statement only in extremely cautious ways, and DO NOT EVER use the Fortran arithmetic IF or the assigned GO TO. (If you find out from some other book what those last two statements are, just be aware that undisciplined use of them can lead to programs that are nearly impossible for a human being to understand. Such programs are likely to contain undetected errors, and it is extremely hazardous to try to modify them.)

Exercise

A student was asked to learn something about the secant method for finding roots, and to modify the program of Figure 5.7 to use it instead of the Newton-Raphson method. What follows is a log of her experiences. You are asked to find and correct the errors in her program, based on the hints that follow and on your own investigations. Your final result is to be a complete, correct program. Wherever you can, you should try to improve on the program shown below, even where it caused no problems as tested.

(If you do run the program to test it, your instructor may be able to supply you with a copy through the computer. The purpose of the exercise is not to test your keyboard skills.)

Our student looked up the secant method in an elementary numerical methods text, and discovered that it works as follows. Given a function $f(x)$ and two points x_{-1} and x_0, we repeatedly apply the iteration formula

$$x_{n+1} = x_n - f(x_n) \frac{x_n - x_{n-1}}{f(x_n) - f(x_{n-1})}$$

The iteration process is stopped when a suitable convergence criterion has been satisfied. The geometrical interpretation is as in this sketch:

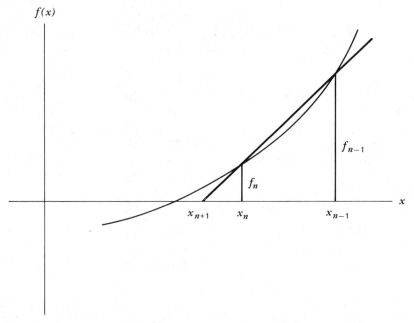

The new approximation is the intersection with the x axis of the secant passing through the points (x_x, f_n) and (x_{n-1}, f_{n-1}). If x_n and x_{n-1} are sufficiently close together, the secant approximates the tangent at x_n. The Newton-Raphson method, by contrast, depends on finding the slope of the tangent by evaluating the derivative at x_n. The secant method, therefore, is useful with functions for which the derivative is difficult to evaluate or perhaps does not even exist. The secant method converges more slowly than the Newton-Raphson method, but more rapidly than the method of bisection (interval-halving).

Our student decided that the modification should not be too difficult, requiring only a change in the iteration formula and maintaining the two previous values instead of just one as in the Newton-Raphson method. (All the changes are in the subroutine SOLVE.) She decided to represent x_{n+1} by XNEW and x_n by X, requiring no program changes, and to represent x_{n-1} by XOLD.

The first version of her program, which we shall charitably assume was written under severe time pressure, is shown in Figure 5.9.

```
*  SUBROUTINE to find root lying between two X values, within EPS
*  Version 3: Uses secant method, for Chapter 5 exercise
*
*********************************************************************
*  THIS PROGRAM CONTAINS DELIBERATE ERRORS
*********************************************************************
*
*  X1 and X2 are used as the two previous approximations
*
*  Returns an error code:
*      IERROR = 0: no errors
*      IERROR = 32: did not converge in 10 iterations;
*                      ROOT has been set to -9E+35
*
*  Variables local to SOLVE:
*      XOLD:    Approximation n-1
*      X:       Approximation n
*      XNEW:    Approximation n+1
*      ITER:    Iteration counter; limit 10
*
*  Subprograms invoked:
*      F:       Value of function
*
        SUBROUTINE SOLVE ( X1, X2, EPS, ROOT, IERROR )

        REAL X1, X2, EPS, ROOT
        INTEGER IERROR
        REAL X, XNEW
        INTEGER ITER
        REAL F,
        INTEGER XOLD

*  Initialize
        ITER = 1
        XOLD = 1X
        X = 2X

*****  REPEAT until converged or too many iterations
  10    CONTINUE

        XNEW = X + F(X) * (X - XOLD) / (F(X) - F(XOLD))
            IF (       (ABS(X - XNEW) .LT. EPS)
     $          .OR. (ITER .GT. 10)             ) GO TO 20
            X = XNEW
            XOLD = X
            ITER = ITER + 1

        GO TO 10
*****  END REPEAT

  20    CONTINUE

        IF ( ITER .GT. 10 ) THEN
            IERROR = 32
            ROOT = -9E+35
        ELSE
            IERROR = 0
            ROOT = XNEW
        END IF

        RETURN

        END
```

Figure 5.9 A student's rewrite of the SOLVE subroutine from
Figure 5.7, modified to use the secant method instead of the Newton-
Raphson method. *This program contains deliberate errors.*

When this program was run, the compiler pointed out three syntactical errors. Presumably they are fairly obvious; if not, your compiler will also complain about them when you try it.

With these errors corrected the compiler gave the comforting report "No diagnostics generated." With a glow of triumph our student ran the program. Joy turned to gloom when this output was produced:

```
 LEFT =  -10.0000000     RIGHT =    10.0000000
 DELTAX =    1.00000000      EPS =  0.999999931E-03
IFY240I VSTAE - ABEND CODE IS: SYSTEM 00CC, USER 0000. SCB/SDWA=00025648.
IFY240I VSTAE - IO HALTED.     PSW=FFF5000C82012A28. ENTRY POINT=00012000.
IFY240I VSTAE - REGS   0-3  7FFFFFFD 00012794 000124DE 00012030
IFY240I VSTAE - REGS   4-7  000126F8 00000000 000124AE 0001278C
IFY240I VSTAE - REGS   8-11 00012210 000121B4 000121BC 800121C8
IFY240I VSTAE - REGS  12-15 00012A94 000129A8 42012850 00012A1C
IFY240I VSTAE - FRGS   0-3  3EFFFFFA 00000900 00000000 00000000
IFY240I VSTAE - FRGS   4-7  00000000 00000000 D9E4D540 40404040
EXPONENT OVERFLOW EXCEPTION AT 012A24
```

This is quite meaningful and helpful, actually, to a person with the appropriate background. But it is gibberish, of course, to any beginning Fortran student. The only consultant on duty was just leaving for class, but on the way out said, "Why don't you put some PRINT statements around whatever you changed, and see if anything looks funny?" She would have preferred an explanation of what "exponent overflow exception" might mean, but decided to take the advice.

In fact, all the changes were in about a dozen lines of SOLVE, and only one of the changes called for any arithmetic. She accordingly placed the statement:

```
PRINT *, ITER, XOLD, X
```

before the assignment statement that computes a new value of XNEW. Running the new program produced this:

```
 LEFT =  -10.0000000     RIGHT =    10.0000000
 DELTAX =    1.00000000      EPS =  0.999999931E-03
           1            -3  -2.00000000
IFY240I VSTAE - ABEND CODE IS: SYSTEM 00CC, USER 0000. SCB/SDWA=00025648.
IFY240I VSTAE - IO HALTED.     PSW=FFF5000C82012A60. ENTRY POINT=00012000.
IFY240I VSTAE - REGS   0-3  7FFFFFFD 00012798 000124DE 00012030
IFY240I VSTAE - REGS   4-7  000126F8 00000000 000124AE 00012790
IFY240I VSTAE - REGS   8-11 00012210 000121B4 000121BC 800121C8
IFY240I VSTAE - REGS  12-15 00012ACC 000129E0 42012884 00012A54
IFY240I VSTAE - FRGS   0-3  3EFFFFFA 00000900 00000000 00000000
IFY240I VSTAE - FRGS   4-7  00000000 00000000 D9E4D540 40404040
EXPONENT OVERFLOW EXCEPTION AT 012A5C
```

Still largely gibberish, of course, but the one line produced by the new PRINT statement tells a lot. Since it was printed only once, just before the gibberish, it told her that the problem developed the very first time the assignment statement was encountered. And the fact that the value of XOLD was printed as in integer pointed out an earlier error that you no doubt saw at once.

On her way to the computer center to fix this one, she ran into the consultant, who took five minutes to give a hint of what this particular machine would do, given a negative integer and told to do real (floating-point) arithmetic with it. You might get a quite different message—or none, just wildly wrong results.

Sadder but wiser, she left the PRINT statement in the program, on the off chance that there might still be problems. Good idea:

```
LEFT =  -10.0000000        RIGHT =    10.0000000
DELTAX =    1.00000000        EPS =   0.999999931E-03
          1  -3.00000000              -2.00000000
          2  -1.20701694              -1.20701694
IFY240I VSTAE - ABEND CODE IS: SYSTEM 00CF, USER 0000. SCB/SDWA=00025648.
IFY240I VSTAE - IO HALTED.      PSW=FFF5000F8201285A. ENTRY POINT=00012000.
IFY240I VSTAE - REGS   0-3  00000002 00012778 000124DE 00012030
IFY240I VSTAE - REGS   4-7  00012030 000127E0 000128C4 000121B4
IFY240I VSTAE - REGS  8-11  00000000 000121B4 000121BC 800121C8
IFY240I VSTAE - REGS 12-15  00012932 000126F8 42012842 00000000
IFY240I VSTAE - FRGS   0-3  00000000 80000000 00000000 00000000
IFY240I VSTAE - FRGS   4-7  00000000 00000000 D9E4D540 40404040
FLOATING-POINT DIVIDE EXCEPTION AT 012856
```

This time there were two lines of output preceding some different gibberish. Actually, not total nonsense: with XOLD and X having the same value the second time, it figures that there would be problems with division by their difference. This pointed to the fact that two statements appear in reverse of the correct order.

With that one corrected, this was the output for the first root, with similar blowups for the others:

```
LEFT =  -10.0000000        RIGHT =    10.0000000
DELTAX =    1.00000000        EPS =   0.999999931E-03
          1  -3.00000000              -2.00000000
          2  -2.00000000              -1.20701694
          3  -1.20701694               8.06608963
          4   8.06608963              15.8598241
          5  15.8598241               23.8057556
          6  23.8057556               33.1973419
          7  33.1973419               45.4679260
          8  45.4679260               62.1312408
          9  62.1312408               85.0336608
         10  85.0336608              116.617355
         11 116.617355               160.209411
NO CONVERGENCE FOR POSSIBLE ROOT NEAR  -3.00000000
```

Now what? It can't be a syntax error, after four clean compiles. The subprograms are probably communicating properly, with the REAL/INTEGER mismatch having been corrected.

There being no more the computer could do, our neophyte programmer was forced to stop and think. Nothing else coming to mind, she decided to compare the assignment statement in detail against the formula in the textbook. But her roommate had the book, so she stopped to rederive the formula from the geometry. (From which she learned some mathematics while she was learning computing. Nice if it always worked that way.) In this way she finally was able to see the error in the assignment statement, which you of course saw long since.

Correcting this last (?) mistake and rerunning led to this output:

```
LEFT =  -10.0000000        RIGHT =    10.0000000
DELTAX =    1.00000000        EPS =   0.100000033E-05
          1  -3.00000000              -2.00000000
          2  -2.00000000              -2.79298210
          3  -2.79298210              -2.92495632
          4  -2.92495632              -2.86912441
          5  -2.86912441              -2.87129688
          6  -2.87129688              -2.87135887
```

```
THERE  IS  A  ROOT  AT   -2.87135887
              1     1.00000000         2.00000000
              2     2.00000000         1.64102554
              3     1.64102554         0.816680431
              4     0.816680431        1.48139762
              5     1.48139762         1.41848182
              6     1.41848182         1.38485145
              7     1.38485145         1.38767719
              8     1.38767719         1.38760471
THERE  IS  A  ROOT  AT    1.38760471
              1     2.00000000         3.00000000
              2     3.00000000         2.31111049
              3     2.31111049         2.45997619
              4     2.45997619         2.54206371
              5     2.54206371         2.52554988
              6     2.52554988         2.52642250
              7     2.52642250         2.52643203
              8     2.52643203         2.52642917
THERE  IS  A  ROOT  AT    2.52643013
              1     6.00000000         7.00000000
              2     7.00000000         6.32581424
              3     6.32581424         6.42314147
              4     6.42314147         6.45999622
              5     6.45999622         6.45726680
              6     6.45726680         6.45731639
              7     6.45731639         6.45732212
THERE  IS  A  ROOT  AT    6.45732116
```

The results have the appearance of being more accurate approximations than those shown in the text. Cursing the fact that what had looked like a 2-hour problem had turned into five hours, spread over three days, our heroine turned in the program and started catching up on her other courses.

You may or may not actually run this program as an exercise. You will learn a good deal if you do, especially if you are using a different computer and/or compiler. If you don't, you are possibly learning somewhat similar things trying to run your own programs. Either way, you are presumably coming to see the wisdom of the following morals of this tale.

1. A clean compile means only that the compiler couldn't find any violations of the rules for writing a legal Fortran program. The program may still have glaring and crippling errors.

2. Some violations of the syntax rules cannot, regrettably, be caught at compile time. Some systems will diagnose them at execution time; others will simply give wrong results or—if you are lucky—give *some* kind of diagnostic message. If there is a warning, it may be incredibly unhelpful—but at least you know there's a problem.

3. Haste makes waste.

4. A great many puzzles can be answered only by seeing values of variables as the computation proceeds. If you don't insert lots of extra PRINT statements into the program in the first place, you will later. Either way, it's simple to take them out when everything seems correct.

5. If the computer has given you weird results, it won't accomplish anything to submit the program again, unchanged, to see if the computer made a mistake. It didn't.

6. The greatest deficiency of computers is that they do what we told them to do, not what we meant to tell them to do.

SIX

FORMATTED INPUT AND OUTPUT

Introduction

Until now we have concentrated our attention on the basics of Fortran: the forms of the most common statements, writing programs in a readable style, and the elements of program organization. All input has been in the free form known as *list-directed* input, and the only nicety for output has been the identification of variables.

Now it is time to take a look at some of the extensive features that Fortran provides for handling input and output in a more sophisticated way. The primary emphasis will be on how to produce more attractive and readable output; list-directed input will still be adequate for most of our continuing work.

Since we wish to concentrate on the simpler output issues and yet combine most of the information about formatted input and output in one place for easier reference use, the chapter has been organized into sections treating the following topics:

- The basic ideas of Fortran format control for the output of real values.
- The fundamentals of the FORMAT statement.
- Format control for the output of integer values.
- Printing large and small numbers with the E format code.
- A more detailed look at the format codes for the output of real and integer values.
- Control of vertical spacing of output.
- Format codes for input of real values.
- Various slightly more advanced topics that are often useful.

All readers should study at least the first four sections now, but the others could be postponed until need for them arises. If you're pressed for time, you could simply read through the examples up through Figure 6.8, about half the chapter.

(One important topic, indexing in the list of an input or output statement, must be postponed until the next chapter, after we have introduced subscripted variables and the DO statement. We shall also postpone discussion of the format codes for data of the types complex, double precision, logical, and character until variables of these types are taken up in Chapter 8.)

The basic ideas of format control for the output of real values

In all of the programs to this point, output has been produced with a PRINT statement that specifies nothing about the format in which the output is to be printed. An asterisk after the word PRINT means that the standard output device (either a printer or a terminal) is to receive the output, and also that the output is to be printed in some standard format chosen by the implementor of the Fortran system. Any character literals in the PRINT statement are printed just as they appear, but numerical values are printed in some standard format.

The format in which output is produced is actually under our total control, if we wish. All that is required is to specify a *format code* corresponding to each value printed. This is an extensive subject, if all of the flexibility that Fortran provides is needed, but simple cases can be handled with a minimum of new information.

For example, let us rewrite the program of Figure 3.19 slightly, so that the values in the two columns are printed with fewer decimal places. Looking at the output in Figure 3.20, we see that the standard format has produced six decimal places in the slimness ratio and five for the safe load. The standard form with your computer and compiler may be different. But, in any event, there will be more digits printed than really make sense. The slimness ratio was being stepped by 0.1, and there can't really be much significance in all those decimals in SAF-LOD. Let's see what is needed to print the slimness ratio with, say, two decimal places and the safe load with one decimal place. This is shown in Figure 6.1, which is a modified version of Figure 3.19.

The only change is in the PRINT statement near the end of the WHILE loop. Following the word PRINT there is now, instead of an asterisk, some information to control the format of printed values. This is called a *format specification*. When included as a part of the PRINT statement as we have done here, it must be enclosed in apostrophes.

Within the parentheses are two *format codes*, also called *edit descriptors*, which specify how the two values (RATIO and SAFLOD) are to be printed. The F11.2 means that the value is to be printed in a *fixed format*, i.e., without an exponent; that the value is to be printed in a field 11 characters wide; and that there are to be two digits after the decimal point. (If the value does not require 11 printing positions, it is *right justified*, i.e., blanks are inserted to the *left* of the value.) This *first* format code goes with the *first* variable, so it is the value of RATIO that will be printed in the form just described. Now comes the *second* format code, the F18.1, which describes the form in which to print the value of the *second* variable, SAFLOD. This time we have said that 18 character positions should be allocated to the value and that it should be printed with one decimal place.

The output of this program is shown in Figure 6.2.

The FORMAT statement

It is not always convenient or even possible to place the format specification information in the PRINT statement itself. In fact, the style just used is applicable

```
*  Column design
*  The program finds the safe loading of a structural column as a
*     function of its slimness ratio, i.e., the ratio of its
*     length to its width
*
*  Version 6: First example of format control of output
*
*  Variables:
*     RATIO:  the slimness ratio of the column
*     TENRAT: 10 times RATIO
*     SAFLOD: the safe load, in pounds per square inch

       REAL RATIO, SAFLOD
       INTEGER TENRAT

*  Print column headings
       PRINT *, ' SLIMNESS RATIO        SAFE LOAD '


*  Give TENRAT its starting value
       TENRAT = 200

*****  WHILE TENRAT is less than or equal to 2000
   10  CONTINUE
       IF ( TENRAT .LE. 2000 ) THEN

*          Convert from TENRAT to RATIO
           RATIO = TENRAT / 10.0

*          Compute safe load from whichever formula applies
           IF ( RATIO .LT. 120.0 ) THEN
               SAFLOD = 1.7E4 - 0.485 * RATIO**2
           ELSE
               SAFLOD = 1.8E4 / (1.0 + RATIO**2 / 1.8E4)
           END IF

           IF ( RATIO .GT. 198.0 ) THEN
               PRINT '(F11.2, F18.1)', RATIO, SAFLOD
           END IF

*          Increment TENRAT
           TENRAT = TENRAT + 1

       GO TO 10
       END IF
*****  END WHILE

       END
```

Figure 6.1 The program of Figure 3.19, modified to produce output with fewer decimal places.

only in rather simple situations. Much more commonly, the formatting information is placed in a separate FORMAT statement.

The question then arises: how do we tell Fortran where the formatting information is? The answer is that the FORMAT statement must always have a statement number; we specify that number in the PRINT statement. Let us see how this works by taking another look at the column design program, as shown in Figure 6.3. The PRINT statement, in the WHILE loop, now contains a statement number instead of the asterisk or the format specification. A PRINT statement,

SLIMNESS RATIO	SAFE LOAD
198.10	5660.0
198.20	5656.1
198.30	5652.2
198.40	5648.3
198.50	5644.4
198.60	5640.5
198.70	5636.6
198.80	5632.7
198.90	5628.8
199.00	5624.9
199.10	5621.0
199.20	5617.1
199.30	5613.3
199.40	5609.4
199.50	5605.5
199.60	5601.6
199.70	5597.8
199.80	5593.9
199.90	5590.1
200.00	5586.2

Figure 6.2 The output of the
program of Figure 6.1.

in this form, means that the output is still to go to the standard output device, whatever that is, but that the formatting information is to come from the FORMAT statement whose statement number is contained in the PRINT.

The FORMAT statement is permitted to appear anywhere in the program after the REAL and INTEGER statements (and certain other *declaratives* that we shall learn about later). The statement number tells Fortran where to find it. When there is no compelling reason to do otherwise, we shall place a FORMAT statement immediately after the input or output statement that refers to it, which makes understanding the program easier.

The choice of a statement number is entirely under your control except that, naturally, there must never be duplicate statement numbers in a program. In the programming style adopted in this book, there are rather few statement numbers in any event. The convention is to use three-digit statement numbers ending in two zeros for FORMAT statements; statement numbers for statements other than FORMAT statements are usually two digits ending in one zero. Statement numbers of each of the two types are assigned in ascending sequence. All such style conventions are at the option of the programmer—unless there are standards for your installation that you are required to follow. And, as with indentation, there are other valid conventions than the statement numbering scheme adopted here. If you don't like this one, invent your own—*and then stick to it*.

The FORMAT statement itself consists of the statement number, the word FORMAT, then the format specification. The format specification is enclosed in parentheses. (Note that, when a format specification appears in a FORMAT statement, *it is not enclosed in apostrophes*. Apostrophes are needed only when the format specification is placed in the PRINT statement.)

The first thing in the format specification is a pair of apostrophes enclosing a blank. This guarantees that the output will be single spaced when it is printed.

```
*  Column design
*  The program finds the safe loading of a structural column as a
*     function of its slimness ratio, i.e., the ratio of its
*     length to its width
*
*  Version 7: First example of the FORMAT statement
*
*  Variables:
*     RATIO:  the slimness ratio of the column
*     TENRAT: 10 times RATIO
*     SAFLOD: the safe load, in pounds per square inch

        REAL RATIO, SAFLOD
        INTEGER TENRAT

*  Print column headings
        PRINT *, ' SLIMNESS RATIO        SAFE LOAD '


*  Give TENRAT its starting value
        TENRAT = 200

*****  WHILE TENRAT is less than or equal to 2000
   10   CONTINUE
        IF ( TENRAT .LE. 2000 ) THEN

*            Convert from TENRAT to RATIO
             RATIO = TENRAT / 10.0

*            Compute safe load from whichever formula applies
             IF ( RATIO .LT. 120.0 ) THEN
                 SAFLOD = 1.7E4 - 0.485 * RATIO**2
             ELSE
                 SAFLOD = 1.8E4 / (1.0 + RATIO**2 / 1.8E4)
             END IF

             IF ( RATIO .GT. 198.0 ) THEN
                 PRINT 100, RATIO, SAFLOD
  100            FORMAT (' ', F10.1, 13X, F5.0)
             END IF

*            Increment TENRAT
             TENRAT = TENRAT + 1

        GO TO 10
        END IF
*****  END WHILE

        END
```

Figure 6.3 Another modified version of the column design program, introducing the FORMAT statement.

We'll learn the details a bit later; for now, let's just get into the habit of always beginning the format specification with apostrophes enclosing a blank.

The first format code, following the spacing control, is F10.1. It is interpreted in the same way as F11.2, which we used before. Next comes a different kind of format code. *It does not correspond to anything in the list of the* PRINT *statement.* The 13X says to put 13 spaces between the field printed as F10.1 and the one printed as F5.0. The combination 13X, F5.0 thus has exactly the same effect as F18.0

would have. (Assuming the result does not exceed 5 digits, of course.) The choice is ours; the X format code sometimes is easier to use or makes the program easier to understand.

How shall we describe the way the list in the PRINT statement "keeps in step" with the format codes, now that there are two variables in the list of the PRINT and three format codes? It's simple if you just remember that there are format codes (such as 13X) that do not correspond to list items in the PRINT statement.

The process is this. The format codes in the format specification are taken from left to right. Those that correspond to list items (like the F10.1) control the format in which values are printed; those that do not (like the 13X) are simply processed as they are encountered.

We shall see as we proceed that there are variations on this theme, but the basic idea carries over to all of the extensions.

The output of the program is shown in Figure 6.4.

Format control for output of integer values

The form of printing the safe load here is not really very attractive: whole-number values are not usually printed with a terminal decimal point. Let's get rid of it. This requires converting the safe load values to Fortran integer form, since real values are *always* printed with a decimal point. To do this, we need only define a new variable of type integer, call it SAFINT, and insert an assignment statement to give it the value of SAFLOD. Then we print SAFINT, instead of SAFLOD. It will be in integer form, the conversion from real to integer form having been made as part of the action of the assignment statement.

Now we need a new format code to describe the integer value to be printed. A format code of I5 says that an integer value is expected and that it should be printed in five printing positions.

The modified program is shown in Figure 6.5 and its output in Figure 6.6.

SLIMNESS RATIO	SAFE LOAD
198.1	5660.
198.2	5656.
198.3	5652.
198.4	5648.
198.5	5644.
198.6	5640.
198.7	5637.
198.8	5633.
198.9	5629.
199.0	5625.
199.1	5621.
199.2	5617.
199.3	5613.
199.4	5609.
199.5	5606.
199.6	5602.
199.7	5598.
199.8	5594.
199.9	5590.
200.0	5586.

Figure 6.4 The output of the program of Figure 6.3.

```
* Column design
* The program finds the safe loading of a structural column as a
*     function of its slimness ratio, i.e., the ratio of its
*     length to its width
*
* Version 8: prints safe load as an integer
*
* Variables:
*     RATIO:  the slimness ratio of the column
*     TENRAT: 10 times RATIO
*     SAFLOD: the safe load, in pounds per square inch
*     SAFINT: the safe load, as an integer

      REAL RATIO, SAFLOD
      INTEGER TENRAT
      INTEGER SAFINT

* Print column headings
      PRINT *, ' SLIMNESS RATIO        SAFE LOAD '


* Give TENRAT its starting value
      TENRAT = 200

***** WHILE TENRAT is less than or equal to 2000
   10 CONTINUE
      IF ( TENRAT .LE. 2000 ) THEN

*         Convert from TENRAT to RATIO
          RATIO = TENRAT / 10.0

*         Compute safe load from whichever formula applies
          IF ( RATIO .LT. 120.0 ) THEN
              SAFLOD = 1.7E4 - 0.485 * RATIO**2
          ELSE
              SAFLOD = 1.8E4 / (1.0 + RATIO**2 / 1.8E4)
          END IF

          IF ( RATIO .GT. 198.0 ) THEN
*             Convert safe load to integer form
              SAFINT = SAFLOD
              PRINT 100, RATIO, SAFINT
  100         FORMAT (' ', F10.1, 12X, I5)
          END IF

*         Increment TENRAT
          TENRAT = TENRAT + 1

      GO TO 10
      END IF
***** END WHILE

      END
```

Figure 6.5 A final modified version of the column design program, introducing the I format code.

Printing large and small numbers with the E format code

Another example will permit us to see the application of the "keeping in step" process in a slightly different way. We can also use the example to introduce a convenient way of printing large and small numbers, the E format code.

SLIMNESS RATIO	SAFE LOAD
198.1	5660
198.2	5656
198.3	5652
198.4	5648
198.5	5644
198.6	5640
198.7	5636
198.8	5632
198.9	5628
199.0	5624
199.1	5621
199.2	5617
199.3	5613
199.4	5609
199.5	5605
199.6	5601
199.7	5597
199.8	5593
199.9	5590
200.0	5586

Figure 6.6 The output of the program of Figure 6.5.

Figure 6.7 is an updated version of the program of Figure 3.25, modified to make the printed output more attractive. The PRINT statement that prints the voltage now has an associated FORMAT statement, the one with statement number 100. The first format code in the FORMAT, after the blank in apostrophes for single spacing, consists of identifying information enclosed in apostrophes. This corresponds to nothing in the list of the PRINT, and is simply transmitted to the output exactly as it stands. The second format code, the E13.5, corresponds to the one variable named in the PRINT. The E format code says to print the number in the form of a fraction between 0.1 and 1.0, times a power of 10. The power of 10 is expressed the same way we are permitted to do it for input: the letter E, followed by a plus or minus sign, followed by a two-digit exponent. The 13 in the format code says to give 13 printing positions to the entire number, and the 5 says to put 5 digits after the decimal point.

Next comes another apostrophe format code that produces the word VOLT preceded by a space. Three more PRINT-FORMAT combinations produce the lines for the resistance, inductance, and capacitance.

Then we have two PRINT statements to produce two lines of column headings; the first line gives the quantities and the second the units. Note that nothing has been done about spacing control here. That is because, with the free-format PRINT (one in which there is simply an asterisk and no format specification and no FORMAT statement number), the compiler automatically inserts a blank to produce single spacing. (At any rate, that is the case with several systems used in preparing programs for this book. Your system might be different.)

Within the WHILE loop, the frequency has been converted to an integer for a more pleasing appearance and then printed under control of an I8 format code.

This program was run with a different set of data from that used in Chapter 3, producing the output in Figure 6.8.

```
* Current in an AC circuit
*
* Modified version to demonstrate output format features
*
* Series circuit contains a voltage source, a resistance,
*     a capacitance, and an inductance; frequency of voltage
*     source is read as data, with a negative value as sentinel
*
* Variables:
*     I: Current, amperes
*     E: Voltage, volts
*     R: Resistance, ohms
*     F: Frequency, hertz
*     L: Inductance, henrys
*     C: Capacitance, farads
*     TWOPI: 2 times pi

      REAL I, E, R, F, L, C, TWOPI
      INTEGER FINT
      TWOPI = 2.0 * 3.1415963

* Read the unchanging parameters

      READ *, E, R, L, C

* Print parameters and column headings

      PRINT 100, E
  100 FORMAT (' ', 'E = ', E13.5, ' VOLT')
      PRINT 200, R
  200 FORMAT (' ', 'R = ', E13.5, ' OHM')
      PRINT 300, L
  300 FORMAT (' ', 'L = ', E13.5, ' HENRY')
      PRINT 400, C
  400 FORMAT (' ', 'C = ', E13.5, ' FARAD')
      PRINT *, '  FREQUENCY    CURRENT'
      PRINT *, '    HERTZ        AMPS'

* Get the first value of the frequency
      READ *, F

***** WHILE frequency is greater than zero
   10 CONTINUE
      IF ( F .GT. 0.0 ) THEN

         I = E / SQRT( R**2 + (TWOPI*F*L - 1.0/(TWOPI*F*C))**2 )
         FINT = F
         PRINT 500, FINT, I
  500    FORMAT (' ', I8, 6X, F7.4)
         READ *, F

      GO TO 10
      END IF
***** END WHILE

      END
```

Figure 6.7 A modified version of the program of Figure 3.25, introducing the E format code.

```
E =    0.40000E+03 VOLT
R =    0.56000E+04 OHM
L =    0.10000E+00 HENRY
C =    0.50000E-08 FARAD
  FREQUENCY    CURRENT
    HERTZ        AMPS
      100       0.0013
      500       0.0063
     1000       0.0126
     3000       0.0386
     5000       0.0619
     6000       0.0689
     7000       0.0714
     8000       0.0702
    10000       0.0625
    15000       0.0435
    20000       0.0325
    50000       0.0128
```

Figure 6.8 The output of
the program of Figure 6.7.

We see that 400 volts has been printed as:

   0.40000E+03

which means 0.40000 times 10^3. The resistance (5600 ohms) is represented in a similar manner. The inductance (0.1 henry) was printed with an exponent of zero, meaning 10 raised to the zero power, which is 1. The capacitance (5 × 10^{-9} farad) was printed as:

   0.50000E-08

which illustrates one advantage of the E format code: without it, we would be counting eight zeros to understand the value of the capacitance. The other advantage, and it is a major one, is that we can provide for the output of values of widely varying sizes in a fixed number of printing positions.

   Now that we have established some of the basic concepts of how the format of printing is controlled, it is time to take a detailed look at the various format codes that are available for specifying the form of output.

Format codes for the output of real values

This section is essentially a summary of information that is at least partly familiar, presented in a form suitable for reference. We can avoid some repetition by starting with some matters that apply equally to both the F and E format codes.

   1. In each case the letter F or E is followed by a number to specify how many printing positions are to be used, a decimal point, and a second number specifying how many digits follow the decimal point in the printed result. We refer to the general format of the F format code, for example, as *Fw.d*, where *w* (width) is the number of printing positions and *d* is the number of decimal places.

2. If the value to be printed does not require w positions, it will be printed at the right of the assigned space (right justified), preceded by blanks.

3. If the value to be printed is too large to fit in w positions, asterisks will be printed to signal the problem.

4. Either of these format codes may be preceded by a *repeat specification*, which is simply an unsigned integer that says how many times the format code should be repeated. (In fact, this is true of any of the format codes that control the transfer of input or output list items.) For example, consider this combination:

```
      PRINT 100, A, B
100   FORMAT (' ', 2F5.2)
```

The values of A and B will both be printed under control of the F5.2 format code.

With this background, describing the two format codes is reasonably straightforward.

■ *Format code F (External Fixed Point).* The general form is $rFw.d$. The r is the (optional) repeat specification. The F indicates conversion between an internal real value and an external number written with a decimal point but without an exponent. The total number of characters in the field (including any sign, the automatic decimal point, and any blanks at the beginning of the field) is w (width). There are d digits after the decimal point. The value of w should always be at least 2 larger than the value of d, to allow space for the decimal point and a possible minus sign.

■ *Format Code E (Floating Point).* The general form is $rEw.d$. As we have seen, the normal form of output with the E format code begins with a zero before the decimal point, followed by d places after the decimal point and an exponent in the form E+nn. The total number of positions, w, includes the sign positions, the leading zero, the decimal point, and the letter E. The value of w should always be at least 6 larger than the value of d, to allow space for the exponent, the decimal point, and a possible minus sign.

The output of integer values

As we have seen in program examples already, printing integer values requires use of the I format code, which is of the general form rIw. Here r is the (optional) repetition code, if present, and w is the number of printing positions.

The control of vertical spacing

Also as seen in the program examples, the first character of each line that is printed is used to control vertical spacing of the output. We have been able to ignore this fact so far because we have put a blank at the beginning of every character literal used for output. This was done to keep things a bit simpler while you were concentrating on other matters. Now that we are taking explicit control of the format of output, we must make provision for handling spacing ourselves.

Failure to take spacing into account can lead to incorrect output and some *very* strange spacing, and there are many occasions when we wish to be able to do things other than single space. Fortunately, the topic is a simple one.

The characters that may be used to control vertical spacing are:

```
Control
Character        Action
_____

Blank            Single space before printing
0 (zero)         Double space before printing
1                Skip to top of next page before printing
+                Suppress spacing (overprint prior line)
```

Putting one of these characters into the output line is a simple matter of beginning the format specification, however that is done, with the appropriate character enclosed in apostrophes. We have already seen this done with the blank for single spacing, and other examples appear in the program at the end of the chapter.

It is good programming practice always to provide explicit spacing control, which we shall do from now on. Several examples appear in the sections that follow.

Format codes for the input of real values

The format codes that we have just considered may all be used for input also. This topic is actually somewhat simpler than the corresponding material on output, since the F and E format codes handle input identically. Furthermore, in all cases leading blanks in the input data are never significant, negative values must be preceded by a minus sign, and positive values may or may not be preceded by a plus sign. For both the F and E format codes, an actual decimal point in the input field overrides the specification that appears in a format code. When it is convenient to provide an actual decimal point in the input data, the number of decimal places can be written as zero in the specification, which is then overridden by the actual decimal point. Finally, both format codes permit, but neither requires, the presence of an exponent in the input field. The exponent may be either a signed integer constant, or the letter E followed by an optionally signed constant. Thus all of the following are permissible and equivalent forms of the exponent, *plus 2:* E+02, E02, E+2, E2, +2.

■ *The F Format Code for input.* The form, as with output, is *Fw.d*, where *w* is the number of characters in the field and *d* is the number of digits after an (assumed) decimal point if the input field does not contain an actual decimal point. For example, suppose that we are reading an input field consisting of the digits:

123456

If this is read under control of F6.1, the value stored would be 12345.6; if read under control of F6.3, the value stored would be 123.456; if read under control of F6.6, the value stored would be 0.123456. On the other hand, if the input field had contained the characters:

123.45

then any of the above format codes would have resulted in the value 123.45 being stored. If we knew that the input field would have a decimal point, we would probably use F6.0.

The E format code for input. The form is *Ew.d*, where *w* is the total width of the field and *d* is the number of places after the (assumed) decimal point. As with the F format code, an actual decimal point overrides the *d* specification.

The I format code for input. The form is *Iw*, where *w* is the width of the field.

Format codes that do not correspond to input or output list items

This heading covers a number of format codes that do not control the formatting of data transmission, but rather deal with such matters as the position of data, character data sent from a format specification directly to the output, and processing more than one card or line with a single READ or PRINT operation. Most of the topics can be treated rather briefly, and are gathered here for easy reference. Most of them are illustrated in the examples at the end of the chapter or in later chapters.

■ *The X format code.* The general form is *nX*, which means to skip *n* printing positions on output or *n* characters on input.

■ *Apostrophe editing.* The apostrophe format code has the form of a character literal, i.e., any collection of characters enclosed in apostrophes. The characters in the literal (not including the apostrophes, of course) are sent to the output device. As always with character literals, an apostrophe within an apostrophe format code is represented by two successive apostrophes.

■ *The slash in a format specification.* The slash format code specifies the end of transmission on the current record. ("Record" is carry-over terminology from an earlier era; it means simply one card or one line of terminal input or output, in our context.) On input, the rest (if any) of the current record is skipped. If more items remain in the list of the READ, transmission picks up with the beginning of the next input record. On output, the current line is transmitted, and, if list items remain, further printing starts a new line.

Repetition of groups of format codes

Just as we can repeat a format code by writing a repetition factor in front of it, so we can repeat a group of format codes by enclosing them in parentheses and writing a group repetition factor in front. For instance, suppose that eight fields in an input record are alternately described by I2 and F10.0. We can write:

```
100 FORMAT (4(I2, F10.0))
```

to get the desired action. We would *not* get the same action with:

```
100 FORMAT (4I2, 4F10.0)
```

which describes a record with four I2 fields, then four F10.0 fields, rather than the desired alternation.

Reversion of control

To this point we have always provided the same number of format codes as list items, with an obvious one-to-one relationship between the list items and those format codes that control format of data transfer. It often turns out to be useful to specify *fewer* format codes than there are list items. When the end of the format specification is reached and more list items remain, format control "reverts," i.e., goes backward, to the format code after the nearest left parenthesis. That can be either the left parenthesis that begins the format specification, or the left parenthesis of a group-repeat if there is one.

The END= feature

We have seen a number of instances where we wanted a program to read data indefinitely, until somehow detecting that the end of data had been reached. Lacking any other means, we have provided a *sentinel* at the end of the data: a zero, a negative value, a very large value, or anything else that could not be a legitimate data value. This has served our purposes, but is sometimes quite inconvenient. The END= feature is available for such situations.

The technique is simply to provide in the READ statement, after the format specification, the characters "END=" followed by a statement number. If the end of the data is detected in executing the READ statement, control will be transferred to the statement specified in the END= option. The system "detects the end of data" when it attempts to read a record after the last one. In other words, everything goes as usual until the system tries to read a "record that isn't there." It is at that point that the statement named in the END= is given control.

A final example

Another version of the ac circuit program will integrate most of the new material that has been presented since the earlier illustrative programs. It also permits illustration of an occasionally useful feature, the *scale factor*. Figure 6.9 is the final revision of this program.

The first change is the inclusion of scale factors with the E and F format codes. With the E format code the use of a 1P scale factor results in moving the decimal point one position to the right and reducing the exponent accordingly. The net effect is that the value represented is the same, but it is a number between 1.0 and 10.0 times a power of 10, rather than a fraction between 0.1 and 1.0 times a power of 10. With the F format code in FORMAT statement 600, the effect of the 3P is to multiply the number being printed by 10^3. This changes a current in amperes into a current in milliamperes, which is a more convenient unit for the values that result from this calculation. The value is "changed" only in terms of how it is printed, of course; the internal stored value is not affected.

FORMAT statement 500 illustrates two of the new features. The zero for carriage control causes double spacing before the printing of the line, which provides a blank line between the four parameters and the first line of the column headings. After the words FREQUENCY and CURRENT have been printed on one line, we want a second line giving the units, HERTZ and MA, for milliamperes. This could be done with a second PRINT and another FORMAT, of course, but

```
* Current in an AC circuit
*
* Final modifed version to demonstrate added I/O features:
*     Scale factor with E and F format codes for output
*     Double spacing with a zero carriage control character
*     The slash in a FORMAT statement
*     The END= option for detecting end of file
*
* Series circuit contains a voltage source, a resistance,
*     a capacitance, and an inductance; frequency of voltage
*     source is read as data, with a negative value as sentinel
*
* Variables:
*     I: Current, amperes
*     E: Voltage, volts
*     R: Resistance, ohms
*     F: Frequency, hertz
*     L: Inductance, henrys
*     C: Capacitance, farads
*     TWOPI: 2 times pi
*     FINT: Frequency, in integer form

      REAL I, E, R, F, L, C, TWOPI
      INTEGER FINT
      TWOPI = 2.0 * 3.1415963

* Read the unchanging parameters

      READ *, E, R, L, C

* Print parameters and column headings

      PRINT 100, E
  100 FORMAT (' ', 'E = ', 1PE13.4, ' VOLT')
      PRINT 200, R
  200 FORMAT (' ', 'R = ', 1PE13.4, ' OHM')
      PRINT 300, L
  300 FORMAT (' ', 'L = ', 1PE13.4, ' HENRY')
      PRINT 400, C
  400 FORMAT (' ', 'C = ', 1PE13.4, ' FARAD')
      PRINT 500
  500 FORMAT ('0', ' FREQUENCY    CURRENT'/' ', '   HERTZ              MA'/)

***** REPEAT until no more data
   10 CONTINUE

      READ (*, *, END=20), F
      I = E / SQRT( R**2 + (TWOPI*F*L - 1.0/(TWOPI*F*C))**2 )
      FINT = F
      PRINT 600, FINT, I
  600 FORMAT (' ', I8, 4X, 3PF6.1)

      GO TO 10
***** END REPEAT

   20 CONTINUE

      END
```

Figure 6.9 A final version of the program shown in Figure 6.7, modified to illustrate several additional input/output features.

this is just the kind of situation where the slash in a FORMAT statement is convenient. It continues the output operation now under way, but specifies that what follows is to go on a new line. Since it is a new line, the next thing that appears should be a carriage control character. And since we do not want a blank line between the two lines of the heading, we provide a blank to specify single spacing. Finally, a slash at the end of the format specification says to go on to another line—but we provide nothing to be printed there. This is a handy way to get a blank line after the second heading line, before the body of the table is printed.

The last new feature that this program illustrates is the END= feature, seen in the READ statement in the REPEAT loop. In order to use it, we must somehow specify the input device, designate a FORMAT statement or indicate that we are using free-format input, write the END= with a statement number, and enclose all this in parentheses. We provide one asterisk to designate the standard input device and a second to specify free-format input. The statement to which we wish to transfer control when the reading operation is complete is the CONTINUE statement after the REPEAT structure, so we write 20, its statement number, after the END=.

This program was run with the same data used before, except that the negative frequency sentinel was deleted from the end. Each time the program found a frequency value, it proceeded with the computation and printing of the current. When a READ operation found that there was no more data, control transferred out of the REPEAT loop and the program terminated. The output produced is shown in Figure 6.10.

The WRITE Statement

The WRITE statement provides an alternative to the PRINT statement, adding certain flexibilities that are occasionally useful. Since you may sometimes need

```
E =     4.0000E+02 VOLT
R =     5.6000E+03 OHM
L =     1.0000E-01 HENRY
C =     5.0000E-09 FARAD

   FREQUENCY     CURRENT
     HERTZ          MA

       100          1.3
       500          6.3
      1000         12.6
      3000         38.6
      5000         61.9
      6000         68.9
      7000         71.4
      8000         70.2
     10000         62.5
     15000         43.5
     20000         32.5
     50000         12.8
```

Figure 6.10 The output of the program of Figure 6.9.

this flexibility or need to be able to read older programs in which the WRITE statement appears, we shall close the chapter with a brief description of it.

The WRITE statement, for the kinds of purposes considered in this chapter, is quite simple. It begins with the word WRITE, then contains a *unit specifier* and a *format specifier* enclosed in parentheses and separated by a comma, and then has an output list of the same type that appears in a PRINT. The "unit specifier" designates the device on which the output is to appear. If we wish to specify the "standard output device," whatever that is, as we have done in all examples to date, we may place an asterisk in this position. Alternatively, unit 6 is by convention the standard output device. An asterisk in the format specifier position designates list-directed output, but it would be unusual in Fortran 77 to do this, since then we are back to the simplicity of the PRINT statement.

Thus, where in the program of Figure 6.9 we had

```
        PRINT 600, FINT, I
  600   FORMAT (' ', I8, 4X, 3PF6.1)
```

we could also have written

```
        WRITE (6, 600) FINT, I
  600   FORMAT (' ', I8, 4X, 3PF6.1)
```

There is no equivalent to the END= feature, in the case of the WRITE statement, since detecting end of file has no application with output.

Finally, the READ statement may also have a unit designator other than the asterisk that we have used to this point. Unit 5 is by convention the standard input device. Thus, the READ statement in the program above, which we wrote as

```
  READ *, E, R, L, C
```

could also have been written as

```
  READ (5, *) E, R, L, C
```

Fortran provides the unit designator primarily for applications in which the input or output does not involve the standard devices, but instead uses tapes or disks. Such applications are assumed to be outside the scope of interest of the typical reader of this book, and are not treated here.

Exercises

*1. Show how the values on each line would be printed under control of the format code at the beginning of the line.

```
I5:    0, 1, 10, -587, 90061, 12345
F7.2:  0.0, 1.0, 16.87, -568.12, 0.04, 12.34
F5.0:  0.0, 1.0, 16.87, -568.12, 0.04, 12.34
E12.4: 0.0, 1.0, 16.87, -568.12, 0.04, 12.34
```

2. Show how the values on each line would be printed under control of the format code at the beginning of the line.

```
I5: 0, 1, 10, -587, 90061, 12347
F6.0: 0.0, 1.0, 23.45, -6.0, 0.0052, -98.71
F8.3: 0.0, 1.0, 23.45, -6.0, 0.0052, -98.71
E14.7: 0.0, 1.0, -1.0, 12.43, 357.91, 0.000001, 10000
```

*3. Here is the beginning of a small program:

```
INTEGER EXP
REAL P, T
EXP = 12
P = 407.8
T = 32.9
WRITE (6, 100) EXP, P, T
```

Show what would be printed if the FORMAT statement were:

a. 100   FORMAT (' ', 'EXP=', I3, '   P=', F6.1, '   T=', F6.1)

b. 100   FORMAT (' ', 'EXPERIMENT NUMBER = ', I3,
   $ '   PRESSURE = ', F6.1, ' TEMPERATURE = ' , F6.1)

c. 100   FORMAT ('1', 'EXPERIMENT NUMBER = ', I3/
   $ '0', ' PRESSURE = ', E10.3, '   TEMPERATURE = ', E10.

4. Consider this small program:

```
INTEGER  READNG
REAL P, V, T
READNG = 61
P = 16.3
V = 4.13
T = 501.6
WRITE (6, 100) READNG, P, V, T
```

Show what would be printed if the FORMAT statement were:

a. 100   FORMAT (' ', 'READING =', I4, '   P =', F7.2,
   $ '   V =', F7.2, '   T =', F7.2)

b. 100   FORMAT (' ', 'READING:', I4, '   PRESSURE = ', F8.3,
   $ '   VOLUME = ', F8.3, '   TEMPERATURE = ', F8.3)

c. 100   FORMAT ('1', 'READING NUMBER: ', I4/
   $ ' ', 'PRESSURE: ', E12.5, ' ATMOSPHERES'
   $ ' ', 'VOLUME: ', E12.5, ' LITERS'/
   $ ' ', 'TEMPERATURE: ', E12.5, ' DEGREES K')

*5. Here is the start of another small program:

```
INTEGER K
REAL A, B
K = 49
A = 67.901
B = -0.38
```

For each of the following lines, show a combination WRITE and FORMAT statements that could produce it.

a. 49   67.901   -0.38

b. K = 49   A= 67.90   B= -0.380

    c. `READING NO: 49   1ST:   67.901   2ND:   -0.380`

    d.
```
67.901   FACTOR 1
-0.380   FACTOR 2
    49   READING
```

6. Same as Exercise 5.

```
INTEGER HOUR, PULSE
REAL TEMP
HOUR = 14
PULSE = 92
TEMP = 101.8
```

    a. `14   92   101.8`

    b. `AT HOUR    14   PULSE WAS   92   AND TEMPERATURE WAS   101.8`

    c. `14 92101.8`

    d.
```
PULSE: 92
TEMP: 101.8
HOUR:   14
```

*7. You are given this start of a program:

```
INTEGER I, J
REAL R, S
I = -16
J = 92017
R = 16.9
S = 437.9
WRITE (6, 100) I, J, R, S
```

For each of the following forms of output, show a FORMAT statement that could produce it.

    a. `-16   92017    16.90   437.90`

    b. `-16   92017   0.169E+02    0.438E+03`

    c. `I=ᵇᵇᵇᵇ -16ᵇ J=   92017 R=   16.9 S= 437.9`

    d.
```
I =     -16
J =   92017
R =    16.9
S =   437.9
```

8. Same as Exercise 7.

```
INTEGER HOUR, MINUTE, SECOND
REAL E, I
HOUR = 09
MINUTE = 43
SECOND = 39
E = 44.9
I = 0.452
WRITE (6, 100) HOUR, MINUTE, SECOND, E, I
```

    a. `94339   44.9 0.452`

    b. `9 43 39 44.900   0.452`

    c. `AT TIME    9 43 39 E =   44.90 I =    0.45`

d. TIME:
     9 HOURS
    43 MINUTES
    39 SECONDS

    VOLTAGE = 0.449E+02
    CURRENT = 0.452E+00

*9. Write READ and FORMAT statements to get four numbers from an input record, such as a card. The numbers are to become the values of four real variables named BOS, PHL, SFO, and ATL. Each number occupies eight character positions, with the first beginning in column 1. Each number contains a decimal point except the last, which is to be treated as though it had two digits after the (nonexistent) decimal point.

10. Same as Exercise 9, with the following changes. The variables are named DET, CLE, ORD, and STL. The first two numbers take 10 character positions each and are to be treated as though they had three digits after the (nonexistent) decimal point. The last two numbers take five character positions each and contain an actual decimal point.

*11. You have a program containing these statements:

```
INTEGER I, J
REAL X, Y
READ (5, 100) I, J, X, Y
```

For each of the following FORMAT statements that might be referenced by the READ, state what the format of the data would have to be for the variables to receive meaningful values.

a.   100   FORMAT (I4, I5, F10.0, F5.2)

b.   100   FORMAT (2I3, 8X, 2F8.0)

c.   100   FORMAT (20X, I1, 10X, I1, 2F10.0)

d.   100 FORMAT (2I5/2F10.0)

12. Same as Exercise 11, with these statements before the FORMAT statement:

```
INTEGER I, J
REAL X, Y, Z
READ (5, 100) I, X, J, Y, Z
```

a.   100 FORMAT (I2, F5.0, I5, F6.1, F7.1, F8.1)

b.   100   FORMAT (I2, F5.0, I5, F6.1, F7.1, F8.1)

c.   100   FORMAT (I4, 6X, F8.2, 6X, I4, 6X, F8.2, 6X, F8.2)

d.   100   FORMAT (I4, F10.0/I4, F10.0/F12.0)

*13. The values of the variables A, B, X, and Z are to be printed on one line. A and B are to be in 12 printing positions each, without exponents, and with four digits after the decimal point. X and Z are to be in 20 positions each, with exponents, and with six digits after the decimal point. Write appropriate statements.

14. Same as Exercise 13, except that an integer K is to be printed in seven positions between A and B, and seven blanks are to be inserted between X and Z.

*15. The FORMAT statement associated with the WRITE has been removed from this program:

```
REAL LABOR, PARTS, ACCESS, OILGRS
REAL SUBTOT, STAX, TOTAL
READ *, LABOR, PARTS, ACCESS, OILGRS
SUBTOT = LABOR + PARTS + ACCESS + OILGRS
STAX = 0.05 * SUBTOT
TOTAL = SUBTOT + STAX
WRITE (6, 100) LABOR, PARTS, ACCESS, OILGRS, SUBTOT, STAX, TOTAL

END
```

When the program (including the FORMAT statement) was run with suitable data, it produced this output:

```
TOTAL LABOR    45.00
TOTAL PARTS    61.12
ACCESSORIES     0.00
OIL, GREASE    14.40

SUBTOTAL      120.52

SALES TAX       6.03

GRAND TOTAL   126.55
```

Write the FORMAT statement.

16. Same as Exercise 15, with this program and output:

```
      REAL PRVBAL, CHARGS, CREDIT, TOTDUE
      INTEGER ACNO1, ACNO2, ACNO3, ACNO4
      READ (5, 100) ACNO1, ACNO2, ACNO3, ACNO4, PRVBAL, CHARGS, CREDIT
100   FORMAT (3I3, I1, 3F7.2)
TOTDUE = PRVBAL + CHARGS - CREDIT
      WRITE (6, 200) PRVBAL, CHARGS, CREDIT, TOTDUE,
     $               ACNO1, ACNO2, ACNO3, ACNO4

      END
```

```
PREVIOUS BALANCE:
    786.43

NEW CHARGES:
    338.55

CREDITS:
    168.64

TOTAL DUE:
    956.34

ACCOUNT NUMBER:
 203 441 820 6
```

*17. Another student has loaned you a deck of cards that contain data in the following fields:

Columns 1–9   Social security number
Columns 10–33   Name

Columns 34–39	Birthdate, as YYMMDD
Columns 40–47	Blank
Columns 48–49	Years of education
Columns 50–51	Occupation code
Columns 52–80	Blank

Write a program that will read such a deck, and, for each card, print the social security number. The required format is three digits, a hyphen, the next two digits, another hyphen, and the last four digits.

18. Modify the program of Exercise 17 so that, along with the social security number, it prints the birthdate. There should be four blank spaces between the two fields, and the birthdate should be printed in the form of the two digits of the month, a slash, the two digits of the day, another slash, and finally the two digits of the year.

19. Modify the program of Figure 3.22 as follows. Place a row of hyphens under the column headings, and then leave a blank line before the beginning of the output. Print the value of X in 11 printing positions, with four digits after the decimal point. Print the value of the polynomial as a floating-point number with five digits after the decimal point, in 15 printing positions. Leave seven blanks between the two columns, as in the output of Figure 3.23.

SEVEN

THE DO STATEMENT AND ARRAYS

Introduction

The Fortran DO statement is one of the most powerful and widely-used features of the language. In one statement it provides for the initialization, incrementing, and testing of a variable, plus repeated execution of a group of statements. Thus it replaces—in those cases where its features permit—the WHILE structure together with the separate initialization and incrementing required when the WHILE is used.

The DO statement finds special usefulness in the processing of data in *arrays*, in which we let one variable name stand for an entire group of related values. In mathematical applications Fortran arrays are the way we handle such things as vectors and matrices, but many other interpretations of arrays are possible.

Since DO statements and arrays are so heavily used and so powerful, we shall provide a number of examples of their application.

Basics of the DO statement

The basic form of the DO statement is:

        DO s, v = el, e2
            .
            .
            .
        s   CONTINUE
or:
        DO s, v = e1, e2, e3
            .
            .
            .
        s   CONTINUE
where:

■ s is the statement number of a statement later in the program, called the *terminal statement*. The statements between the DO statement and the terminal statement are called the *range* of the DO statement. The comma after the statement number is optional; since some systems do not permit it, we shall not use it.

■ v is a variable of type integer, real, or double precision (see Chapter 8). It is called the *index variable* or the DO *variable*.

■ e_1, e_2, and (if present) e_3 are expressions of type integer, real, or double precision. e_1 is called the *initial parameter*, e_2 is called the *terminal parameter*, and e_3 is called the *incrementing parameter*.

In the simplest case, where $e_2 > e_1$ and e_3 is positive, the action of the DO statement is quite simple. The DO variable, v, is given the value of the initial parameter e_1, and then the range is executed repeatedly; after each execution the variable is incremented by the value of the incrementing parameter e_3 (and by 1 if e_3 is absent), repeated execution stops when the incremented value is greater than the value of the terminal parameter e_2.

Some examples

This may sound more complicated than it is. Consider a simple example in which we wish to read one value from each of 10 input records, form their sum, and print it. This program fragment will do that:

```
      REAL X, SUM
      SUM = 0.0
      DO 10 I = 1, 10
         READ *, X
         SUM = SUM + X
10    CONTINUE
      PRINT *, 'SUM = ', SUM
```

The DO statement says to execute the range 10 times, with I taking on the values from 1 to 10 in succession. With SUM having been initialized to zero, after the ten data values have been added to it, its contents will be their sum, which is printed when repetition stops.

In the preceding example the DO loop was used merely to count the desired number of repetitions. More commonly the DO variable is either itself involved in the computation in the range, or controls the accessing of array elements as we shall see later. For an example of use of the DO variable in the computation, suppose we have forgotten the formula for the sum of the first N integers and want to compute it, with N read as data:

```
      INTEGER K, N, SUM
      READ *, N
      SUM = 0
      DO 30 K = 1, N
         SUM = SUM + K
30    CONTINUE
      PRINT *, 'THE SUM OF THE FIRST ', N, ' INTEGERS IS ', SUM
```

Here, as K takes on the successive values from 1 to N, each of its values is added to the integer variable SUM, which was initialized to zero.

Or suppose we wanted the sum of the squares of the odd integers not greater than a value N read as data:

```
       INTEGER K, N, SUMOSQ
       READ *, N
       SUMOSQ = 0
       DO 10 K = 1, N, 2
          SUMOSQ = SUMOSQ + K**2
10     CONTINUE
       PRINT *, 'SUM OF SQUARES OF THE ODD INTEGERS NOT GREATER THAN'
       PRINT *, N, ' IS', SUMOSQ
```

This time, the DO variable K is given the initial value 1 and then incremented by 2 after each execution of the range, until the incremented value is greater than N. Thus, if the value of N read from the card is 3, the range will be executed with K equal to 1 and then 3. If N is 5, the successive values of K will be 1, 3, and 5. Suppose N is 6; the successive values of K will also be 1, 3, and 5, since a further incrementation produces 7, which is greater than the terminal value. If the value of N is 1 or 2, the range will be executed once, with the DO variable equal to 1.

And if the value of N should be zero or negative, the range will not be executed at all. This is because a test is made at the beginning of execution of the DO statement to determine how many executions of the range are specified by the DO parameters. This is called the *iteration count*. The range is not executed at all unless the iteration count is greater than zero.

We thus see that the DO statement is a WHILE loop, in the sense that the test for completion is made at the beginning of the loop, and that the body of the loop is not executed at all if the test is satisfied the first time. It will perhaps be uncommon to take deliberate advantage of this feature, but it is important to know how it operates.

(Some older versions of Fortran always executed the range at least once, even if the DO parameters were such as to logically specify zero. Some programmers took deliberate, but unwise, advantage of this quirk, leading to programs that will not work correctly if compiled by a compiler adhering to the new standard. It is possible to specify to our Fortran compiler that it should interpret DO statements so as to maintain compatibility with the old standard on this matter. We shall have no occasion to use this most awkward feature, which is mentioned in passing only to show how difficult it is to remove bad features from established languages.)

It is permissible for the "incrementing" parameter to be negative, in which case the DO takes on the value from e_1 down to e_2 in steps of e_3. This is useful, for example, in working backward through arrays of variables.

The DO **variable can be real, too**

The examples so far have all had integer DO variables. For an example using a real variable, let us modify the program of Figure 4.2 so that it utilizes a DO statement to run through the values of X, rather than a loop implemented with an IF statement. Figure 7.1 is the modified program.

Other than some minor changes to take advantage of what we learned in Chapter 6 about output formatting, the changes are near the end of the main program. The DO statement combines the actions of initializing X, which was done with a preliminary assignment statement before, the incrementing of X near the end of the loop, and the testing to determine when to terminate rep-

```
*  A modified version of the program of Figure 4.2:
*     Uses a DO loop to generate the 11 values of X
*     Produces formatted output
*
*  Variables:
*     LEFT:   Left end of interval
*     RIGHT:  Right end of interval
*     DELTAX: (RIGHT - LEFT)/10
*     X:      Real variable used in DO loop

       REAL LEFT, RIGHT, DELTAX, ZERVAL
       REAL POLY4F

*  Read LEFT and RIGHT, print column headings, then compute
*     and print the value of the polynomial for 11 equally spaced
*     intervals from LEFT to RIGHT, inclusive

       READ *, LEFT, RIGHT

       PRINT *, '          X          POLYNOMIAL'
       DELTAX = (RIGHT - LEFT) / 10.0

*  Run through all the values of X
       DO 10 X = LEFT, RIGHT, DELTAX
          PRINT 100, X, POLY4F(X)
  100     FORMAT (' ', 2F12.4)
   10  CONTINUE

       END

*  A FUNCTION subprogram to evaluate a certain fourth-degree polynomial

       REAL FUNCTION POLY4F(X)
       REAL X
       POLY4F = 2.0*X**4 - 15.0*X**3 - 2.0*X**2 + 120.0*X - 130.0
       RETURN
       END
```

Figure 7.1 A modified version of the program of Figure 4.2. This version generates the values of X using a DO statement.

etition. This version of the program is not only shorter, but once one is familiar with the operation of the DO statement, easier to understand.

It should also be noted that since the operation of the DO statement is to check its parameters and determine the number of repetitions *before* carrying out the range even once, it would be possible to omit the test to determine whether the value of LEFT is less than the value of RIGHT. If LEFT should be greater than RIGHT, the DO statement would not execute the range at all. But of course that is not fully equivalent to what we had before, since the DO version would give one execution of its range for equal values of LEFT and RIGHT, and in any event there would be no error message.

The output of the program when run with appropriate data is shown in Figure 7.2.

A few rules

The DO statement is subject to a few simple rules.

X	POLYNOMIAL
6.4570	-0.1169
6.4571	-0.0806
6.4572	-0.0439
6.4573	-0.0078
6.4574	0.0288
6.4575	0.0647
6.4576	0.1050
6.4577	0.1387
6.4578	0.1753
6.4579	0.2117
6.4580	0.2520

Figure 7.2 The output of the program of Figure 7.1 when run with a suitable set of data.

1. It is permissible for the range of one DO (which we will call the *outer* DO) to contain one or more other DOs (which we call the *inner* DOs. The DO statements are then said to be *nested*. When this is done, it is required that all statements in the range of an inner DO also be in the range of an outer DO. It turns out that this is only common sense, as will presumably be obvious from the examples that follow, and we will not be tempted to violate the rule. However, various kinds of programming mistakes—incorrect statement numbers, for example—will lead to diagnostic error messages giving notice of unintentional violations of this rule.

2. The last statement in the range of a DO must not be any of the statements having to do with transfer of control: GO TO, another DO, etc. We shall also not be tempted to violate this rule, since the terminal statement of the range will always be a CONTINUE statement, which avoids all such worries, as well as making program modification easier.

3. No statement within the range of a DO may alter any of the parameters of the DO. This is a rule that we might be tempted to break, on the assumption that the termination test would be based on the modified values. But that is not how the statement works. Rather, the number of iterations is computed when the DO is first encountered, and so changes in the DO parameters would simply not be noticed.

4. It is not permissible to transfer control into the range of a DO other than by executing the DO itself.

A puzzle solved with nested DOs

The use of nested DO statements is illustrated in a program to solve a simple puzzle: Find all three-digit numbers that are equal to the sum of the cubes of their digits. The program in Figure 7.3 involves a nest of three DO statements to run through the 900 three-digit decimal numbers. With HUNDRD, TENS, and UNITS having been declared as integer variables, the first DO statement says that the statements in its range should be carried out with HUNDRD taking on the values from 1 through 9. (The first digit must be nonzero to produce a three-digit number.)

```
*  The "Stickler Problem" -- from
*  Forsythe, Keenan, Organick, and Stenberg,
*  Computer Science: a First Course,
*  Fortran Language supplement, 2nd edition, page 226
*
*  Problem: to find all three-digit decimal numbers
*     that are equal to the sum of the cubes of their digits.

       INTEGER HUNDRD, TENS, UNITS

       PRINT 100
   100 FORMAT (' ', 'THE FOLLOWING NUMBERS ARE EQUAL TO'/
      $            ' ', 'THE SUM OF THE CUBES OF THEIR DIGITS'/)

*  Run through the numbers from 100 to 999
       DO 30 HUNDRD = 1, 9
          DO 20 TENS = 0, 9
             DO 10 UNITS = 0, 9
                IF (      (100*HUNDRD + 10*TENS + UNITS)
      $            .EQ. (HUNDRD**3 + TENS**3 + UNITS**3) ) THEN
                   PRINT 200, HUNDRD, TENS, UNITS
   200             FORMAT (' ', 3I1)
                END IF
    10       CONTINUE
    20    CONTINUE
    30 CONTINUE

       END
```

Figure 7.3 A program using three nested DO statements to solve a puzzle: find all three-digit numbers that are equal to the sum of the cubes of their digits.

The first statement within the range of the first DO is another DO. This means that after HUNDRD has been set to 1, TENS will be given all the values from zero to nine. The first statement in the range of the second DO is another DO, giving UNITS the values from zero to nine. The numbers will therefore be generated in the order 100, 101, 102, 103, etc., up through 999.

For each combination of three values, the value of the corresponding three-digit number is computed and compared with the sum of the cubes. If the two are equal, the number is printed. Here is the output when the program was run:

```
THE FOLLOWING NUMBERS ARE EQUAL TO
THE SUM OF THE CUBES OF THEIR DIGITS

153
370
371
407
```

Exercises

1. State in a few words what the following programs would do.

*a.
```
          INTEGER I
          DO 10 I = 1, 11, 2
             WRITE (6, 100) I, I**3
   100       FORMAT (' ', 2I6)
    10    CONTINUE

          END
```

b.
```
        INTEGER K, SUM
        DO 20 K = 21, 31, 2
            SUM = SUM + K
    20  CONTINUE

        WRITE (6, 100) SUM
    100 FORMAT (' ', I5)

        END
```

*c.
```
        INTEGER I
        REAL X, SUM

        SUM = 0.0
        DO 10 I = 1, 10000
            READ (*, *, END=20) X
            SUM = SUM + X
    10  CONTINUE

    20  CONTINUE
        WRITE (6, 100) SUM
    100 FORMAT (' ', E14.7)

        END
```

d.
```
        INTEGER N1, N2, K
        N1 = 0
        N2 = 0
        DO 10 I = 1, 10000
            READ (*, *, END=20) K
            IF ( K .EQ. 1 ) THEN
                N1 = N1 + 1
            END IF
            IF ( K .EQ. 2) THEN
                N2 = N2 + 1
            END IF
    10  CONTINUE

    20  CONTINUE
        WRITE (6, 100) N1
    100 FORMAT (' ', 'THERE WERE ', I3, ' 1''S')
        WRITE (6, 200) N2
    200 FORMAT (' ', 'THERE WERE ', I3, ' 2''S')

        END
```

*2. Identify the errors in these program segments.

a.
```
        DO I = 1, 10
            PRINT *, I, I**4
    10  CONTINUE
```

b.
```
        DO 10 I = 1, 10
            PRINT *, I, I**5
            I = I + 1
    10  CONTINUE
```

c.
```
        DO 10 I = 1, 10
            DO 20 J = 1, 20
                PRINT *, I, J, I**2 + J**2
    10      CONTINUE
    20  CONTINUE
```

d.
```
        READ *, N
        DO 10 I = 1, N
            IF ( N .GT. 10 ) THEN
                N = 10
            END IF
            PRINT *, I, I**6
    10  CONTINUE
```

*3. Write a program that will produce a conversion table from temperatures in Centigrade degrees to Fahrenheit, according to the formula from Chapter 2:

$F = 1.8 C + 32.0$

for all Centigrade temperatures from 0 to 100 in steps of 1 degree. There will be no input to the program.

4. Write a program that reads a card giving two integer values. These are to be the beginning and ending temperatures for a conversion table like that of the previous exercise.

*5. Write a program to print 20 lines, each line giving one of the numbers from 1 to 20, together with its square root.

6. Write a program to print 20 lines, each line giving one of the integers from 50 to 69, together with its square and square root.

*7. Write a program to read a deck of cards, each card containing a real quantity in columns 1–8. Print a line giving the quantity, for all cards in which the number read is greater than 100.

8. Extend the program of Exercise 7 so that it prints a total of the amounts printed. (This means not to include in the total the data values of 100 or less.)

*9. The formula

$$Y = 41.298\sqrt{1 + x^2} + x^{1/3}e^x$$

is to be evaluated for

$x = 1.00, 1.01, 1.02, \ldots, 3.00$

Each $x-y$ pair is to be printed on a separate line, with E20.6 format codes. Write a program using a DO loop to carry out this computation.

10. The formula

$$z = \frac{e^{ax} - e^{-ax}}{2} \sin (x + b) + a \log \frac{b + x}{2}$$

is to be evaluated for all combinations of

x: 1.0(0.1)2.0
a: 0.10(0.05)0.80
b: 1.0(1.0)10.0

where x: 1.0(0.1)2.0 means $x = 1.0, 1.1, 1.2, \ldots, 2.0$, and so on. For each combination of x, a, and b (there are 1650 combinations) a line giving x, a, b, and z is to be written. Write a program containing three DO loops to carry out this computation.

*11. A deck of cards is punched with real numbers in columns 1 to 8, with a decimal point. The numbers should be all positive except for a zero sentinel, but it is thought that there may be quite a few error cards with negative numbers on them. Write a program that will read such a deck and print all the positive numbers together with the card numbers from which they came. Upon detecting the sentinel, print a count of the number of error cards.

12. Modify the program of Exercise 11 as follows. The number from every card (except the sentinel) is to be printed. Every line should have a line number, with the negative numbers printed in a middle column and the positive ones in a right-hand column.

*13. Write a program to determine the largest factorial you can compute on your computer, both as an integer and as a real. With enough knowledge of how your computer stores numbers, this can be determined by analysis, of course. But this exercise asks you to write a program that keeps computing factorials until something "blows up," or until you otherwise establish that the answers are wrong.

   Use two **FUNCTION** subprograms, one for the integer value and one for the real, so that you will have them for possible later use.

14. The value of e, the base of the natural logarithms, is given by

$$1 + \frac{1}{1!} + \frac{1}{2!} + \frac{1}{3!} + \cdots$$

Taking terms through $1/10!$ gives an approximation that is good to six decimals. Write a program using a DO loop and your integer subprogram for the factorial from Exercise 13, to compute and print an approximation to e.

15. The approximation to e can be obtained with far less computation by factoring the series, which requires only that we decide in advance how many terms to use. If we take five terms, we have:

$$1 + \frac{1}{1}\left(1 + \frac{1}{2}\left(1 + \frac{1}{3}\left(1 + \frac{1}{4}\left(1 + \frac{1}{5}\right)\right)\right)\right)$$

Write a program that computes and prints an approximation to e using the factored series, with terms through $1/10!$.

16. The series for e^x is:

$$e^x = 1 + \frac{x}{1!} + \frac{x^2}{2!} + \frac{x^3}{3!} + \cdots$$

Taking terms through $x^{15}/15!$ will give eight decimal digit accuracy for $-2 < x < 2$. Write a program to do this, using the factored series.

17. Write a program to find all integer solutions to the equation $a^2 + b^2 = c^2$, with $a < b$ and $c < 50$. Try to minimize the amount of computation, taking advantage of whatever you know about number theory or—what is the same thing in this case—the geometry of right triangles. See if you can find a way to do the job with only two DO loops, that is, having given values to a and b, simply test whether the sum of their squares is a perfect square and print it if so.

A first look at arrays

Suppose we needed to write a program to read and process 100 numbers, perhaps computing their mean and median. Using the techniques we know so far, it would be necessary to invent 100 data names. All 100 would have to appear in READ statements, and all 100 would have to appear in statements specifying the computation. Doing the job this way is tedious, wasteful, and error-prone. Worse, it is inflexible. If the program had to be able to accept a *variable* number of data points, rather than a fixed number known in advance, many kinds of programs couldn't be written at all. This challenge is answered in Fortran by the use of *arrays*.

A Fortran array is a collection of values having a single name. We specify which of the values we wish to refer to by writing a *subscript* in parentheses after the name. Thus, where in usual mathematical notation we might write x_1 or σ_2, in Fortran we could write X(1) or SIGMA(2). A subscript can be any integer-valued expression, so besides constants as in the examples just shown, we may write things like Z(I) or LINK(K − 1). And, by giving the variables in the subscript expressions appropriate values, we can process a sequence of values in an array.

Arrays are not limited to the linear forms of these examples, but may, for instance, be two-dimensional, as in a table. If we have a table named, let us say, DETERM, which has five rows and five columns, we may specify the entry in the second row and fourth column by writing DETERM(2, 4).

And how does Fortran know that we wish for a variable name to refer to a whole set of values rather than just one, as we have done until now? And how does it know how many elements there are in the array? And how does it know whether we mean a linear array, a table, or a higher dimensional array? All of these questions are answered by an *array declaration*. The simplest way to provide an array declaration is to add it to a type declaration. For example, if we want X to be the name of a linear array having 20 real elements, we can write:

```
REAL X(20)
```

The fact that the type declaration is REAL means that all of the elements of X are of type real; the fact that the name is followed by parentheses giving information about the array size and shape makes X the name of an array rather than the name of a simple variable; the fact that there is just one number within parentheses means that X is a one-dimensional array; and the 20 means that there are 20 elements. The declaration:

```
INTEGER MATRIX (10, 20)
```

would define an array of elements of type integer, having 10 rows and 20 columns for a total of 200 elements.

The simple forms of array declarations illustrated here assume subscripts that run from one up to the size given in the declaration. With X declared by REAL X(20), for example, the array elements would be numbered from 1 to 20. It is also possible to specify what range of subscripts we wish, by writing *subscript bounds*. If we wanted the 20 elements of X to be numbered from zero to 19, for example, we could write:

```
REAL X(0:19)
```

Or if we wanted the rows of a table named CONDUC to be number from 1 to 5 but the five columns numbered from 5 to 10, we could specify:

```
REAL CONDUC (5, 5:10)
```

The ability to specify subscript ranges that do not start at 1 is a notational convenience. In familiar mathematical notation, for example, the coefficients of a polynomial are numbered so that the subscript and the exponent are the same in each term; this makes the constant term a_0, and it makes a program clearer if the array containing the coefficients can be numbered accordingly.

Some examples

Suppose that the ordinates of five points on a curve are represented by the five elements of an array named Y, which appears in the declaration:

```
REAL Y(5)
```

If we need to obtain the average of the first three elements and assign it as the new value of a variable named AVER1, we can write:

```
AVER1 = ( Y(1) + Y(2) + Y(3) ) / 3.0
```

If we want to replace the second element by the average of the first three (including the second), we can write:

```
Y(2) = ( Y(1) + Y(2) + Y(3) ) / 3.0
```

If the spacing between the x values corresponding to these values is given by the value of a variable named H, we can approximate the area under the curve represented by these values, according to the trapezoidal rule, from:

```
TRAP = 0.5 * H * ( Y(1) + 2.0*(Y(2) + Y(3) + Y(4)) + Y(5) )
```

Processing arrays with the DO statement

The ability to process the elements of an array with the DO statement is one of the most powerful features of Fortran. Let us take a simple example to start.

Suppose we wish to read twenty values of an array named X from twenty input records, and, before proceeding to other computations, compute and print the sum and average (arithmetic mean) of the elements. This program fragment does that:

```
      REAL X(20), SUM, AVERAG
      INTEGER I
      SUM = 0.0
      DO 10 I = 1, 10
         READ *, X(I)
         SUM = SUM + X(I)
10    CONTINUE
      AVERAG = SUM / 10.0
      PRINT *, 'SUM =', SUM, ' AVERAGE = ', AVERAG
```

The REAL statement defines X as a one-dimensional array of twenty elements, all of type real. The DO statement causes the READ statement to be executed twenty times, with the values read from twenty input records going into the elements of X in succession from element 1 to element 20. As each value is read, it is immediately added to SUM, which was initialized to zero. When the DO loop terminates, the average is printed along with the sum.

In truth, everything done by this program fragment could be done without arrays; the values could simply be summed as they are read, assuming we have no further need for the individual values themselves. Let us add to the requirement that we are to compute the standard deviation of the ten values, which we obtain from the formula:

$$\text{standard deviation} = \sqrt{\frac{\sum_{i=1}^{n} (X_i - \overline{X})^2}{N}}$$

$$\text{where } \overline{X} = \text{mean} = \frac{\Sigma X_i}{N}$$

We need accordingly to follow the fragment above with statements that compute the difference between each element and the mean, square it, and add to the previous value of the sum of the squares. When this loop is completed we can divide by 20, take the square root, and print the result. This fragment, following the one shown above, will do the job:

```
      SUMSQ = 0.0
      DO 20 I = 1, 20
         SUMSQ = SUMSQ + (X(I) - AVERAG)**2
   20 CONTINUE
      STDDEV = SQRT(SUMSQ/20.0)
      PRINT *, 'THE STANDARD DEVIATION IS ', STDDEV
```

Suppose we wanted to find the largest and smallest of the elements in the array:

```
      BIG = X(1)
      SMALL = X(1)
      DO 30 I = 2, 20
         BIG = AMAX1(BIG, X(I))
         SMALL = AMIN1(SMALL, X(I))
   30 CONTINUE
      PRINT *, 'LARGEST = ', BIG, ';SMALLEST =', SMALL
```

AMAX1 is a function that returns the largest of its two or more real arguments, and AMIN1 is similar for minimum values. The process is to assume that the first element is both the largest and the smallest—which of course won't turn out to be the final result unless all the numbers are equal. Then we test the values of BIG and SMALL against all of the other 19 elements in succession, replacing the value of SMALL whenever we find an element that is smaller than what is in SMALL and replacing the value of BIG whenever we find an element that is larger than what is in BIG.

Exercises

*1. What would the following program do?

```
      REAL X(50), DIFFX(49)
      INTEGER I

      DO 10 I = 1, 50
         READ *, X(I)
   10 CONTINUE
```

```
      DO 20 I = 1, 49
         DIFFX(I) = X(I+1) - X(I)
20    CONTINUE

      END
```

2. What would the following program do?

```
      REAL X(50), DIFFX(49)
      INTEGER I

      DO 10 I = 1, 50
         READ *, X(I)
10    CONTINUE

      DO 20 I = 1, 49
         DIFFX(I) = X(I+1) - X(I)
20    CONTINUE

      PRINT *, ' THE SECOND DIFFERENCES OF THE X VALUES'
      DO 30 I = 1, 48
         WRITE (6, 100) I, DIFFX(I+1) - DIFFX(I)
100      FORMAT (' ', I3, E17.7)
30    CONTINUE

      END
```

Note: The remaining exercises in this set are to give you some practice in programming with subscripted variables, in most case in conjunction with the **DO** statement. You are not asked to write complete programs, but do include **REAL** or **INTEGER** statements for all variables.

*3. An array named A contains 10 elements. Write separate program segments to accomplish the following.

    a. Place the product of the first and second elements in PROD.

    b. Replace the third element by the average of the first, third, and fifth elements.

    c. If the last element is zero or positive, do nothing, but if it is negative reverse its sign.

    d. Replace every element by two times itself, using a **DO** loop.

4. An array named B contains 20 elements. Write separate program segments to accomplish the following:

    a. Divide the fourth element by the sum of the fifth and sixth elements, and place the result in ABC.

    b. Replace the last four elements by zero, without using a loop.

    c. If the tenth element is greater than TEST, replace the tenth element by the average of the ninth and eleventh elements.

    d. Replace very element in the array by zero, using a **DO** loop.

*5. Two one-dimensional arrays named A and B each contain 30 elements. Write a program segment using a **DO** statement to compute

$$D = \left[\sum_{i=1}^{30} (A_i - B_i)^2 \right]^{1/2}$$

6. Two arrays named A and B each have 15 elements. Form the sum of the products of corresponding elements, take the square root of the result, and assign its value to ANORM.

*7. Two one-dimensional arrays named R and S have 40 elements each. The number of elements containing valid data is given by the value of previously computed integer variable M. Compute the first M elements of an array named T, which also has 40 elements, according to

$$T(i) = R(i) + S(i)$$
$$i = 1, 2, \ldots, M$$

If R, S, and T are thought of as vectors, this is the operation of vector addition

8. Two one-dimensional arrays, A and B, have 18 elements each. N is an integer, the value of which does not exceed 18. Compute

$$C = \sum_{k=1}^{N} A_k B_k$$

If A and B are thought of as vectors, then C is their scalar (or inner) product.

*9. The two arrays named A and B have 15 elements each. If the element from A is greater than the corresponding element from B, for all 15 pairs, write a line that says OK; otherwise write a line that says NO WAY.

10. A two-dimensional array named AMATR contains 10 rows and 10 columns. A one-dimensional array named DIAG contains 10 elements. Write a program segment to compute the elements of DIAG from

$$DIAG(I) = AMATR(I, I)$$
$$I = 1, 2, \ldots, 10$$

*11. A one-dimensional array named M contains 20 integers. Write a program segment using a DO statement to replace each element by itself, multiplied by its element number. In other words, replace m_i by $i \cdot m_i$, $i = 1, 2, \ldots,$ 20.

12. A one-dimensional array named F contains 50 elements. Each of the elements, except the first and last, is to be replaced by

$$F_i = \frac{F_{i-1} + F_i + F_{i+1}}{3}$$

This is an example of techniques for *smoothing* experimental data to reduce the effect of random errors.

*13. A one-dimensional array named B contains 50 elements. Place the largest of these elements in BIGB and the element number of BIGB in NBIGB.

*14. A two-dimensional array A contains 15 rows and 15 columns. A one-dimensional array X contains 15 elements. Compute the 15 elements of a one-dimensional array B according to

$$B_i = \sum_{j=1}^{15} A_{ij}X_j \qquad i = 1, 2, \ldots, 15$$

This can be viewed as multiplication of a matrix and a vector.

15. Three two-dimensional arrays A, B, and C have 15 rows and 15 columns each. Given the arrays A and B, compute the elements of C from

$$C_{ij} = \sum_{k=1}^{15} A_{ik}B_{kj} \qquad i, j = 1, 2, \ldots, 15$$

This is matrix multiplication.

A first program using sorting to process exam grades

Many of the ideas that have been presented so far can be illustrated by a program to process the grades received by students on an examination. For concreteness we shall suppose that the requirements are as follows.

There are no more than 50 students in the class. Each person taking the exam received a whole-number grade and is identified by a five-digit number called the student ID. The grades are entered into the computer in the order in which the instructor graded them, and are therefore not in any regular sequence for either grade or ID. The instructor wishes to post a grade report giving all the grades in order by student ID, and also giving a summary containing the highest grade and the lowest grade, each accompanied by the ID of the student receiving it, together with the arithmetic mean of the grades and the median grade. (The median grade is the one that separates all the grades into two groups: half higher and half lower.)

This task requires that the data be *sorted* twice. That is, the grades must first be put into ascending sequence on the student ID's, each ID accompanied by the grade which that student received. Then the grades must be sorted into ascending sequence on the grades so that the program can determine the high, low, and median.

The final version of the program will use a SUBROUTINE to do the sorting, but in order to study the sorting method a little more easily we shall first present a version that sorts on grades only and also prints both arrays after each step of the sorting process. See Figure 7.4.

There are two type-statements here, one for the arrays and one for the non-subscripted variables. This separation is not necessary; all the variables could have been listed in one type-statement, or there could have been eight type-statements.

The first executable statement is a DO that reads the data and computes the sum of the grades. Recall that we are doing no error checking in this program. If there should be more than 50 data records, the program would process the first 50 and give no error indication. Assuming that there are 50 or fewer, the DO loop will be exited with K having a value one greater than the number of records read. K is greater than N because the index of the DO loop is incremented before the END= option detects the end of file condition.

After computing the value of N, we compute the mean (average) grade for later printing.

```
* A program to process a set of exam grades and student ID's
* Program finds mean, median, highest and lowest grades
* Grades and ID's are sorted by method of exchange sorting
*
* Variables:
*     GRADE:   Exam grade, integer array of 50 elements
*     ID:      Student identification, integer array of 50 elements
*     N:       Number of grades; determined by the program
*              (There are assumed to be at least two but no more than 50;
*                  no error checking is performed.)
*     I:       Index of outer sorting loop
*     J:       Index of inner sorting loop
*     K:       Auxiliary variable used for various subscripts
*     SUM:     Auxiliary variable in finding mean
*     MEAN:    Arithmetic mean of grades
*
* (Separate variables not needed for high, low, and median:
*     high will be in GRADE(N) after sorting and low in GRADE(1);
*     median is taken as "middle" entry in table,
*     with no separate handling of odd vs. even number of entries.)
*
* Subprogram:
*     SWAP:    Exchanges two integer arguments

      INTEGER GRADE(50), ID(50)
      INTEGER N, I, J, K, SUM, MEAN

* Read the grades and ID's; sum the grades
      SUM = 0.0
      DO 10 K = 1, 50
         READ (*, *, END = 20) GRADE(K), ID(K)
         SUM = SUM + GRADE(K)
   10 CONTINUE

   20 CONTINUE

* How many records did we read?
      N = K - 1

* Compute arithmetic mean
      MEAN = SUM / N

* Sort the elements of both arrays into ascending sequence
*     on the values of the elements of GRADE

      DO 40 I = 1, N-1
         DO 30 J = I+1, N
            IF ( GRADE(I) .GT. GRADE(J) ) THEN
               CALL SWAP ( GRADE(I), GRADE(J) )
               CALL SWAP ( ID(I), ID(J) )
            END IF
*           Print the arrays so we can watch sorting process
*           Note use of implied DO indexing
            PRINT 100, I, J, (GRADE(K), ID(K), K = 1, N)
  100       FORMAT (' ', 'I =', I3, '  J =', I3/50(' ', 2I7/))
   30    CONTINUE
   40 CONTINUE

* Print the results
      PRINT 200, GRADE(N), ID(N)
  200 FORMAT (' ', 'HIGH GRADE WAS:  ', I4, '  BY', I7, '  GOOD WORK!')
      PRINT 300, GRADE(1), ID(1)
  300 FORMAT (' ', 'LOW GRADE WAS:   ', I4, '  BY', I7, '  (SEE ME)')
      PRINT 400, MEAN
  400 FORMAT (' ', 'MEAN GRADE WAS:  ', I4)
      PRINT 500, GRADE((N+1)/2)
  500 FORMAT (' ', 'MEDIAN GRADE WAS:', I4)

      STOP

      END
```

```
*  A SUBROUTINE to exchange two integer quantities

      SUBROUTINE SWAP ( ARG1, ARG2 )

      INTEGER ARG1, ARG2, TEMP

      TEMP = ARG1
      ARG1 = ARG2
      ARG2 = TEMP

      RETURN

      END
```

Figure 7.4 A program that sorts input data into ascending sequence, then prints various statistics. Note nested DO statements, and implied DO indexing in the first PRINT statement.

Now comes the sorting loop that is the heart of the program. Let us follow the logic of the method in the abstract, then see it in operation with sample data.

The outer DO loop starts its index, I, at 1. Control passes to the inner DO, which runs its index, J, through all the values from one more than I up through and including N. The first grade is therefore compared with every other grade in the GRADE array. The first comparison is thus between GRADE(1) and GRADE(2). If the first element is less than or equal to the second element, we do nothing, but if the first element is greater than the second, we interchange the first and second elements of GRADE, and likewise interchange the first and second elements of ID. The interchanges are carried out in the program by a simple subroutine named SWAP.

After comparing the first and second elements of GRADE and swapping if necessary to make sure that the first element of GRADE is the smaller, we compare the first and third elements of GRADE and swap them if necessary. The first element of GRADE in this second comparison may well be the one that was initially in the *second* element of GRADE, if an interchange was performed the first time. Next we compare the first element of GRADE with the *fourth* element of GRADE, etc., until we have compared the first element of GRADE with the last element of GRADE, in every case exchanging the corresponding values of GRADE and ID whenever necessary to get the smaller of the two GRADE values in the first position.

The result of this process will be to guarantee that the smallest element in the GRADE array is in element number 1. Next we compare the *second* element of GRADE with the third, fourth, etc., through the last. This will assure that the second smallest value of GRADE is in element number 2. We continue in this way, comparing successive elements of GRADE with all following elements, exchanging as necessary, until we have compared the next-to-last and the last.

The net result of the entire process is to place the elements of GRADE into ascending sequence, and to keep the corresponding elements of ID in proper relation to the elements of GRADE.

The program specifies this processing with two nested DO statements, the first of which runs its DO variable from 1 up to one less than the number of elements, and the second of which runs its DO variable from I+1 up to the number of elements.

This first version of the program has been set up with a PRINT statement within the nested DO statements so that we can watch the sorting process. This PRINT statement employs implied DO indexing to print the first N values of each of the two arrays. The technique is to use parentheses to enclose the array names and the indexing information. The FORMAT statement uses the slash to separate records, which in this case causes the printing of the values of I and J on a separate line from the values of the arrays. The FORMAT statement also utilizes the group repeat so that when control reaches the end of the FORMAT statement, control reverts to the beginning of the group repeat, not to the beginning of the FORMAT statement. With a slash at the end of the FORMAT statement, each pair of values from the GRADE and ID arrays will be printed on a separate line and there will be a blank line after the last pair.

It may help understand this PRINT-FORMAT combination to realize that its operation is identical to this sequence of statements:

```
        PRINT 100, I, J
  100   FORMAT (' ', 'I =', I3, '   J =', I3)
        DO 10 K = 1, N
          PRINT 200, GRADE(K), ID(K)
  200     FORMAT (' ' 2I7)
   10   CONTINUE
        PRINT *, ' '
```

The final PRINT is necessary to get the effect of the slash—the insertion of a blank line into the output—after printing the last pair of values.

With the first N values in both arrays now in ascending sequence on the grades, we can print the summary statistics. The highest grade is the one in GRADE(N) and the lowest is the one in GRADE(1). The mean we already have, and the median is the grade is the "middle" of the GRADE array. "Middle" is placed in quotes because it is ambiguous: the "middle" of a table having an odd number of values is clear enough, but what do we do when the number of values is even? When it matters, statisticians take the average of the two middle values as the median in the latter case, and we could certainly arrange for the program to do that as well. All that would be required would be a test to determine whether N is odd or even, which can readily be done with the techniques sketched in Exercise 3 on page 175.

We choose here not to go to this trouble. The "middle" element is taken to be the one identified by the subscript value $(N+1)/2$, which will be the middle element if N is odd and the smaller of the two elements in the middle of the table if N is even.

The work of the program is now complete. The SUBROUTINE SWAP is as we saw it in Chapter 4, except that this time its arguments are of integer type.

The program was run with this data:

```
    84 11111
    77 22222
    37 33333
    91 44444
    56 55555
```

The student ID's were chosen so as to facilitate study of the process of exchanging that is to be seen in the output of Figure 7.5.

Recall that the printing is done *after* the exchange, if any, of array elements. Thus, the five values of GRADE and ID are first printed after element 1 and 2 of each have been swapped. The second time through the inner loop, with I equal to 1 and J equal to 3, the element in GRADE(1) is the one that originally

```
I =    1   J =    2
        77    22222
        84    11111
        37    33333
        91    44444
        56    55555

I =    1   J =    3
        37    33333
        84    11111
        77    22222
        91    44444
        56    55555

I =    1   J =    4
        37    33333
        84    11111
        77    22222
        91    44444
        56    55555

I =    1   J =    5
        37    33333
        84    11111
        77    22222
        91    44444
        56    55555

I =    2   J =    3
        37    33333
        77    22222
        84    11111
        91    44444
        56    55555

I =    2   J =    4
        37    33333
        77    22222
        84    11111
        91    44444
        56    55555

I =    2   J =    5
        37    33333
        56    55555
        84    11111
        91    44444
        77    22222

I =    3   J =    4
        37    33333
        56    55555
        84    11111
        91    44444
        77    22222

I =    3   J =    5
        37    33333
        56    55555
        77    22222
        91    44444
        84    11111

I =    4   J =    5
        37    33333
        56    55555
        77    22222
        84    11111
        91    44444

HIGH GRADE WAS:     91  BY  44444  GOOD WORK!
LOW GRADE WAS:      37  BY  33333  (SEE ME)
MEAN GRADE WAS:     69
MEDIAN GRADE WAS:   77
```

Figure 7.5 The output of the sorting program
when run with the data shown above.

was in GRADE(2), which is simply part of the way the method works. An exchange is needed this time, too. This exchange, as it happens, puts the smallest grade in GRADE(1), but the program of course doesn't "know" that, and proceeds to test GRADE(1) against GRADE(4) and GRADE(5).

At this point we can be absolutely certain of one thing: the smallest grade is in GRADE(1). We accordingly ignore it in further processing, next establishing that the second smallest grade is in GRADE(2), which, as it turns out, requires two exchanges. Getting the next larger grade into GRADE(3) requires only two comparisons and, as it happens, one exchange. Getting the next larger grade into GRADE(4) takes only one comparison and (as it happens) an exchange. Nothing more need be done; the largest grade is now also in GRADE(5).

The full grade processing program

Now that we have seen how the sorting method works and studied the other new aspects of the program, let us extend it to do the full task set out at the beginning of the previous section.

Since we need to sort twice, it seems natural to turn the sorting portion of the program into a SUBROUTINE so that it need appear only once in the program. The task of the main program will then be to read the data, compute the mean, call the sort routine twice, and print the results. (In "real life" the program would probably also print the data as read, before any sorting, so that the instructor could check the correctness of data input, but that would add no educational value for our purposes.)

The program is shown in Figure 7.6. The first part of the program is nearly the same as before, the only difference being that since I and J are no longer needed in the main program they have been removed from the type-statement. Now, after computing the mean, there is a CALL SORT statement in which the arguments are, in order, ID, GRADE, and N. (Observe that when the name of an entire array is being passed to a subprogram no subscripts are written.) The SUBROUTINE named SORT has been written so that it sorts on the elements of the *first* of its arguments, so this CALL means to sort both arrays on the ID's.

A PRINT statement using implied DO indexing, combined with a suitable FORMAT statement, now prints both column headings and the grades in ID sequence.

A second CALL in which the first argument is GRADE now sorts the data into ascending grade sequence, after which the printing is as before.

The SUBROUTINE to do the sorting must be able to sort any two integer arrays of up to 50 elements, so it would be highly misleading to use GRADE and ID as the dummy argument names. A and B have been chosen. Note that both must be declared as arrays *within the subprogram*. Without such a declaration, the compiler would simply report a long series of error messages of the general form "You have used a subscript on a nonsubscripted variable," except that the messages would probably be less explicit than that. We see that statement numbers 10 and 20 have been used here as well as in the main program. This does not constitute duplicate use of the statement numbers, since the main program and the subprogram are fundamentally distinct program segments. The only way they "know anything about each other" is through the arguments in the CALL statement. (There are in fact other ways to provide this communication, as we

```
*  A program to process a set of exam grades and student ID's
*  It first sorts both arrays on ID's and prints both
*  It then finds mean, median, highest and lowest grades
*  Grades and ID's are sorted by method of exchange sorting
*
*  Variables:
*       GRADE:   Exam grade, integer array of 50 elements
*       ID:      Student identification, integer array of 50 elements
*       N:       Number of grades; determined by the program
*                (There are assumed to be at least two but no more than 50;
*                   no error checking is performed.)
*       K:       Auxiliary variable used for various subscripts
*       SUM:     Auxiliary variable in finding mean
*       MEAN:    Arithmetic mean of grades
*
*  (Separate variables not needed for high, low, and median:
*       high will be in GRADE(N) after sorting and low in GRADE(1);
*       median is taken as "middle" entry in table,
*       with no separate handling of odd vs. even number of entries.)
*
*  Subprogams:
*       SORT:    Sorts two integer arrays of N elements on first array
*       SWAP:    Exchanges two integer arguments; used by SORT

         INTEGER GRADE(50), ID(50)
         INTEGER N, K, SUM, MEAN

*  Read the grades and ID's; sum the grades
         SUM = 0.0
         DO 10 K = 1, 50
            READ (*, *, END = 20) GRADE(K), ID(K)
            SUM = SUM + GRADE(K)
   10    CONTINUE

   20    CONTINUE

*  How many records did we read?
         N = K - 1

*  Compute arithmetic mean
         MEAN = SUM / N

*  Sort the elements of both arrays into ascending sequence
*       on the values of the elements of ID
         CALL SORT ( ID, GRADE, N )

*  Print ID's and grades
         PRINT 100, (ID(K), GRADE(K), K = 1, N)
  100    FORMAT (' ', '    ID    GRADE'//50(' ', 2I7/))

*  Sort the elements of both arrays into ascending sequence
*       on the values of the elements of GRADE
         CALL SORT ( GRADE, ID, N )

*  Print the results
         PRINT 200, GRADE(N), ID(N)
  200    FORMAT (' ', 'HIGH GRADE WAS:   ', I4, '  BY', I7, '  GOOD WORK!')
         PRINT 300, GRADE(1), ID(1)
  300    FORMAT (' ', 'LOW GRADE WAS:    ', I4, '  BY', I7, '  (SEE ME)')
         PRINT 400, MEAN
  400    FORMAT (' ', 'MEAN GRADE WAS:   ', I4)
         PRINT 500, GRADE((N+1)/2)
  500    FORMAT (' ', 'MEDIAN GRADE WAS:', I4)

         END
```

Figure 7.6 An extended version of the sorting program of Figure 7.4. This program sorts data on both input arrays, for different purposes.

```
*  A SUBROUTINE to sort the elements of two arrays into ascending
*     sequence on the values of the first one.
*
*  Arrays are assumed to have a maximum of 50 elements each;
*     N, the actual number of elements, is a third argument
*
*  Arguments:
*     A:        Array on which both arrays are sorted
*     B:        Array of same size as A
*     N:        Actual number of elements in A and B
*
*  Variables local to subprogram:
*     I:        Index of outer DO
*     J:        Index of inner DO

      SUBROUTINE SORT ( A, B, N )

      INTEGER A(50), B(50), N
      INTEGER I, J

      DO 20 I = 1, N-1
         DO 10 J = I+1, N
            IF ( A(I) .GT. A(J) ) THEN
               CALL SWAP ( A(I), A(J) )
               CALL SWAP ( B(I), B(J) )
            END IF
10       CONTINUE
20    CONTINUE

      RETURN

      END

*  A SUBROUTINE to exchange two integer quantities

      SUBROUTINE SWAP ( ARG1, ARG2 )

      INTEGER ARG1, ARG2, TEMP

      TEMP = ARG1
      ARG1 = ARG2
      ARG2 = TEMP

      RETURN

      END
```

Figure 7.6 (*cont'd*) An extended version of the sorting program of Figure 7.4. This program sorts data on both input arrays, for different purposes.

shall see in Chapter 9, but we choose to use these other facilities with great caution.)

This version of the program was run with a longer set of data, producing the output of Figure 7.7.

Exercises

*1. How would the program of Figure 7.4 operate if the second CALL SWAP were omitted?

ID	GRADE
10926	66
11227	62
11452	87
21895	67
23415	88
31805	66
36173	88
38097	60
49002	78
50020	61
55909	59
57244	75
58290	41
62210	73
65109	83
72231	93
77004	91
89077	47

```
HIGH GRADE WAS:     93  BY  72231   GOOD WORK!
LOW GRADE WAS:      41  BY  58290   (SEE ME)
MEAN GRADE WAS:     71
MEDIAN GRADE WAS:   67
```

Figure 7.7 The output of the program of Figure 7.6 when run with a new set of data.

2. How would the program of Figure 7.4 operate if, in the subroutine SWAP, TEMP were declared to be real instead of integer?

*3. Modify the program of Figure 7.4 so that, if the number of grades is even, the average of the two middle grades is used as the median. [*Hint:* The MOD function returns the remainder after dividing its first (integer) argument by the second. Thus, MOD(N, 2) will be zero if N is even and 1 if N is odd.]

4. Modify the program of Figure 7.6 to incorporate the following changes.

■ Print the raw grades, that is, the grades as read, before sorting.

■ Incorporate the change in computing the median, as in Exercise 3.

■ Modify the SORT subroutine so that it takes a fourth argument, which specifies whether the sort is to be in ascending or descending sequence. Then change the second sort so that it sorts the data in descending sequence on the grades and prints the data in this order. (This will require minor changes in the statements for printing the highest and lowest grades.)

*5. A biology student is working with several hundred experimental insects, which fall into 20 classes numbered from 1 to 20. Part of the experiment requires recording the weight of each insect and producing a table showing the average weight of each class.

You are given a deck of cards, each card containing a class number in columns 1–2 and a weight in grams in columns 3–8 with a decimal point. Write a program that will read each card, up to a zero sentinel, counting the number in each class and adding up their weights. When the sentinel is detected, compute and print all the average weights, along with identifying class numbers.

6. A psychology student has accumulated some hundreds of test scores, each of which falls into one of 30 groups. He wants to produce a report showing the highest and lowest scores in each group. Make reasonable assumptions about card format and a sentinel, and write a program to produce such a result. The output is to consist of 30 lines, each of which contains a group number, the number of items in that group, and the lowest and highest scores for that group.

7. The sorting method used here requires making identical exchanges on both of the arrays, to get the data into sequence on the values in one of them. If each record consisted of many fields instead of just two, this would involve a great deal of data movement. We can eliminate this costly internal data movement by taking advantage of the fact that a Fortran subscript can itself be a subscripted variable, as follows.

We establish a new one-dimensional integer array named NELMNT, having as many elements as the data being sorted. We shall assume that the data is being sorted on the contents of ID. At the beginning of the program, we load each element of NELMNT with its element number, so that NELMNT(I) = I. Then during the process of comparing elements of the ID array we make three changes. First, whenever it has been established that two elements, call them J and K, should be exchanged, we in fact do nothing at all with the ID array elements, but instead exchange elements J and K of NELMNT. Second, whenever we wish to refer to an element of ID in the testing, the subscript of ID becomes the appropriate subscripted value of NELMNT. For instance, the comparison of two elements of ID might involve a statement like:

```
IF ( ID(NELMNT(J)) .GT. ID(NELMNT(K)) ) THEN
```

Third, during the comparison phase, we do nothing at all with the array GRADE.

When we have gone through a complete set of passes of the ID array, carrying out all exchanges of NELMNT that may be required, the elements of NELMNT will specify the order in which the elements of ID and GRADE should be picked up to place them in ascending sequence on the elements of ID. For example, this statement would print the largest ID and the corresponding grade:

```
PRINT (6, 100) ID(NELMNT(N)), GRADE(NELMNT(N))
```

The Gauss-Seidel method for solving simultaneous equations

We turn now to a larger problem, one that nicely illustrates the use of arrays and the DO statement in a standard numerical method. The application is the solutions of a system of simultaneous linear algebraic equations, which is required in many scientific and engineering uses of computers. Several common methods are available for finding the solution; we shall employ one that is fast and accurate when it applies. The limitation is given below.

A simple way to get started is to describe the method in terms of a system of three equations in three unknowns:

$$a_{11}x_1 + a_{12}x_2 + a_{13}x_3 = a_{14}$$
$$a_{21}x_1 + a_{22}x_2 + a_{23}x_3 = a_{24}$$
$$a_{31}x_1 + a_{32}x_2 + a_{33}x_3 = a_{34}$$

Suppose we make guesses at the values of x_2 and x_3—it doesn't matter whether they are good guesses; zeros will work. Then we solve the first equation for x_1, writing a prime to indicate that this is a new approximation:

$$x_1' = \frac{a_{14} - a_{12}x_2 - a_{13}x_3}{a_{11}}$$

Now using this new value for x_1 and the initial guess for x_3, we solve the second equation for x_2:

$$x_2' = \frac{a_{24} - a_{21}x_1' - a_{23}x_3}{a_{22}}$$

Finally, using the new approximations to x_1 and x_2, we solve the third equation for x_3:

$$x_3' = \frac{a_{34} - a_{31}x_1' - a_{32}x_2'}{a_{33}}$$

This process of computing a new value for each of the variables constitutes one *iteration*. Now we perform another iteration, always using the most recently computed value of each variable. If the system of equations satisfies certain conditions that will be stated shortly, the approximations will eventually converge to a solution of the system of equations, regardless of how good or bad the initial guesses were. This is called the *Gauss-Seidel iteration method* for solving a system of simultaneous equations.

A *sufficient* condition for convergence is that the main diagonal term in each row dominate the other coefficients in that row, which means that:

$$|a_{ii}| > \sum_{j \neq i} |a_{ij}| \qquad \begin{aligned} i &= 1, 2, \ldots, n \\ n &= \text{number of equations} \end{aligned}$$

We emphasize that this condition is *sufficient*. *Necessary* conditions are considerably less stringent, but unfortunately are beyond the scope of this text. In other words, there are systems of equations in which the main diagonal terms do not all dominate their rows, yet the Gauss-Seidel method does converge to a solution. Furthermore, there are methods for greatly accelerating the convergence, which are heavily used in the numerical solution of certain partial differential equations. An example appears in Chapter 9.

We shall present a program that applies the Gauss-Seidel method to any system of equations having up to 50 equations. The program will include the reading of coefficients and constant terms, which could total as many as 2550. The program performs a certain amount of data validation as the input is read. A test for convergence determines when to stop iterating, with a maximum number of permitted iterations read as part of the data, to prevent endless iteration in the case of a system that cannot converge. Zeros are used as the starting guesses.

A pseudocode of the organization and operation of the program to be written is shown in Figure 7.8. The program will consist of a main program and two subroutines. The main program begins by initializing all the arrays to zero; this is all it takes to specify zeros as the initial guesses and to make it unnecessary

---

A program to solve simultaneous equations by the Gauss-Seidel method:

```
Main program:
Initialize all arrays to zero
CALL READER to get and validate data
IF any errors THEN
    PRINT error message
    STOP
END IF
DO WHILE no more than maximum allowed iterations
    CALL ITRATE to compute one new approximation
    PRINT latest approximation to solution
    IF convergence achieved, STOP
END WHILE
IF no convergence, PRINT error message and STOP

READER subroutine:
READ parameters: N, max iterations, convergence test,
    and biggest allowable coeficient or constant term
Initialize error count to zero
REPEAT until end of file
    READ I (row number), J (column number), and a data value
    IF subscript out of range or value too big THEN
        PRINT error message, with I, J, and data value
        Add 1 to error count
    ELSE
        Store data value
    END IF
END REPEAT

ITRATE subroutine:
Performs one iteration of the Gauss-Seidel method
Stores new approximations to the unknowns,
    and returns maximum residual, RESID, which is the maximum
    difference between two successive approximations to an unknown

RESID <-- 0.0
For each row I:
    Sum the off-diagonal terms
    Compute new X(I) and store temporarily
    Update RESID if necessary
    Store the new X(I)
```

Figure 7.8 Pseudocode of a method for solving a system of simultaneous equations by the Gauss-Seidel iteration method.

to read the zero coefficients. Next the main program calls a subroutine that reads the data and makes certain rudimentary validity checks; one of its output parameters is the number of data errors it detected. Solution is not attempted unless this number is zero. The main program next repeatedly calls the subroutine that computes one new approximation, up to a maximum number that was specified in the input. After each iteration it prints the latest approximate solution, so we can watch convergence, and stops if convergence has been achieved. If convergence is not reached in the maximum specified iterations, an error message is printed.

The subroutine to read the data is as described already. The validity checking is simply to make sure that the subscripts are within allowable limits, and that the data value read is not greater (in absolute value) than the biggest permissible, which is a parameter read at the beginning of the subroutine.

As noted, the "iteration" subroutine actually computes just one new approximation to the solution, that is, a new set of values for all of the variables. The repeated calling of the subroutine is handled by the main program. The subroutine ITRATE, therefore, has a relatively simple task. It first sets to zero a variable which at the end of the process will contain the largest residual, i.e., the largest difference (in absolute value) between two successive approximations to any unknown. Then it computes a new set of approximations, one row at a time, always using the latest values of all unknowns. As each new unknown value is computed, it is first stored in a temporary location for use in the residual calculation, then placed in its array location.

The program is shown in Figure 7.9. Since it follows the pseudocode closely, we need only point out some of the programming considerations. Observe that the array that holds the coefficients and constant terms, A, is declared as having 50 rows and 51 columns, and that this is done in the main program and in both subroutines. Similarly with the array for the unknowns, X. It is important to realize that since A and X are arguments of the subroutines, declaring them in the subroutines does not cause the allocation of additional storage. This is not merely a matter of saving storage space, but also of proper communication among the various routines: it is of course essential that A should mean the same thing wherever it appears in a program, and similarly with X.

This way of doing things, reserving space for the maximum size arrays and then using only a small portion of it, naturally wastes considerable storage space for small systems. We shall see in Chapter 9 how this could be avoided.

Observe in READER the use of the END= option on the READ statement, which is a natural way to handle a situation where we have no idea how much data there may be, even after we know the size of the system, because only the non-zero values are required to be provided. Notice also how the subscripts are checked against the array size implied by the value of N. It would not do, of course, to check against the maximum size of the array.

The iteration subroutine, ITRATE, follows the pseudocode very closely. The most direct way to get the off-diagonal terms in each row is to provide two DO statements, one of which runs from the first term in the row up to the one before the diagonal, and another that runs from the term after the diagonal to the end of the row. And these DO statements work correctly for the first row, where there is no term before the diagonal, and the last, where there is no term after it. Look at the first DO, for example. When I is 1, specifying the first row, which has no term before the diagonal, I−1 will be zero. This means that the range of the DO will not be executed at all for the first row. Similarly for the last row.

This program was run with data specifying the following system of equations:

$$6x_1 \quad\quad + \quad x_3 = \quad 9$$
$$x_1 \; -5x_2 \quad\quad = \; -9$$
$$2x_2 + 9x_3 = \quad 31$$

It produced this output:

```
1    1.500000  2    2.099999  3    2.977776
1    1.003704  2    2.000740  3    2.999835
1    1.000027  2    2.000005  3    2.999998
```

```
***********************************************************
* The Gauss-Seidel method for solving simultaneous equations
***********************************************************
*
* This program solves a system of N equations in N unknowns;
*    N (not greater than 50) is read as input.
* Only the nonzero elements need be read.
* Each data record contains one element, with its row and column number.
* The program reads the following parameters prior to reading
*    the coefficients and constants:
*
*    N:      The number of equations in the system
*    MAXIT:  The maximum number of iterations to be permitted
*    EPS:    The convergence criterion
*    BIGGST: The maximum size (in absolute value) to be permitted
*            of any coefficient or constant term
*
* The coefficients are stored in a 50 by 51 two-dimensional array,
*    with the constant term for equation N in column N+1 of its row.
*
* All input is checked for validity, even if errors are found;
*    no solution is attempted with invalid data.
*
* Other variables:
*    I:      Row index in various places
*    J:      Column index in various places
*    ERRORS: Error count returned by subprogram READ
*    ITER:   Iteration count; tested against MAXIT
*
* Subprograms:
*    READER: Reads data, validity tests it, returns error count
*    ITRATE: Performs one iteration of the Gauss-Seidel method

      REAL A(50, 51), X(50), EPS
      INTEGER N, I, J, MAXIT, ERRORS, ITER

* Initialize the arrays to zero
      DO 20 I = 1, N
         X(I) = 0
         DO 10 J = 1, N+1
            A(I, J) = 0
   10    CONTINUE
   20 CONTINUE

* Get the data and a count of errors
      CALL READER (A, N, EPS, MAXIT, ERRORS)

      IF ( ERRORS .NE. 0 ) THEN
         PRINT 100, ERRORS
  100    FORMAT ('0', I5, ' ERRORS FOUND - SOLUTION NOT ATTEMPTED')
         STOP
      END IF

* Call the solution subroutine, ITRATE, up to MAXIT times.
* Note that in this developmental version of the program,
*    output is handled by ITRATE so we can watch convergence.
      DO 30 ITER = 1, MAXIT
         CALL ITRATE (A, X, N, RESID)
*        print latest approximation to solution
         PRINT 200, (I, X(I), I = 1, N)
  200    FORMAT (' ', 6(I3, F12.6))
*        check for convergence:
         IF ( RESID .LT. EPS ) STOP
   30 CONTINUE

*     if next statement ever reached, there was no convergence:
      PRINT 300, MAXIT
  300 FORMAT ('0', 'NO CONVERGENCE IN', I5, ' ITERATIONS')

      END
```

Figure 7.9

```
***********************************************************************
* A SUBROUTINE to read the data and count errors
***********************************************************************
      SUBROUTINE READER (A, N, EPS, MAXIT, ERRORS)

      REAL A(50, 51), EPS, BIGGST, TEMP
      INTEGER N, I, J, MAXIT, ERRORS

* Read the parameters
      READ *, N, MAXIT, EPS, BIGGST

* Initialize the error count
      ERRORS = 0
* Read the elements of the array, with checking.

***** REPEAT until end of file
   10 CONTINUE
         READ (*, *, END = 20) I, J, TEMP
         IF (          (I .LT. 1)
     $        .OR. (I .GT. N)
     $        .OR. (J .LT. 1)
     $        .OR. (J .GT. N+1)
     $        .OR. (ABS(TEMP) .GT. BIGGST) ) THEN
            ERRORS = ERRORS + 1
            PRINT 100, I, J, TEMP
  100       FORMAT (' ', 'ERROR IN ELEMENT WITH I = ', I2, ' J = ', I2,
     $                   ' VALUE = ', F10.6)
         ELSE
            A(I, J) = TEMP
         END IF
      GO TO 10
***** END REPEAT

   20 CONTINUE

      RETURN

      END

***********************************************************************
* The SUBROUTINE to iterate for a solution
***********************************************************************
      SUBROUTINE ITRATE (A, X, N, RESID)

      REAL A(50, 51), X(50), RESID, SUM, TEMP
      INTEGER N, I, J

*     initialize the residual, i.e., the difference between
*     successive values of variables:
      RESID = 0

*     index I selects a row:
      DO 30 I = 1, N

*        next statement is executed once per row:
         SUM = 0

*        get sum of terms in row I, excluding diagonal:
         DO 10 J = 1, I-1
            SUM = SUM + A(I, J) * X(J)
   10    CONTINUE
         DO 20 J = I+1, N
            SUM = SUM + A(I, J) * X(J)
   20    CONTINUE

*        compute a new approximation to variable X(I):
         TEMP = (A(I, N+1) - SUM) / A(I, I)

*        at end of sweep, the following statement will have put
*        largest residual in RESID:
         RESID = AMAX1 (RESID, ABS(TEMP - X(I)))

*        store new approximation to X(I):
         X(I) = TEMP
   30 CONTINUE

*     one sweep has now been completed

      RETURN

      END
```

Figure 7.9 (cont'd) A program for solving a system of simultaneous equations by the Gauss-Seidel iteration method, corresponding to the pseudocode of Figure 7.7.

We observe fairly rapid convergence to within the given convergence criterion (0.01) of the exact answer of (1, 2, 3).

The program was next run with intentionally erroneous data for a system of three equations and a maximum data value of 100:

```
ERROR IN ELEMENT WITH I =  0 J =  2 VALUE =   -1.060000
ERROR IN ELEMENT WITH I =  6 J =  3 VALUE =    2.669000
ERROR IN ELEMENT WITH I =  1 J =  0 VALUE =   -0.119000
ERROR IN ELEMENT WITH I =  2 J =  1 VALUE =  222.000000
ERROR IN ELEMENT WITH I =  2 J =  2 VALUE = **********

    5  ERRORS FOUND - SOLUTION NOT ATTEMPTED
```

The asterisks indicate a value that was not only too large to pass the validity test, but was also too large to print in the space available.

Finally, the program was run with data specifying this system:

$$
\begin{aligned}
x_1 + x_2 + x_3 \qquad\qquad &= \quad 2 \\
x_1 + 3x_2 - x_3 + x_4 \qquad &= -12 \\
x_1 - 3x_2 + 7x_3 \qquad\qquad &= \quad 30 \\
x_1 + 2x_2 \qquad\qquad - 4x_4 &= -13
\end{aligned}
$$

It produced this output:

```
1     2.000000   2    -4.666666   3     1.428574   4   1.416667
1     5.238092   2    -5.742060   3    -0.420061   4   1.688493
1     8.162121   2    -7.423553   3    -2.393860   4   1.578753
1    11.817413   2    -9.263341   3    -4.748892   4   1.572685
1    16.012222   2   -11.444595   3    -7.481493   4   1.530758
1    20.926086   2   -13.979441   3   -10.673795   4   1.491802
1    26.653229   2   -16.939590   3   -14.396922   4   1.443512
1    33.336502   2   -20.392303   3   -18.740906   4   1.387974
1    41.133209   2   -24.420685   3   -23.808807   4   1.322960
1    50.229492   2   -29.120407   3   -29.721375   4   1.247169

  NO CONVERGENCE IN    10   ITERATIONS
```

This system does have a solution (1, -2, 3, -4) but the main diagonal terms are not, in all cases, larger in absolute value than the sum of the absolute values of the other terms in their rows, and the method diverges.

A discussion of the choice in methods for solving simultaneous equations is well beyond the scope of this text. The message of this final example is simply that, although the Gauss-Seidel method has decided advantages in the cases to which it applies, it does not always apply.

Exercises

*1. Modify the READER subroutine in Figure 7.9 to include a test that the system of coefficients satisfies the sufficient condition for convergence of the Gauss-Seidel method, i.e., that:

$$|a_{ii}| > \sum_{i \neq j} |a_{ij}| \qquad i = 1, 2, \ldots, n$$

2. Continuing with Exercise 1, a slightly less stringent condition actually suffices. The greater than sign can be replaced by greater than or equal, so long as in at least one equation there is strict inequality. Modify READER to make this test.

*3. Modify the program of Figure 7.9 so that it prints the solution only at the end, if and when convergence has been achieved.

*4. A solution to the following specialized system of equations is to be found:

$$
\begin{aligned}
a_{11}x_1 &= b_1 \\
a_{21}x_1 + a_{22}x_2 &= b_2 \\
a_{31}x_1 + a_{32}x_2 + a_{33}x_3 &= b_3
\end{aligned}
$$

$$
a_{n1}x_1 + a_{n2}x_2 + a_{n3}x_3 + \cdots + a_{nn}x_n = b_n
$$

First write a program to solve this system on the assumption that the a's are contained in a two-dimensional array that will have zeros for the missing elements. This is a moderately simple program: first solve for x_1, substitute this result into equation 2, and so on.

The difficulty, from the standpoint of computer solution, is that there is a great deal of wasted space in the array, which uselessly restricts the size of the system that can be solved. Devise a method of storing the coefficients in a *one*-dimensional array and write a program to find the unknowns. Assume that it must be possible to handle a maximum of 100 equations in 100 unknowns. The actual number of equations is given by the value of N.

5. Same as Exercise 15, except that the system of equation is

$$
a_{1,\,1}x_1 + \cdots + a_{1,n-2}x_{n-2} + a_{1,n-1}x_{n-1} + a_{1,n}x_n = a_{1,n+1}
$$

$$
\begin{aligned}
a_{n-2,n-2}x_{n-2} + a_{n-2,n-1}x_{n-1} + a_{n-2,n}x_n &= a_{n-2,n+1} \\
a_{n-1,n-1}x_{n-1} + a_{n-1,n}x_n &= a_{n-1,n+1} \\
a_{n,n}x_n &= a_{n,n+1}
\end{aligned}
$$

6. Given a two-dimensional array named C, with 10 rows and 11 columns, compare $C(1, 1)$ with all other elements in the first column, looking for the element with the largest absolute value; make the value of L equal to the row number of the element in column 1 with the largest value. If at the end of these operations $L = 1$, do nothing more; otherwise exchange the elements in row 1 with the elements in row L, whatever it is.

7. One series for inverse sine is

$$
\sin^{-1}x = x + \frac{x^3}{2 \cdot 3} + \frac{1 \cdot 3x^5}{2 \cdot 4 \cdot 5} + \frac{1 \cdot 3 \cdot 5x^7}{2 \cdot 4 \cdot 6 \cdot 7} + \cdots
$$

This can be evaluated directly, with a loop developing one term at a time, or the series can be rewritten as:

$$
\sin^{-1}x = x\left(1 + \frac{1x^2}{2}\left(\frac{1}{3} + \frac{3x^2}{4}\left(\frac{1}{5} + \frac{5x^2}{6}\left(\frac{1}{7} + \cdots\right)\right)\right)\right)
$$

Evaluating the rewritten series, which involves very much less arithmetic for the same number of terms, requires determining in advance how many terms are to be computed, since the nested parentheized expression must be evaluated from the inside out.

Write a program, using a **DO** loop to generate the integers needed in the nested version, to evaluate this series. Let $x = 0.5$, evaluate the series, and multiply the result by 6.0 to provide an estimate for the value of π.

8. The Gauss-Seidel program of Figure 7.9 was run with data for this system:

$$3x_1 + x_2 + x_3 + x_4 = 3$$
$$x_1 + 4x_2 - x_3 - x_4 = -10$$
$$x_1 + 2x_3 - x_4 = 5$$
$$x_1 + x_2 + 2x_4 = -3$$

This was the output:

1	1.000000	2	-2.750000	3	2.000000	4	-0.625000
1	1.458333	2	-2.520833	3	1.458333	4	-0.968750
1	1.677083	2	-2.796875	3	1.177083	4	-0.940104
1	1.853298	2	-2.904079	3	1.103299	4	-0.974609
1	1.925130	2	-2.949110	3	1.050130	4	-0.988010
1	1.962330	2	-2.975052	3	1.024830	4	-0.993639
1	1.981287	2	-2.987524	3	1.012537	4	-0.996881
1	1.990623	2	-2.993741	3	1.006248	4	-0.998441
1	1.995311	2	-2.996876	3	1.003124	4	-0.999218
1	1.997656	2	-2.998437	3	1.001563	4	-0.999609
1	1.998827	2	-2.999218	3	1.000782	4	-0.999804
1	1.999413	2	-2.999609	3	1.000391	4	-0.999902
1	1.999706	2	-2.999804	3	1.000196	4	-0.999951
1	1.999853	2	-2.999902	3	1.000097	4	-0.999976
1	1.999927	2	-2.999950	3	1.000049	4	-0.999988

Observe that after the first few lines, each variable approaches its final value from the same direction, slowly arriving at a stable value. With such a convergent system, it can materially speed up the convergence to *accelerate* the process, as follows.

At the point where a new approximation to the ith variable has been computed but not yet stored (we called it x_i'), replace the old value of that variable, x_i, not with x_i' but with $x_i + \omega(x_i' - x_i)$. Assuming that ω is greater than 1.0, this will move the new approximation beyond what it would otherwise have been, in other words, accelerating the convergence.

The Gauss-Seidel method is almost always used with acceleration because the reduction in the number of iterations is almost always worth the trouble, and sometimes results in dramatic savings.

Even more interesting, in some cases, is the fact that a divergent system can sometimes be made to converge by *decelerating* the process, that is, by using a value of $\omega < 1.0$.

Modify the program of Figure 5.12 to read a value of OMEGA along with the other parameters, and experiment with values of OMEGA to study the effect on convergence rate. It might be interesting, for example, to prepare a plot of the number of iterations required for convergence, as a function of OMEGA, for values of OMEGA from 1.0 to 1.9 in steps of 0.05.

The rate of convergence as a function of ω depends very strongly on the coefficients of the system.

EIGHT

DOUBLE PRECISION, COMPLEX, LOGICAL, AND CHARACTER VARIABLES

Introduction

So far we have used variables, constants, and operations that were either of type integer or of type real. These are adequate for many scientific and engineering applications, but are quite restrictive in others. In this chapter we shall investigate four additional types of variables and associated constants, operations, and functions:

■ Double precision, with which we can avoid, or at least postpone, the problems created by the finite precision with which quantities are necessarily represented in a digital computer.

■ Complex, which permits such operations as finding the complex roots of a polynomial, solving systems of simultaneous equations having complex coefficients, etc. Many problems in electrical engineering, in particular, are most conveniently formulated in terms of complex variables.

■ Logical, which permits simple formulation of problems in which variables have only the "values" true and false. This facility is useful for storing the result of a decision, and for any problem in which variables are inherently limited to two values.

■ Character, which permits convenient use of Fortran in a wide variety of applications in which some or all of the data is not numerical at all, but rather, is symbolic.

For each of the four we shall consider the basic concepts and definitions, then see a case study that illustrates the ideas in practice.

The chapter will not be of equal interest to all readers. Complex variables are used more heavily in certain engineering applications, and logical variables have limited application. Some skimming is in order, depending upon interests. All readers should look at the material on double precision and character variables, however. Double precision must often be resorted to in order to sidestep roundoff errors; the case study in this section is highly instructive. And character variables will often be a useful adjunct of other operations, such as identifying output—not to mention the many applications in which processing of symbolic data is at the heart of the work.

Double precision constants, variables, and operations

A double precision number is one that is represented and used more or less like a real number but has more digits. In some computers, though not all, there are in fact twice as many digits, as suggested by the name. It is probably safe to say *at least* twice as many; one important class of computers has about two and one-third times as many digits in a double precision number as in a real number. In both cases a number is represented in *floating point,* that is, as a fraction times a number base raised to a power.

The extra digits of precision in the representation of numbers are most commonly used to guard against the damaging effects of rounding errors in long sequences of arithmetic operations that combine very large and very small numbers in addition or subtraction. We shall examine some examples of this need later.

A double precision constant is written in exponent form like a real constant, but with a D in place of the E. The following are acceptable double precision constants:

```
1.5D0
5.0D4
5.0D-4
1.2345678923456D0
```

A double precision variable is one that has been named in a DOUBLE PRECISION type statement. The initial-letter naming convention has no application here because a variable is never considered to be double precision unless it has been so declared.

Arithmetic expressions are formed in double precision according to the same rules that govern the hierarchy of operators when using real expressions.

Double precision operands may be combined in arithmetic expressions with real (single precision) and integer operands. In such cases the real and/or integer operands are first converted to their double precision equivalents and the arithmetic is done in double precision.

It is permissible for an assignment statement to have an expression of type integer, real, or double precision on the right of the equal sign and a variable of some other type on the left. When this happens the arithmetic is done according to the usual rules governing expressions, and the result is converted to the type of the variable on the left just before the value is stored.

In the following examples, assume that variables beginning with I are of type integer, those beginning with R are of type real, and that those beginning with D are of type double precision. Then all of the following are legitimate assignment statements:

```
D1 = D2*D3 + (D4 - 8.76543231D1) / D5
D1 = 4.0*D2 - D3/1.1D0
D1 = R1 + D1 + R2
R1 = (D1*D2 + D3*D4) / (D1*D5 - D3*D6)
```

```
D1 = R1 + 2.0
D1 = (I1 - 8) * I2
D1 = D2**2
RTRAP = (R1/3) * (RA + 2.0*DSUM + RB)
```

Observe in the first example that the characters D1 appear both as a variable name and as the exponent in a constant; the Fortran compiler has no difficulty distinguishing the two uses. Notice in the second example that a simple-appearing constant is written as double precision. This is worth doing in a binary or hexadecimal machine because the decimal number 1.1 has no exact equivalent in binary or hexadecimal. The double precision approximation will naturally be closer to the true value than the single precision. In the third example both additions will be done in double precision. In the fourth example all arithmetic will be done in double precision and the result converted to real form before being stored as the new value of R1.

The situation in the last example is a little more complicated. The 3 will first be converted to real form and the division done in real. Then 2.0 will be converted to double precision form and the multiplication done in double precision. Next RA and RB will be converted to double precision and the additions done in double precision. Next the quotient will be converted to double precision and the final multiplication done in double precision. Finally, the result will be converted to single precision and stored as the new value of RTRAP. (A "smart" compiler would probably anticipate the need for converting the quotient to double precision and do it at the earlier stage.)

This last example is a realistic one, illustrating one common application of double precision variables. DSUM is assumed to contain the sum of a great many real values; in the course of summing them, much significance can be lost in the course of adding small numbers to large. The use of double precision prevents this loss, even though all the other quatitites in its use are of single precision (real) length. See pp. 230–231 in the following chapter for a program that illustrates this phenomenon.

Built-in double precision functions are provided for the standard mathematical functions. The naming system is straightforward: the names of all double precision functions begin with D. Thus the double precision square root is DSQRT and the double precision cosine is DCOS. A complete list is provided in the Appendix, where it will be noted that there are a few special-purpose functions involving double precision. For instance, the function SNGL ("single") takes a double precision argument and supplies as the function value the real representation of the most significant part of the argument. Most double precision functions, contrary to this example, take one or more double precision arguments and return a double precision function value.

One new feature is involved in the input and output of double precision values: the D format specification. This may be used to make it explicit that double precision quantities are involved; the main result is to print double precision values with a D instead on an E in the exponent. On the other hand, the new field specification is not really needed, because the E field specification will also produce a D exponent when the value being printed is double precision. Input is simple: just put a D exponent on the value, and any of the field specifications F, E, or D will accept the value correctly. Output without an exponent can be handled with an F field specification, by simply providing more space.

Error analysis in a sine series evaluation

The Taylor series for the sine:

$$\sin x = x - \frac{x^3}{3!} + \frac{x^5}{5!} - \frac{x^7}{7!} + \cdots$$

is usually described as valid for any finite angle, and the truncation error is said to be less in absolute value than the first term neglected. That is, we can get as accurate an approximation to the sine as we need simply by taking many terms and stopping after finding that they have gotten sufficiently small. These statements are "mathematically" true, which is to say that they are true in some ideal universe in which all quantities are represented exactly. In the nonideal universe of digital computers with finite representation of numbers, they are decidedly not true. Actually the Taylor series is useless for large angles. Fortunately, there are easy ways around the problem, and the built-in sine functions have no problem with large angles.

This case study takes a direct approach to the sine series, to show one circumstance in which double precision can be useful. By a "direct" approach to the sine series we mean to start with the first term and keep computing and summing terms until finding one that is less than, say, 10^{-8}. The sum of the series should then, according to the standard presentation in calculus, be within 10^{-8} of the correct value of the sine.

The program requires a strategem to avoid producing intermediate results too large to be represented in the computer. The largest angle we shall consider will be about 70 radians; if we were to try to raise 70 to the large powers that will be required, we should greatly exceed the sizes permitted of real and double precision values in most computers. Therefore, we shall take a different approach, that of computing each term from the one preceding, which is also faster. The relationship for computing a term from the previous one is not complicated. Having the first term, x, we can get the next term by multiplying by $-x^2$ and dividing by 2×3. Having the second term, we can get the third by multiplying by $-x^2$ and dividing by 4×5. In short, given any term we can get the next one by multiplying by $-x^2$ and dividing by the product of the next two integers.

Figure 8.1 is a pseudocode of the algorithm just described. We wish to try the method out on a variety of angles, so the entire computation is enclosed in a REPEAT UNTIL loop that terminates on detecting the end of the data.

The program in Figure 8.2 fills in the details of the program design. After reading an angle in degrees and converting it to radians, there are a few initialization operations before going into the WHILE loop. The first term is just X, so the sum of the series as we prepare to enter the WHILE loop is also X. To get the next term, which requires dividing by the product of 2 and 3, we set DENOM equal to 3.0. Finally, it saves a bit of time to precompute x^2. The rest of the program follows the pseudocode and the previous discussion closely.

Figure 8.3 shows the output when this program was run with a set of input values that were ± 30° plus multiples of 360°. The result in every case should therefore have been either +0.5 or −0.5. We see that the program gets the sine of 30° to as much precision as is possible in the computer used, but things begin to deteriorate rather badly for larger angles. For 1110° and larger values, the results are nonsense. Let us see how this happened.

```
Sine computation for the Taylor Series:

REPEAT until end of data
    READ angle in degrees, convert to radians
    Compute first term of series
    SUM <-- first term
    WHILE term > 1E-8 (absolute value)
        Compute next term from previous term
        Add term to SUM
        Prepare for computing next term
    END WHILE

    Compute sine from SIN function
    PRINT angle, sine from series, sine from SIN
END REPEAT
```

Figure 8.1 Pseudocode of a method for finding the sine of an angle from its Taylor Series, using a recursion relation to compute successive terms.

A modified version of the program, not shown, was used to produce the values of all of the terms and the partial products, including first term, which is just x. Figure 8.4 shows this output for an angle of 30°. The fifth term in the series, the one with x^9, is evidently less than 10^{-8}, and the result is indeed "exactly" 0.5—at least to the precision possible in the computer used.

Figure 8.5 shows similar output for an angle of 1110°, where there are very serious problems. The main culprit here is the impossibility of accurately representing the largest terms. Observe the term for which DENOM was 19, for instance. Following our instructions in the FORMAT statement, the program has dutifully printed 15 digits in the value—but the rightmost 7 digits are zeros, which is suspicious to say the least. In the computer used a real (single precision) number has the equivalent of slightly less than seven decimal digits of significance, so of the 15 digits printed the rightmost 8 are indeed meaningless. The value shown is therefore potentially in error by as much as 10. The sum of the terms is very much larger that than, of course, but when the sum has been reduced back to a small value, the significance lost earlier can never be recovered.

If we really wanted to compute sines, we obviously would not proceed in this manner. Since the sine function is periodic with a period of 2π, it is a simple matter to reduce the angle to a value less than 2π, below which our method is seen to be almost tolerable. In fact, it is not much extra trouble to reduce the angle to a value less than $\pi/2$, simply keeping track of the quadrant of the original angle so the correct sign can be attached to the result.

However, there are often situations in which such techniques are not available. Let us proceed with our example to see how much can be accomplished using double precision in an attempt to deal with the problem. Figure 8.6 presents our program modified to use double precision quantities in most places, permitting us to examine some typical double precision programming examples.

The REAL statement has been replaced by a DOUBLE PRECISION statement, which as with all type-statements must appear before the first executable statement. The FORMAT for the heading line has been modified to provide more space, and the D field specification has been used in reading—although the program would have worked correctly without this latter change. The conver-

```
* A program to compute the sine function from
*    the Taylor Series expansion
* Terms are computed from a recursion relation, until finding
*    one that is smaller in absolute value than 1E-8
*
* Single precision version
*
* Variables:
*     DEGREE: The angle, in degrees
*     X:      The angle, in radians
*     XSQ:    The square of the angle in radians
*     DENOM:  Largest integer in denominator of current term
*     TERM:   Current term of Taylor Series
*     SUM:    Sum of Taylor Series so far
*     COMPAR: The sine as computed by the builtin SIN function

      REAL DEGREE, X, XSQ, DENOM, TERM, SUM, COMPAR

* Print column headings
      PRINT 100
 100  FORMAT ('1', ' DEGREE', 10X, 'X', 9X,
     $     'SERIES SINE      FUNCTION SINE'/)

***** REPEAT until end of file
  10  CONTINUE
         READ (*, *, END = 999) DEGREE
*        convert to radians
         X = DEGREE / 57.29578
*        initialize
         TERM = X
         SUM = X
         DENOM = 3.0
         XSQ = X**2

***** WHILE TERM > 1E-8 (absolute value)
  20     CONTINUE
         IF ( ABS(TERM) .GT. 1E-8 ) THEN

*            get new term from previous, add to sum, and increment DENOM
             TERM = - TERM * XSQ / (DENOM * (DENOM - 1.0))
             SUM = SUM + TERM
             DENOM = DENOM + 2.0

         GO TO 20
         END IF
*****    END WHILE

* Get value of sine from built-in SIN function for comparison
         COMPAR = SIN(X)

* Print results, go around again
         PRINT 200, DEGREE, X, SUM, COMPAR
 200     FORMAT (' ', F6.0, 3F16.7)
         GO TO 10
***** END REPEAT

 999  CONTINUE

      END
```

Figure 8.2 The single precision version of a program for computing a sine from the Taylor series, corresponding to the program design shown in Figure 8.1.

DEGREE	X	SERIES SINE	FUNCTION SINE
30.	0.5235988	0.5000000	0.4999999
390.	6.8067837	0.5000133	0.4999996
750.	13.0899696	0.5055103	0.5000001
1110.	19.3731537	-11.2978935	0.4999991
1470.	25.6563263	891.7993160	0.4999881
1830.	31.9395142	639408.0000000	0.4999903
2190.	38.2227020	***************	0.4999925
2550.	44.5058899	***************	0.4999948
-30.	-0.5235988	-0.5000000	-0.4999999
-390.	-6.8067837	-0.5000133	-0.4999996
-750.	-13.0899696	-0.5055103	-0.5000001

Figure 8.3 The output of the program of Figure 8.2. The asterisks represent values that were too large to print in the specified field width.

sion from degrees to radians uses the quotient of 180 and π, a table giving that quotient to 16 digits not being handy. The value of π was written with a D0 exponent to guarantee that it would be interpreted as a double precision constant. Most compilers these days will do the required division of two constants at compile time. We note that the two functions used in the program—absolute value and sine—have been converted to their double precision forms. Finally, observe the F field specification in writing the results. This form may be used with double precision values as well as real.

Figure 8.7 shows the output of this program when it was run with the same values as before, plus a few even larger angles. We see that the additional precision has indeed alleviated the problem. The sine of 1110°, which before was meaningless, is now accurate to nine digits. The algorithm does not break down completely until 2550°, by which time the single precision version was giving utter nonsense.

Still, the problem is *postponed*, not *solved*. Eventually even double precision cannot hold enough significance to avoid the fundamental problem of finite representation of rational numbers. The example was chosen deliberately to try to make this point: double precision is often a help, but it can't solve all problems. Furthermore, it should not be used as a way of avoiding thinking: the proper solution in this case is to use some elementary mathematics, as described earlier, to make the problem go away entirely!

DENOM	TERM	SUM
1.	0.5235988	0.5235988
3.	-0.0239246	0.4996742
5.	0.0003280	0.5000021
7.	-0.0000021	0.5000000
9.	0.0000000	0.5000000

Figure 8.4 The first four terms of the Taylor series for the sine of an angle of 30°.

DENOM	TERM	SUM
1.	19.3731537	19.3731537
3.	-1211.8515600	-1192.4782700
5.	22741.5312000	21549.0508000
7.	-203222.0000000	-181672.9370000
9.	1059347.0000000	877674.0620000
11.	-3614480.0000000	-2736805.0000000
13.	8696041.0000000	5959236.0000000
15.	-15541847.0000000	-9582611.0000000
17.	21445376.0000000	11862765.0000000
19.	-23534656.0000000	-11671891.0000000
21.	21030944.0000000	9359053.0000000
23.	-15599421.0000000	-6240368.0000000
25.	9757921.0000000	3517553.0000000
27.	-5216996.0000000	-1699443.0000000
29.	2411375.0000000	711932.0000000
31.	-973155.0620000	-261223.0620000
33.	345874.3750000	84651.3125000
35.	-109086.6870000	-24435.3750000
37.	30737.4531000	6302.0781200
39.	-7784.3046900	-1482.2265600
41.	1781.4609400	299.2343750
43.	-370.2189940	-70.9846191
45.	70.1768188	-0.8078003
47.	-12.1825523	-12.9903526
49.	1.9440212	-11.0463314
51.	-0.2861285	-11.3324594
53.	0.0389657	-11.2934933
55.	-0.0049241	-11.2984171
57.	0.0005790	-11.2978373
59.	-0.0000635	-11.2979002
61.	0.0000065	-11.2978935
63.	-0.0000006	-11.2978935
65.	0.0000001	-11.2978935
67.	-0.0000000	-11.2978935

Figure 8.5 The first 34 terms of the Taylor series for the sine of 1110°.

The D field specification makes it possible to see how big the values were that in Figure 8.7 could not be printed in the space available. Figure 8.8 is the output when the 3F23.14 was changed to 3D23.14.

Exercises

*1. Consider the following system of equations:

$$140679x + 556685y = 146710$$
$$81152x + 321129y = 84631$$

If we write:

$$ax + by = c$$
$$dx + ey = f$$

```
* A program to compute the sine function from
*     the Taylor Series expansion
* Terms are computed from a recursion relation, until finding
*     one that is smaller in absolute value than 1D-17
*
* Double precision version
*
* Variables:
*     DEGREE: The angle, in degrees
*     X:      The angle, in radians
*     XSQ:    The square of the angle in radians
*     DENOM:  Largest integer in denominator of current term
*     TERM:   Current term of Taylor Series
*     SUM:    Sum of Taylor Series so far
*     COMPAR: The sine as computed by the builtin SIN function

      DOUBLE PRECISION DEGREE, X, XSQ, DENOM, TERM, SUM, COMPAR

* Print column headings
      PRINT 100
 100  FORMAT ('1', ' DEGREE', 14X, 'X', 16X,
     $    'SERIES SINE', 11X, 'FUNCTION SINE'/)

***** REPEAT until end of file
  10  CONTINUE
          READ (*, *, END = 999) DEGREE
*         convert to radians
          X = DEGREE / (180.0/3.141592653589793D0)
*         initialize
          TERM = X
          SUM = X
          DENOM = 3.0
          XSQ = X**2

***** WHILE TERM > 1D-17 (absolute value)
  20      CONTINUE
          IF ( DABS(TERM) .GT. 1D-17 ) THEN

*             get new term from previous, add to sum, and increment DENOM
              TERM = - TERM * XSQ / (DENOM * (DENOM - 1.0))
              SUM = SUM + TERM
              DENOM = DENOM + 2.0

          GO TO 20
          END IF
*****     END WHILE

* Get value of sine from built-in DSIN function for comparison
          COMPAR = DSIN(X)

* Print results, go around again
          PRINT 200, DEGREE, X, SUM, COMPAR
 200      FORMAT (' ', F6.0, 3F23.14)
      GO TO 10
***** END REPEAT

 999  CONTINUE

      END
```

Figure 8.6 The program of Figure 8.2 modified to use double precision.

DEGREE	X	SERIES SINE	FUNCTION SINE
30.	0.52359877559830	0.50000000000000	0.50000000000000
390.	6.80678408277788	0.50000000000000	0.50000000000000
750.	13.08996938995747	0.50000000000101	0.50000000000000
1110.	19.37315469713705	0.50000000051688	0.49999999999999
1470.	25.65634000431664	0.49999970301821	0.49999999999999
1830.	31.93952531149623	0.50029132120488	0.49999999999999
2190.	38.22271061867581	0.63373514786285	0.49999999999999
2550.	44.50589592585540	179.81746342879037	0.49999999999999
2910.	50.78908123303498	49902.58337544370990	0.49999999999999
3270.	57.07226654021457**********************		0.49999999999999
3630.	63.35545184739416**********************		0.49999999999999
3990.	69.63863715457374**********************		0.49999999999999
-30.	-0.52359877559830	-0.50000000000000	-0.50000000000000
-390.	-6.80678408277788	-0.50000000000000	-0.50000000000000
-750.	-13.08996938995747	-0.50000000000101	-0.50000000000000

Figure 8.7 The output of the program of Figure 8.6.

DEGREE	X	SERIES SINE	FUNCTION SINE
30.	0.52359877559830D+00	0.50000000000000D+00	0.50000000000000D+00
390.	0.68067840827779D+01	0.50000000000000D+00	0.50000000000000D+00
750.	0.13089969389957D+02	0.50000000000101D+00	0.50000000000000D+00
1110.	0.19373154697137D+02	0.50000000051688D+00	0.49999999999999D+00
1470.	0.25656340004317D+02	0.49999970301821D+00	0.49999999999999D+00
1830.	0.31939525311496D+02	0.50029132120488D+00	0.49999999999999D+00
2190.	0.38222710618676D+02	0.63373514786285D+00	0.49999999999999D+00
2550.	0.44505895925855D+02	0.17981746342879D+03	0.49999999999999D+00
2910.	0.50789081233035D+02	0.49902583375444D+05	0.49999999999999D+00
3270.	0.57072266540215D+02	-0.12789275840736D+08	0.49999999999999D+00
3630.	0.63355451847394D+02	0.28150790008241D+09	0.49999999999999D+00
3990.	0.69638637154574D+02	0.11883695457511D+14	0.49999999999999D+00
-30.	-0.52359877559830D+00	-0.50000000000000D+00	-0.50000000000000D+00
-390.	-0.68067840827779D+01	-0.50000000000000D+00	-0.50000000000000D+00
-750.	-0.13089969389957D+02	-0.50000000000101D+00	-0.50000000000000D+00

Figure 8.8 The output when the program of Figure 8.6 was changed to use a D field specification for output.

a solution is given by:

$$x = \frac{ce - bf}{ae - bd}$$

$$y = \frac{af - cd}{ae - bd}$$

Write two programs to solve this system, one using real variables throughout and the other using double precision. Why should there by any difference in the results, considering that in a machine that can hold at least six decimal digits of precision in a real quantity the numbers in the equations can all be represented exactly?

2. Write a program, using double precision throughout, to evaluate and print these values:

$$D1 = (10^9 + \sin 2 - 10^9) \cdot 10^7$$
$$D2 = (10^9 - 10^9 + \sin 2) \cdot 10^7$$

Why is there a difference? Is not addition commutative?

3. Consider the following program.

```
REAL SUM
REAL LOG2
DOUBLE PRECISION H
DOUBLE PRECISION X
INTEGER I

READ *, N

H = 1.0/N
X = 1.0 + H
SUM = 0.0
DO 10 I = 1, N-1
    SUM = SUM + 1.0/X
    X = X + H
10  CONTINUE

LOG2 = (H/2)*(1.0 + 2.0*SUM + 0.5)
WRITE (6, 100) N, LOG2
100  FORMAT (' ', 'WITH N = ', I6, '    THE APPROX TO LOG(2) = ', F14.7)

END
```

This program uses the trapezoidal rule to evaluate the integral:

$$\int_1^2 \frac{dx}{x} = \ln 2$$

The line segments of the trapezoidal rule approximation always lie above the curve of the reciprocal; so with exact computations the approximation should always be larger than ln 2. Run the program with values of n in the range of 100 to 2000 to see what happens. Then change SUM from real to double precision and run again. Can you explain the difference?

4. Considering all the examples where we have seen the effect of the finite precision of number representation in digital computers, it seems intuitively unlikely that, even with double precision, we should be able to get exact equality in the identify:

$$\sin^2 x + \cos^2 x = 1$$

Write a program to see if can find values of x for which the relationship does not hold. If you find a region where it diverges more than can be explained by a variation in the last digit of the sine and cosine functions, explore nearby values. You may have discovered a bug in the sine and cosine routines in your compiler. The vendor will appreciate hearing from you if so. (But don't write unless you can really *prove* an error.)

Complex constants, variables, and expressions

A complex quantity in Fortran is an ordered pair of Fortran real quantities, the first representing the real part of the complex number and the second the imaginary part. We are able to write constants that represent complex numbers; we can specify, using a COMPLEX type-statement, that a variable represents the real and imaginary parts of a complex number; we have functions available for carrying out various mathematical functions on complex quantities. The capability provided by Fortran complex operations is a great savings in programming effort in some kinds of applications.

A complex constant consists of a pair of real constants enclosed in parentheses and separated by a comma. The following are examples of complex constants and their meanings:

```
(2.0, 3.0)          2.0 + 3.0i
(2.E5, 2.5E4)       200,000 + 25,000i
(1.075, -0.653)     1.075 - 0.653i
(1.0, 0.0)          1.0   (pure real)
(0.0, 5.0)          5.0i  (pure imaginary)
```

A Fortran complex variable is one that has been named in a COMPLEX statement. A complex variable is stored within the computer as two real quantities.

The five familiar arithmetic operations are all defined for operations on complex quantities, although, of course, the computer must do all the manipulations necessary to separate the complex operations into action on the real and imaginary parts. These may be reviewed, for reference:

$$(a + bi) + (c + di) = (a + c) + (b + d)i$$
$$(a + bi) - (c + di) = (a - c) + (b - d)i$$
$$(a + bi)*(c + di) = (ac - bd) + (ad + bc)i$$
$$(a + bi)/(c + di) = \frac{(ac + bd)}{c^2 + d^2} + \frac{(bc - ad)}{c^2 + d^2} i$$
$$(a + bi)**n$$

A meaning has not been shown for exponentiation because the method of raising a complex number to a power depends on the size of the exponent. For small integer powers, an actual multiplication is used. In other cases the complex logarithm and exponential functions are used, according to the relation:

$$(a + bi)^n = e^{n \ln (a + bi)}$$

The exponent may itself be complex.

Fortran provides functions for computing the exponential function, logarithm, sine, cosine, and square root of a complex number. The argument and function value are both complex in these cases. The complex absolute value function supplies a real value from a complex argument. The first letter of all complex-valued functions is C, which simplifies remembering them. The complete list may be found in the Appendix.

Four other functions that are provided for manipulating complex variables find heavy use. REAL supplies the real part of its complex argument and AIMAG supplies the imaginary part of its complex argument, in both cases as a Fortran real number. The function CMPLX takes two real arguments, separated by a comma, and supplies as the function value the value of the complex number composed of the two values. For instance, we could write:

```
Z = CMPLX(A, B)
```

A and B would have to be Fortran real variables and Z a Fortran complex variable. The result of the statement would be to assign the value of A as the real part of Z, and B as its imaginary part. Since the arguments of CMPLX can be any real-valued expressions, we might also write:

```
Z2 = CMPLX(0.0, 4.0*OMEGA + 3.56)
```

This would create in Z a number having zero as its real part and the value of the second expression as its imaginary part.

The function CONJG takes a complex argument and supplies the complex conjugate of the number as the function value:

```
CONJG(a + bi) = (a - bi)
```

Input and output of complex numbers is fairly simple. Whenever a complex variable appears in the list of an input or output statement, Fortran expects to transmit two real values. For list-directed input we provide the two real values in the form of a Fortran complex constant, that is, as two real quantities enclosed in parentheses and separated by a comma. When a FORMAT statement is involved we must provide two format codes for each complex quantity. In that case the two real values do not have to be enclosed in parentheses and no comma is required. For instance, if H is real and Z is complex, we could write:

```
        PRINT 100, H, Z
100    FORMAT (' ', F10.0, F10.3, F12.5)
```

The F10.0 would be associated with the value of H, the F10.3 with the real part of Z, and the F12.5 with the imaginary part of Z. Formatted output requires two format codes for each complex value; the real and imagainary parts are printed as two real values, without the parentheses or the separating comma.

Complex and real Fortran quantities may be mixed in an expression. In such a case the Fortran reals are taken to be pure reals, mathematically; the result is always complex. Thus, if Z has been declared to be complex, we may write expressions such as 4.0*Z or Z − H, where H is real. Mixing of complex and integer quantities is permitted, but it would be uncommon to do so except for raising a complex quantity to an integer power. Mixing of complex and double precision quantities is not permitted.

Figure 8.9 is a program to illustrate operations with complex quantities. The terminal session when the program was run is shown in Figure 8.10. We see that for output with a FORMAT statement, two format codes were provided for each complex quantity. In most cases the same format code was used for the real and imaginary parts, but this is not required. The square root example demonstrates that complex expressions may appear in the list of an output statement, just as with other types of quantities. The next example shows the mixing of real and complex quantities in an expression and the use of a complex constant. The final example illustrates the use of the CMPLX function.

```
* A program to illustrate Fortran complex operations
*
      COMPLEX A, B, Z, R

      READ *, A, B
      PRINT 100, A, B
100   FORMAT ('0', 'A = ', 2F5.1, '  B = ', 2F5.1)

      Z = A + B
      PRINT 200, Z
200   FORMAT ('0', 'ADDITION:       ', 2F8.2)

      Z = A * B
      PRINT 300, Z
300   FORMAT ('0', 'MULTIPLICATION:', 2F8.2)

      Z = CEXP(A)
      PRINT 400, Z
400   FORMAT ('0', 'EXPONENTIAL:    ', F11.5, F12.6)

      PRINT 500, CSQRT(A)
500   FORMAT ('0', 'SQUARE ROOT:    ', F11.5, F12.6)

      Z = 2.0 * A + (10.0, 20.0)
      PRINT 600, Z
600   FORMAT ('0', 'EXAMPLE 5:      ', 2F8.2)

      R = CMPLX(1.0, -2.0)
      Z = A - R**2
      PRINT 700, R, Z
700   FORMAT ('0', 'EXAMPLE 6:      ', 4F5.0)

      END
```

Figure 8.9 A demonstration program to show some Fortran complex operations.

```
19.14.12 >run c8cpx
EXECUTION:
>(1,2) (3,4)

A =    1.0  2.0  B =    3.0  4.0

ADDITION:           4.00    6.00

MULTIPLICATION:    -5.00   10.00

EXPONENTIAL:       -1.13120    2.471726

SQUARE ROOT:        1.27202    0.786151

EXAMPLE 5:         12.00   24.00

EXAMPLE 6:          1.   -2.    4.    6.
```

Figure 8.10 The output of the program of Figure 8.9.

Computing the frequency response of a servomechanism

The following practical engineering calculation exhibits many of the ideas about Fortran complex operations that we have been studying.

The *transfer function* of a certain servomechanism is given by:

$$T(j\omega) = \frac{K(1 + j0.4\omega)(1 + j0.2\omega)}{j\omega(1 + j2.5\omega)(1 + j1.43\omega)(1 + j0.02\omega)^2}$$

(In electrical engineering j is used for the imaginary unit, i being reserved for current.)

Without attempting to present the complete theory, which would be out of place here, we may characterize a transfer function as follows. Consider a "black box" with two input terminals and two output terminals. An input signal of a certain frequency and amplitude is applied to the input terminals. The signal appears at the output terminals with the same frequency, but with its amplitude increased or decreased, and with its phase shifted forward or backward. In complex variable language, the increase or decrease is given by the magnitude of the transfer function, which is computed by the CABS function. The phase shift is given by the arctangent of the quotient of the imaginary part of the transfer function and its real part.

Both the magnitude change and the phase shift depend strongly on the frequency of the applied signal. We shall develop a program, therefore, to display this dependence. Such information is important in designing the servomechanism for stability, among other reasons.

The program design shown in Figure 8.11 works as follows. We read data including the amplification factor, K, and parameters defining the range of frequencies desired: the starting and stopping frequencies and a multiplicative increment. The input data is checked to make sure that it is reasonable. For each frequency value we are to compute T as a complex number, find its magnitude and phase, and print all of these values.

The program in Figure 8.12 carries all this out in a way that offers no new material except the complex operations.

The type-statements establish all of the variables as either real or complex.

```
Servomechanism computation with complex quantities:

READ parameters, including frequency limits and increment
Check data and stop with message if not valid
Give frequency its starting value
WHILE frequency < stopping value
    Compute intermediate factors
    Compute transfer function, T
    Compute magnitude and phase of transfer function
    PRINT frequency and results
    increment frequency
END WHILE
```

Figure 8.11 Program design for a servomechanism frequency response calculation.

```
      * A program to compute the frequency response of a servomechanism
      * Uses COMPLEX constants, variables, operations, and functions
      *
      * Input variables (all real):
      *     K:       Amplification factor
      *     FIRST:   Starting frequency
      *     LAST:    Stopping frequency
      *     INC:     Multiplicative frequency increment, > 1.0
      *
      * Output variables:
      *     OMEGA:   Frequency, radians/sec, real
      *     ABSVAL:  Magnitude of transfer function, real
      *     PHASE:   Phase angle of transfer function, real
      *     T:       Transfer function, complex
      *
      * Intermediate variables, all complex
      *     N1:      First factor (after K) in numerator
      *     N2:      Second factor in numerator
      *     D1:      First factor in denominator
      *     D2:      Second factor in denominator
      *     D3:      Third factor in denominator
      *     D4:      Fourth factor in denominator
      * These factors are set up using the CMPLX function, which converts
      *     from the form of two REAL values, representing the real and
      *     imaginary parts of the complex numbers, to the form of one
      *     Fortran COMPLEX number

            REAL K, FIRST, LAST, INC, OMEGA, ABSVAL, PHASE
            COMPLEX T, N1, N2, D1, D2, D3, D4

      *Read parameters, validate
            READ *, K, FIRST, LAST, INC
            IF (        (FIRST .GE. LAST)
           $     .OR. (INC    .LE. 1.0 ) ) THEN
               PRINT *, 'IMPOSSIBLE DATA; PROGRAM NOT EXECUTED'
               STOP
            END IF

      * Print amplification factor and identifications
            PRINT 100, K
       100  FORMAT ('1', 20X, 'SERVOMECHANISM FREQUENCY RESPONSE'/
           $     '0', '         AMPLIFICATION FACTOR = ', F12.5///
           $     ' ', 8X, 'OMEGA', 7X, 'T REAL', 8X, 'T IMAG', 7X,
           $     'ABS VALUE', 6X, 'PHASE'//)

      * Set frequency to starting value
            OMEGA = FIRST

      ***** WHILE OMEGA <= LAST
       10   CONTINUE
            IF ( OMEGA .LE. LAST ) THEN

               N1 = CMPLX(1.0, 0.4 * OMEGA)
               N2 = CMPLX(1.0, 0.2 * OMEGA)
               D1 = CMPLX(0.0, OMEGA)
               D2 = CMPLX(1.0, 2.5 * OMEGA)
               D3 = CMPLX(1.0, 1.43 * OMEGA)
               D4 = CMPLX(1.0, 0.02 * OMEGA)
               T = K * N1 * N2 / (D1 * D2 * D3 * D4**2)

      *        get complex absolute value, = magnitude of output
               ABSVAL = CABS(T)

      *        use two-argument arctangent function to get phase,
      *        retaining quadrant information; convert to degrees
               PHASE = 57.29578 * ATAN2 (AIMAG(T), REAL(T))

      *        print results; note that complex value needs two format codes
               PRINT 200, OMEGA, T, ABSVAL, PHASE
       200     FORMAT (' ', 5E14.4)

      *        increment frequency
               OMEGA = INC * OMEGA

            GO TO 10
            END IF
      ***** END WHILE

            END
```

Figure 8.12 A program for a servomechanism frequency response
calculation.

Since all the values that are read are real, we have no special actions there, and the validation is straightforward.

Within the WHILE loop we begin by assembling the real and imaginary parts of N1, the first complex factor in the numerator of the transfer function. This and the following five statements would, of course, make no sense if the variables had not been named in COMPLEX type-statements. D1 is the $j\omega$ factor in the denominator. Pure imaginaries like this must be set up as complex numbers, whereas as pure reals can simply be written as Fortran reals. We see this in the expression for the transfer function, which contains both the real variable K and the six complexes.

The absolute value is found with the CABS function. The phase is a little more complicated. Mathematically, the arctangent is a many-valued function: there are infinitely many angles having the same tangent. The writers of the Fortran arctangent function therefore had to make some assumptions regarding which range of angles will be regarded as the principle values. The function ATAN accepts a single argument and provides a function value that is in the range of $-\pi/2$ to $\pi/2$. The function ATAN2 accepts two arguments, which are taken to be, respectively, the ordinate and abscissa of a point in any of the four quadrants. The function provides the angle whose tangent is the quotient of the first argument divided by second. With this much information, the function can distinguish the four quadrants, and return a result in the range of $-\pi$ to π.

The result, for either function, is an angle in radians. We convert it to degrees by multiplying by $180/\pi$.

Printing the results requires only remembering to provide two format codes corresponding to the complex variable T.

Incrementing the frequency is done by multiplying it by INC, whereas in previous illustrations we have added an increment.

The results when this program was run with suitable data are shown in Figure 8.13.

(To servo fans: Notice that for frequencies out through about 20 radians/sec the input is amplified, but after that attenuated. The phase angle starts out near $-90°$, moves to nearly $-180°$, moves across the real axis to values a little less than $+180°$, then back down again, then up again. This behavior for larger frequencies is of considerable interest to the servomechanism engineer, who is concerned with the stability of the device represented by the transfer function.)

Exercises

*1. Find any syntax errors in the following program.

```
        COMPLEX I, Z1, Z2, Z3

        READ ( 5, 100) Z1, Z2
100     FORMAT (4F5.0)

        I = (0.0, 1.0)
        PRINT *, (-I)**I
        IF ( Z1 .GT. Z2 ) THEN
            PRINT *, ' FIRST INPUT VALUE IS LARGER'
```

SERVOMECHANISM FREQUENCY RESPONSE

AMPLIFICATION FACTOR =   900.00000

OMEGA	T REAL	T IMAG	ABS VALUE	PHASE
0.2000E-01	-0.3024E+04	-0.4483E+05	0.4493E+05	-0.9386E+02
0.3000E-01	-0.3012E+04	-0.2974E+05	0.2989E+05	-0.9578E+02
0.4500E-01	-0.2986E+04	-0.1961E+05	0.1984E+05	-0.9866E+02
0.6750E-01	-0.2929E+04	-0.1276E+05	0.1309E+05	-0.1029E+03
0.1012E+00	-0.2807E+04	-0.8062E+04	0.8537E+04	-0.1092E+03
0.1519E+00	-0.2562E+04	-0.4783E+04	0.5426E+04	-0.1182E+03
0.2278E+00	-0.2127E+04	-0.2498E+04	0.3281E+04	-0.1304E+03
0.3417E+00	-0.1502E+04	-0.1027E+04	0.1820E+04	-0.1456E+03
0.5126E+00	-0.8502E+03	-0.2761E+03	0.8939E+03	-0.1620E+03
0.7689E+00	-0.3838E+03	-0.2582E+02	0.3847E+03	-0.1762E+03
0.1153E+01	-0.1493E+03	0.1183E+02	0.1498E+03	0.1755E+03
0.1730E+01	-0.5622E+02	0.5083E+01	0.5645E+02	0.1748E+03
0.2595E+01	-0.2226E+02	-0.5438E+00	0.2227E+02	-0.1786E+03
0.3892E+01	-0.9517E+01	-0.2082E+01	0.9742E+01	-0.1677E+03
0.5839E+01	-0.4422E+01	-0.1938E+01	0.4828E+01	-0.1563E+03
0.8758E+01	-0.2265E+01	-0.1396E+01	0.2661E+01	-0.1483E+03
0.1314E+02	-0.1297E+01	-0.8659E+00	0.1559E+01	-0.1463E+03
0.1971E+02	-0.8062E+00	-0.4416E+00	0.9192E+00	-0.1513E+03
0.2956E+02	-0.4925E+00	-0.1460E+00	0.5137E+00	-0.1635E+03
0.4434E+02	-0.2562E+00	0.6127E-02	0.2563E+00	0.1786E+03
0.6650E+02	-0.1020E+00	0.4045E-01	0.1097E+00	0.1584E+03
0.9976E+02	-0.3090E-01	0.2633E-01	0.4060E-01	0.1396E+03
0.1496E+03	-0.7663E-02	0.1115E-01	0.1353E-01	0.1245E+03
0.2245E+03	-0.1691E-02	0.3892E-02	0.4243E-02	0.1135E+03
0.3367E+03	-0.3516E-03	0.1242E-02	0.1291E-02	0.1058E+03
0.5050E+03	-0.7109E-04	0.3805E-03	0.3871E-03	0.1006E+03
0.7575E+03	-0.1419E-04	0.1144E-03	0.1153E-03	0.9707E+02

Figure 8.13 The output of the program of Figure 8.12.

```
        END IF
        Z3 = (2.0, 3.0)Z1
        Z4 = 2.0*Z1 - Z2/2
        WRITE (6, 200) Z1, Z2, Z3, Z4
200     FORMAT (' ', 2F10.7, 4E16.7, F12.5, F13.6)

        END
```

*2. A student was asked to write a program to find the solution to this system of equations:

$$(2 + 3i)x + (4 - 2i)y = (5 - 3i)$$
$$(4 + i)x + (-2 + 3i)y = (2 + 13i)$$

Does this program do that?

```
        COMPLEX A, B, C, D, E, F, X, Y, DENOM

        A = (2.0, 3.0)
        B = (4.0, -2.0)
        C = (5.0, -3.0)
        D = (4.0, 1.0)
        E = (-2.0, 3.0)
        F = (2.0, 13.0)
```

```
      DENOM = A*E - B*D
      X = (C*E - B*F)/DENOM
      Y = (A*F - C*D)/DENOM
      WRITE (6, 100) X, Y
100   FORMAT (' ' 4F12.7)
      END
```

3. Write a program, using complex variables throughout, to solve the quadratic equation:

$$ax^2 + bx + c = 0$$

where a, b, and c are complex. Use the familiar formula:

$$x = \frac{-b \pm \sqrt{b^2 - 4ac}}{2a}$$

Run the program with data specifying:

$a = 1 + i$
$b = 2 + 3i$
$c = -7 + i$

Run it again, but with this data:

$a = 1$
$b = -5$
$c = 6$

Is it any surprise that a program using complex operations can get correct results with real data? How would you phrase your answer mathematically?

*4. Write a program, using complex variables where appropriate, to compute and print z and e^z for a succession of values along a vertical line through $(1, 0)$, such as $1 - 5i, 1 - 4i, 1 - 3i, \ldots, 1 + 4i, 1 + 5i$. This will give a numerical demonstration how the complex exponential function maps a vertical line.

5. Same as Exercise 4, except the line runs through the origin and makes a 45° angle with the x axis; so the values might be $-5 - 5i, -4 - 4i, -3 - 3i, \ldots, 4 + 4i, 5 + 5i$.

6. Same as Exercise 4, except use 20 points equally spaced on a circle of radius 1 with its center at the origin.

7. Write a program to provide a numerical demonstration that the identify

$$\sin^2 x + \cos^2 x = 1$$

holds for complex variables.

8. Modify the program of Figure 8.12 to use this transfer function:

$$T(j\omega) = \frac{100}{j\omega(1 + 0.25j\omega)(1 + 0.0625j\omega)}$$

9. Modify the program of Figure 8.12 to use this transfer function:

$$T(j\omega) = \frac{1260}{j\omega(1 + 0.25j\omega)(1 + 0.001j\omega) + \dfrac{20(j\omega)^4}{(1 + 0.5j\omega)(1 + 0.4j\omega)}}$$

*10. Think of a friend who has had algebra at some point, but didn't really take to it. Suppose you wanted to illustrate what is meant by "elegance" in mathematics by explaining the equation

$$e^{i\pi} + 1 = 0$$

which combines the five most fundamental entities in mathematics in one beautifully simple equation. Can you think of any way to use a computer to help in the demonstration?

Logical constants, variables, and expressions

A Fortran logical quantity is one that can take on only the values "true" and "false." The representation of these "values" inside the computer is up to the writer of the Fortran compiler, and is of no interest to us.

A *logical constant* is either of the following:

```
.TRUE.
.FALSE.
```

Note that, as with the relational operators, the periods are required to distinguish from variables that might be named TRUE or FALSE.

A *logical variable* is a variable that has been declared in a LOGICAL type-statement. A logical variable can take on only the values .TRUE. and .FALSE..

A *logical assignment statement* has the form:

```
variable = expression
```

in which the variable on the left is a *logical* variable and the expression on the right is a logical expression.

A logical expression might be as simple as a single relational expression, such as I .EQ. 20. Thus if L1, L2, and so on, are logical variables, we may write logical assignment statements like these:

```
L1 = .TRUE.
L2 = .FALSE.
L3 = A .GT. 25.0
L4 = I .EQ. 0
L5 = L6
```

The first two are the equivalent for logical variables of the form:

```
variable = constant
```

L3 will be set to .TRUE. if the value of the real variable A is greater than 25.0, and to .FALSE. otherwise. Likewise, L4 will be set to .TRUE. if the value of I is in fact equal to zero and to .FALSE. otherwise. L5 will be given the same truth value as L6 currently has.

We have seen that a *relational expression* is one that compares integer, real, or double precision variables by using the relational operators .LT., .LE., .EQ., .NE., .GT., and .GE.. A *logical expression* combines logical values and/or relational expressions by using the *logical operators* .AND., .OR., and .NOT.. Thus we can write logical assignment statements like these:

```
L1 = D .LT. EPS .OR. ITER .GT. 20
L2 = D .GE. EPS .AND. ITER .LE. 20
L3 = BIG .GT. TOLER .OR. SWITCH
```

L1 will be given the value .TRUE. if *either* or *both* of the relations is true, and .FALSE. otherwise. L2 will be given the value .TRUE. if and only if *both* relations are satisfied, and .FALSE. otherwise. L3 will be given the value .TRUE. if the relation is satisfied, if the value of the logical variable SWITCH is .TRUE., or both.

The logical operator .NOT. reverses the truth value of the expression that it precedes. For instance, consider the statement:

```
L = A .GT. B .AND. .NOT. SWITCH
```

if SWITCH is .TRUE., then .NOT. SWITCH is .FALSE., and if SWITCH is .FALSE., then .NOT. SWITCH is .TRUE.. Thus LI will be given the value .TRUE. if the value of A is greater than the value of B *and* the value of SWITCH is .FALSE..

In the absence of parentheses, the hierarchy of logical operators is .NOT., .AND., and .OR.. For instance, take this statement, in which all the variables are logical:

```
L = A .AND. B .OR. C .AND. D .AND. .NOT. D .AND. E
```

The .NOT. is performed first, then the three .AND.s, then the .OR.. It is much better to use extra parentheses, however, than to rely on these rules, if there is any doubt about the meaning of an expression.

Input or output of logical quantities is not common, but when needed either can be handled with the L format code, which basically operates with the letters T and F.

Logical variables may be used for a variety of purposes. Aside from some rather specialized applications, their most common usefulness is in "storing" the result of a logical test, which is what we shall illustrate in the case study for this material.

The use of a logical variable in the method of bubble sorting

The task is to modify the sorting program of the previous chapter so that it examines the elements in a different sequence, and which takes advantage of any correct ordering of the data to reduce the effort required.

You will recall that the sorting method before was to compare the first element with all the others, exchanging them as necessary; then compare the second element with all following; then compare the third with all following, etc. A problem is that the method cannot recognize initial ordering. In fact, if the elements were already in sequence to begin with, the method would still make all the same comparisons, although no exchanges would be necessary.

The method to be used here, which is called *bubble sorting*, proceeds as follows. We first compare the first element with the second and exchange if necessary. Now compare the second with the third, the third with the fourth, etc., until we have compared the N − 1st with the Nth. This constitutes one pass, and is guaranteed to place the largest element in the Nth position. Now we repeat the entire process, which is guaranteed to place the next-to-largest element in the N − 1st position. After N − 1 passes, the array is assured of being in ascending order. The advantage is this: we don't necessarily have to make N − 1 passes. If we ever complete a pass without having to make any exchanges, we know we are done. An array that was initially already in sequence would require only one pass to discover that fact.

Figure 8.14 shows just the sorting loop portion of the program, which is otherwise unmodified from that of Figure 7.4. The basic idea is that we shall continue making passes through the GRADE array until discovering at the beginning of a new pass that no exhanges were required on the previous pass; we then know that the job is finished. The way the program "knows" whether any exchanges were required is that at the beginning of each pass it sets to .FALSE. a logical variable that records whether there were any exchanges. This variable is named ANYEXC, for "any exchanges?" It would, of course, have to appear in a new type-statement at the beginning of the program:

```
LOGICAL ANYEXC
```

In other words, the program assumes the best, namely, that no exchanges will be necessary. Then, if at any point the IF statement determines that an exchange of two elements is necessary, the variable is set to .TRUE., and the pass continues. When the test is next made at the beginning of the WHILE loop, the answer will be "yes, there was at least one exchange on the previous pass." Again hoping for the best, the program sets ANYEXC back to .FALSE. and goes through another pass.

We do have to be sure that the program does make at least one pass through the data, so just before going into the WHILE loop we set ANYEXC to .TRUE. to force at least one pass through the data. After that, the decision is up to the program, depending on what it discovers about the data.

```
* Sort the elements of both arrays into ascending sequence
*      on the values of the elements of GRADE

* Set ANYEXC ("any exchanges?") to .TRUE. to force
*      one execution of WHILE loop
       ANYEXC = .TRUE.

***** WHILE any exchanges
   30  CONTINUE
       IF ( ANYEXC ) THEN

           ANYEXC = .FALSE.
           DO 40 I = 1, N-1
              IF ( GRADE(I) .GT. GRADE(I+1) ) THEN
                 CALL SWAP ( GRADE(I), GRADE(I+1) )
                 CALL SWAP ( ID(I), ID(I+1) )
                 ANYEXC = .TRUE.
              END IF
   40      CONTINUE
*          Print the arrays so we can watch sorting process
*          Note use of implied DO indexing
           PRINT 100, (GRADE(K), ID(K), K = 1, N)
  100      FORMAT (50(' ', 2I7/))

           GO TO 30
       END IF
***** END WHILE
```

Figure 8.14 The central sorting loop of Figure 7.4, modified to use bubble sorting.

As the program is shown here, there is one serious inefficiency. After the first pass, the element in GRADE(N) never again needs to be moved; it is in its final position, and therefore need never be involved in comparisons, either. Similarly, after the second pass, the last *two* elements are in their final position and could be ignored thereafter. The program does not take advantage of this fact; modifying it to do so would not be difficult.

This program was run with the same data as used to illustrate the earlier program, namely:

```
84 11111
77 22222
37 33333
91 44444
56 55555
```

The output is shown in Figure 8.15. We see that on the first pass the 84 got moved down toward its final position, and the 56 that was at the bottom moved up one position toward its eventual position near the top. The 91 is now at the bottom, where it belongs. After the second pass the 84 and the 91 are properly located. The 56 has moved up only one place. The third pass moves the 56 up to its final position and in fact the job is done, but the program doesn't know that without going through one more pass in which no exchanges are needed.

The way the 56 had to "bubble up" through the array to find its final position is the basis of the name for the method.

```
77    22222
37    33333
84    11111
56    55555
91    44444

37    33333
77    22222
56    55555
84    11111
91    44444

37    33333
56    55555
77    22222
84    11111
91    44444

37    33333
56    55555
77    22222
84    11111
91    44444

HIGH GRADE WAS:      91   BY   44444   GOOD WORK!
LOW  GRADE WAS:      37   BY   33333   (SEE ME)
MEAN GRADE WAS:      69
MEDIAN GRADE WAS:    77
```

Figure 8.15 The output of Figure 8.14, run with the data shown above.

Here is another set of data, involving the same grades but in a different initial sequence:

```
77 10000
37 20000
56 30000
91 40000
84 50000
```

(The student ID's have been changed to avoid confusion.) At a quick glance it would not appear that this set is significantly "better ordered" than the other, although admittedly the concept of "good ordering" takes some effort to define precisely.

The output is shown in Figure 8.16, where we see that exactly one pass was required (not counting the final pass needed to establish that the sorting is finished.) Can you see what it was about the input data that made this possible?

```
            37    20000
            56    30000
            77    10000
            84    50000
            91    40000

            37    20000
            56    30000
            77    10000
            84    50000
            91    40000

HIGH GRADE WAS:      91   BY   40000   GOOD WORK!
LOW GRADE WAS:       37   BY   20000   (SEE ME)
MEAN GRADE WAS:      69
MEDIAN GRADE WAS:    77
```

Figure 8.16 The output of the program of Figure 8.14, when run with data that could be sorted in only one pass.

Exercises

1. What would be the output of this program?

```fortran
LOGICAL L1, L2
REAL X

L1 = .TRUE.
L2 = .FALSE.
X = 1.0

IF ( L1 ) THEN
    PRINT *, ' ANYTHING THAT EVALUATES TO A LOGICAL VALUE OK HERE'
END IF
IF ( L2   .OR. X .GT. 0.5 ) THEN
    PRINT *, ' THIS COMBINATION PERFECTLY OK'
END IF
IF ( L1 .OR. L2 ) THEN
    PRINT *, ' WHAT OTHER CHOICES ARE THERE?'
END IF
IF ( L1 .AND. L2 ) THEN
    PRINT *, ' GET ME A NEW COMPILER, QUICK!'
END IF

END
```

2. How would the program of Figure 8.14 operate if, within the WHILE loop, the variable ANYEXC were not set to .FALSE.?

*3. Describe the output of this program, and suggest a suitable punishment for any programmer who would create such a monstrosity.

```
LOGICAL TRUE, FALSE

READ *, TRUE, FALSE

IF ( .TRUE. .OR. FALSE ) THEN
    FALSE = .TRUE.
END IF

      WRITE (6, 100) TRUE, FALSE
100   FORMAT (' ', L1, 3X, L1)

      END
```

Character constants, variables, and operations

A *character constant* in Fortran is any collection of the characters from the Fortran character set, enclosed in apostrophes. (A character constant is also called a *character literal*.)

Here are some examples of character constants:

```
'CAT'
'CATS AND DOGS'
'FRIDAY, JULY 30, 1982'
'SEAN T. O''REILLY'
'1234'
'FLOW RATE '
' LITRES/SEC'
```

Blanks are significant in character constants; said otherwise, "blank" is a character in the Fortran character set.

The length of a character constant is the number of characters it contains (including any blanks, of course). The enclosing apostrophes are not considered to be part of the constant and are not counted in determining its length. If the constant contains two adjacent apostrophes to represent one actual apostrophe, that counts as one character.

A *character variable* is one that has been declared in a CHARACTER type-statement. Every character variable has a length attribute associated with it. If the CHARACTER type-statement contains no length information, the length of the character declared in it is one. Lengths may be declared either for all of the variables in a CHARACTER statement by following the word CHARACTER with an asterisk and the number of characters, or by following the name of a variable in a CHARACTER statement with an asterisk and the number of characters. For example, consider these type-statements:

```
CHARACTER X, Y
CHARACTER*10, A, B, C
CHARACTER D, E*2, F*200
```

The variables X, Y, and D would have length one, A, B, and C would have length 10, E would have length 2, and F would have length 200.

The elements of an array can be of character type. The declaration of such a variable requires that we specify both the dimension information (how many dimensions and the range of each subscript), and the length attribute. The length specified is that of each element in the array. Consider these declarations:

```
CHARACTER*8 UNITS(20)
CHARACTER PIECES(8, 8)*6
```

The first type-statement says that UNITS is a one-dimensional CHARACTER array having 20 elements, each of length 8. (These might be identifications for output of some kind.) The second says that PIECES is a two-dimensional array having eight rows and eight columns, and each element has six characters. (This might have something to do with a chess program.)

There are two ways to operate on character items. One is with the *concatenation* operator and the other with the use of *substrings*.

The concatenation operator is two successive slashes. It combines two character items into one. The length of the new item is the sum of the lengths of the two operands.

The expression created by concatenation may be used wherever a character constant or variable may be used.

A character assignment statement is one such place. It is of the usual form:

```
variable = expression
```

The "variable" must have been declared in a character type-statement, and the "expression" must be of type CHARACTER. A character expression may be a constant, variable, array reference, FUNCTION reference, a substring reference, in every case of type CHARACTER. Or, a character expression may any two of the above joined together with the concatenation operator.

A *substring* is any one or more characters of a string, up to and including the full string. The characters in the substring are designated by writing two *substring expressions*, separated by a colon, within parentheses following the name of the variable. The substring expressions, which must be integer-valued, specify the first and last character positions of the substring. Suppose, for example, that a program contained these statements:

```
CHARACTER C1*8

C1 = 'ABCDEFGH'
```

Then a reference to C1(2:4) would get the characters BCD, C1(8:8) would get the character H, and C1(1:8) would obtain the entire string.

The concatenation and substring operations are illustrated in the example program of Figure 8.17. We see that C2 and C5 are defined as having three characters, C1 has four, C3 seven, C4 twelve, and DAYABB ("day abbreviation") has five. DAY is a character array having seven elements each of length nine.

The first two statements assign character constants to character variables, and the next two illustrate concatenation. The PRINT statement will show the results of these operations, with one blank between each pair of output items. The next statement asks for characters three through five of the variable C4. The next statement prints this result and also demonstrates that a substring reference may be placed directly in the list of an output statement. The FORMAT statement that is referenced by this PRINT illustrates the use of the A format code. This is

```
* A program to illustrate elementary character operations
*
      CHARACTER*3 C2, C5
      CHARACTER C1*4, C3*7, C4*12, DAYABB*5
      CHARACTER*9 DAY(7)
      INTEGER I

      C1 = 'ABCD'
      C2 = 'EFG'
      C3 = C1 // C2
      C4 = C3 // 'HIJKL'
      PRINT *, C1, ' ', C2, ' ', C3, ' ', C4
      C5 = C4(3:5)
      PRINT 100, C5, C3(4:7)
 100  FORMAT (' ', A3, 2X, A4)

* Read the full names of the days of the week
      DO 10 I = 1, 7
         READ 200, DAY(I)
 200     FORMAT (A9)
  10  CONTINUE

      PRINT *, 'THE WEEKDAYS ARE:'

      DO 20 I = 1, 5
         DAYABB = '(' // DAY(I)(1:3) // ')'
         PRINT 300, I, DAY(I), DAYABB
 300     FORMAT (' ', 'DAY ', I1, ' IS ', A9, A7)
  20  CONTINUE

      END
```

Figure 8.17 A program to illustrate certain operations on character variables, especially the concatenation and substring operations.

of the form, *Aw*, where *w* is the width of the field. If the length of the output is shorter than *w*, the characters are printed at the *left* of the *w* character positions, followed by blanks. If the output item is longer, only the leftmost *w* characters are printed, with no error message.

Input of character data can be handled without a **FORMAT** statement, i.e., in list directed form; in that case the **CHARACTER** data must be enclosed in apostrophes. With formatted input apostrophes are not needed, and the *Aw* format code specifies the width of the input field.

This is illustrated in the reading loop to obtain the names of the days of the week.

Now comes a loop to print the weekdays, with their day number (Monday = 1), the full name, and the three-letter abbreviation enclosed in parentheses. The most interesting statement is the one to get **DAYABB**. We see that a left parenthesis is concatenated to a substring of **DAY**, which is concatenated to a right parenthesis. The formatted **PRINT** statement references a **FORMAT** statement that obtains spacing between the various output items through a combination extra spaces within literals and a **A7** format code for an item having only five characters.

The output when this program was run is shown in Figure 8.18.

Another illustration will give a glimpse of what can be done with **CHARACTER**

```
ABCD EFG ABCDEFG ABCDEFGHIJKL
CDE  DEFG
THE WEEKDAYS ARE:
DAY 1 IS MONDAY      (MON)
DAY 2 IS TUESDAY     (TUE)
DAY 3 IS WEDNESDAY   (WED)
DAY 4 IS THURSDAY    (THU)
DAY 5 IS FRIDAY      (FRI)
```

Figure 8.18 The output of the program of Figure 8.17, when run with the data shown in the text.

variables and the substring operation. We will read a set of names that are in the form of last name followed by a comma, followed immediately (no intervening space) by the first name. The name in this form is a maximum of 31 characters long. The last name has at most 20 characters, the first name has at most 10, and the comma takes one position. We are to print the name as read and also the first and last names separately. The main task is to locate the comma and use its location to pick up the first and last names for printing. Figure 8.19 is a program to do this.

We see that the READ references a FORMAT statement and uses the END= option to detect the end of data. A DO loop now runs through the name looking for a comma, inspecting each character in turn, using the substring operation to compare successive characters with a character constant consisting of a comma in apostrophes. If the DO is satisfied normally, it did not find a comma, which is flagged as an error condition.

If the IF statement causes an escape from the DO, K now points at the comma. The last name then runs from character 1 up through character $K-1$, and the first name runs from chracter $K+1$ through the end. No special action is taken if the last name is longer than 20 characters or the first name is longer than 10; in either case, excess characters at the right will be dropped.

The program was run with this data:

```
SMITH,ROBERT
JOHNSON,ALBERT
O'REILLY,SEAN
O'REILLY III,SEAN
JONES,SAM
ST. GEORGE,CHEVALIER
MCCRACKEN,DANIEL D.
JONES,SAM
JONESSAM
JONES,
,GEORGE
```

Observe that formatted CHARACTER input does not have to be enclosed in apostrophes, and that an embedded apostrophe does *not* have to be represented by two apostrophes.

When the program was run with this data, it produced the output shown in Figure 8.20. We see that the program had no difficulty with "normal" names, that blanks within the first or last names caused no problem, and that a name

```
* A program to illustrate substring operations on names
* Input:
*     NAME:    A name of the form SMITH,ROBERT or JOHNSON,ALBERT
*              31 characters, with trailing blanks
*
* Output:
*     FIRST:   First name, 10 characters
*     LAST:    Last name, 20 characters
*
* An error indication is given if the NAME does not contain a comma

      CHARACTER NAME*31, LAST*20, FIRST*10
      INTEGER K

***** REPEAT until end of data
  10  CONTINUE

          READ (*, 100, END = 40) NAME
 100      FORMAT (A)

* Locate the comma, if there is one
          DO 20 K = 1, 31
              IF ( NAME(K:K) .EQ. ',' ) GO TO 30
  20      CONTINUE

* If this statement is reached, DO terminated without finding a comma
          PRINT *, 'THIS NAME DID NOT CONTAIN A COMMA; NOT PROCESSED:'
          PRINT *, NAME
          GO TO 10

  30      CONTINUE

* K now points at the comma
* Use substring operations to get first and last names
          LAST = NAME(1:K-1)
          FIRST = NAME(K+1:31)
          PRINT *, NAME, ' ', FIRST, ' ', LAST

      GO TO 10
***** END REPEAT

  40  CONTINUE

      END
```

Figure 8.19 A program to process names, written in the form of last-comma-first, into first name and last name.

without a comma was correctly flagged as being in error. A name with a comma but no following first name produced a first name that was blank, which might seem reasonable; the problem specifications did not say what to do with such a case. A name that began with a comma led to trouble, however. The problem is that with the comma in the first position, the value of K was 1, and $K-1$ was zero. Since there is no such position, the object program flagged the error, the took the "standard corrective action." What the standard corrective action might be, could be looked up in an appropriate manual. But the real solution to such a situation, if the program is expected to be able to handle it, would be to put an explicit test for the condition into the program.

```
SMITH,ROBERT                      ROBERT     SMITH
JOHNSON,ALBERT                    ALBERT     JOHNSON
O'REILLY,SEAN                     SEAN       O'REILLY
O'REILLY III,SEAN                 SEAN       O'REILLY III
JONES,SAM                         SAM        JONES
ST. GEORGE,CHEVALIER              CHEVALIER  ST. GEORGE
MCCRACKEN,DANIEL D.               DANIEL D.  MCCRACKEN
JONES,SAM                         SAM        JONES
THIS NAME DID NOT CONTAIN A COMMA; NOT PROCESSED:
JONESSAM
JONES,                                       JONES

IFY197I CMOVE - INVALID SOURCE LENGTH FOR CHARACTER MOVE.

TRACEBACK OF CALLING ROUTINES; MODULE ENTRY ADDRESS=00012000
  CMOVE#  (000142B8) CALLED BY MAIN     (00012000) AT ISN  **    OFFSET (000272)
     ARGUMENT LIST AT (000120FC)
        ARGUMENT  1 AT(0001211C)=D1D6D5C5(HEX); 'JONE'; -774449723;-0.247688E+2
        ARGUMENT  2 AT(00012098)=00000014(HEX); '....';          20;+0.000010E-7
        ARGUMENT  3 AT(00012130)=6BC7C5D6(HEX); '.GEO'; 1808254422;+0.467149E+5
        ARGUMENT  4 AT(80012168)=00000000(HEX); '....';           0;+0.000000E+0
  MAIN    (00012000) CALLED BY (OP/SYS)

STANDARD CORRECTIVE ACTION TAKEN, EXECUTION CONTINUING.
GEORGE                            GEORGE     JONES

SUMMARY OF ERRORS - ERROR NUMBER    ERROR COUNT
                     197                 1
```

Figure 8.20 The output of the program of Figure 8.19, when run with the data shown in the text.

Sorting by the insertion method

For a final example of a program using CHARACTER data, let us take one last look at sorting methods.

The instructor in Chapter 7 who needed student exam grades and ID's sorted, now wants to add student names to the input, so that results can be sorted into name order for easier entry into the gradebook. This gives us a chance to look at character operations in a different light, and to see another—and very effective—sorting method.

Insertion sorting proceeds as follows. We compare the first and second items. If they are already in correct sequence, nothing need be done; if they are out of sequence, they are exchanged. *Relative to each other*, those two elements are now in correct final position. They may end up at the end of the list or with many other elements between them, but they will never need to be exchanged with each other. Stated otherwise, we have a two-element subset of the initial list that is properly ordered. We now take up the third item in the original list and decide where it belongs in a subset of three. If it is larger than the element now in position two, nothing more need be done. But if it is smaller than element two, those two must be exchanged and we then ask whether the subset of three is now in correct sequence. How would we know that? By comparing what is now element two with element one, and exchanging if necessary. With whatever exchanges were necessary, if any, we now know that relative to each other the first three elements of the (possibly reordered) list are in correct sequence. Now we take up the fourth element and through a similar process decide where to "insert" it in a subset of four. Then the fifth, the sixth, etc.

One of the things that makes this method efficient is that as we seek to find the correct place to insert the next element in the list, we can stop making comparisons as soon as we find that it has been exchanged into the position where it belongs. Sorting a list that is initially in correct sequence therefore is very fast, and the method takes full advantage of whatever ordering exists in the input.

One detail must be considered. What happens when an element that is being inserted into the new subset belongs at the head of the list? How do we know that we have reached the beginning? One answer is to provide a test of the element number against which the candidate is being tested, and stop the process when that element number has been reduced to one. A simpler and faster alternative is to provide a special "sentinel" element at the beginning of the list, one with contents that are smaller than anything that could appear in the list, namely, blanks.

A program to do this job is shown in Figure 8.21. NAME is declared to be a CHARACTER array having 51 elements—numbered from 0 to 50—each with 10 characters. There are now two subprograms, since we need to be able to swap two character arguments as well as two integer arguments, and there is no practical way to have one Fortran SUBROUTINE do both. The reading loop is similar to earlier versions, except that, of course, it now includes reading of the names.

Before going into the sorting process we set up the "buffer" value of blanks in element zero of the NAME array. The character constant used for this purpose need not have ten blanks, since a character move of a shorter item to a longer is always "padded" with blanks.

The sorting loop can be viewed in this light. The subscript I picks an element that is to be inserted into its proper place in a subset of the full list. The subscript J works its way backward from that element, looking for the proper place to insert it. This "looking" is done by a process of comparing and exchanging if necessary, until finding that the element identified by J is larger than the element identified by $J-1$, at which the proper spot has been found.

The sorting loop includes a PRINT statement so that we can watch this happening.

This data was used to demonstrate the program:

```
73 12345 'SOONG'
69 32167 'MARTIN'
55 98145 'PETERS'
46 11927 'JONES'
81 73410 'O''BRIEN'
87 74519 'MARTINEZ'
```

Its output is shown in Figure 8.22. The first line shows that SOONG and MARTIN were exchanged, putting them into their final relative position. The second line shows that PETERS was moved into the proper position in a subset of three. The next three lines show JONES moving to her or his position in a subset of four, which happens to be the final position—but the program doesn't know that and doesn't care. Inserting O'BRIEN into correct position requires two

```
* A program to do insertion sorting on character data
* Extends the scope of the sorting program in Chapter 8 to include
*     student name, and sort on it
*
* The method of insertion sorting is used, with a "buffer"
*     value of blanks in array element zero
*
* Variables:
*     NAME:     Student name, CHARACTER*10 array of 50 elements
*     GRADE:    Exam grade, integer array of 50 elements
*     ID:       Student identification, integer array of 50 elements
*     N:        Number of grades; determined by the program
*               (There are assumed to be at least two and not more than 50
*                no error checking is performed.)
*     I:        Index of outer sorting loop
*     J:        Index of inner sorting loop
*     K:        Auxiliary variable used for various subscripts
* Subprograms:
*     ISWAP:    Exchanges two integer arguments
*     CSWAP:    Exchanges two CHARACTER*10 arguments

      CHARACTER*10 NAME(0:50)
      INTEGER GRADE(50), ID(50)
      INTEGER N, I, J, K

* Read the grades, ID's, and names
      DO 10 K = 1, 50
          READ (*, *, END = 20 ) GRADE(K), ID(K), NAME(K)
   10 CONTINUE

   20 CONTINUE

* How many records did we read?
      N = K - 1

* Put "buffer" value of blanks in element zero of NAME
      NAME(0) = ' '

* Sort the records into ascending sequence on the names
      DO 50 I = 2, N
          J = I
*****     WHILE NAME(J) is smaller than the one before it
   30     CONTINUE
          IF ( NAME(J) .LT. NAME(J-1) ) THEN

              CALL CSWAP ( NAME(J), NAME(J-1) )
              CALL ISWAP ( GRADE(J), GRADE(J-1) )
              CALL ISWAP ( ID(J), ID(J-1) )
              PRINT 100, (NAME(K), K = 1, N)
  100         FORMAT (' ', 7A10)
              J = J - 1

          GO TO 30
          END IF
*****     END WHILE

   40     CONTINUE

   50 CONTINUE

* Print data in alphabetical order of names
      PRINT 200, (NAME(K), ID(K), GRADE(K), K = 1, N)
  200 FORMAT ('1', '  NAME            ID      GRADE'//
     $       50(' ', A, 2I7/))

      END
```

Figure 8.21

```
* A SUBROUTINE to exchange two CHARACTER*10 arguments

      SUBROUTINE CSWAP ( ARG1, ARG2 )
      CHARACTER*10 ARG1, ARG2, TEMP

      TEMP = ARG1
      ARG1 = ARG2
      ARG2 = TEMP

      RETURN

      END

* A SUBROUTINE to exchange two integer arguments

      SUBROUTINE ISWAP ( ARG1, ARG2 )

      INTEGER ARG1, ARG2, TEMP

      TEMP = ARG1
      ARG1 = ARG2
      ARG2 = TEMP

      RETURN

      END
```

Figure 8.21 (*cont'd*) A program to sort student names using the method of insertion sorting.

moves. Finally, MARTINEZ has to be inserted between MARTIN and O'BRIEN, which takes three exchanges.

As it happens, insertion sorting is a good choice of method for files up to the range of perhaps 500 to 1000 records, details depending of factors outside the scope of this book.

```
MARTIN    SOONG     PETERS    JONES     O'BRIEN   MARTINEZ
MARTIN    PETERS    SOONG     JONES     O'BRIEN   MARTINEZ
MARTIN    PETERS    JONES     SOONG     O'BRIEN   MARTINEZ
MARTIN    JONES     PETERS    SOONG     O'BRIEN   MARTINEZ
JONES     MARTIN    PETERS    SOONG     O'BRIEN   MARTINEZ
JONES     MARTIN    PETERS    O'BRIEN   SOONG     MARTINEZ
JONES     MARTIN    O'BRIEN   PETERS    SOONG     MARTINEZ
JONES     MARTIN    O'BRIEN   PETERS    MARTINEZ  SOONG
JONES     MARTIN    O'BRIEN   MARTINEZ  PETERS    SOONG
JONES     MARTIN    MARTINEZ  O'BRIEN   PETERS    SOONG
```

```
     NAME         ID      GRADE

JONES         11927       46
MARTIN        32167       69
MARTINEZ      74519       87
O'BRIEN       73410       81
PETERS        98145       55
SOONG         12345       73
```

Figure 8.22 The output of the program of Figure 8.21 when run with the data shown in the text.

Exercises

*1. Identify two syntactic errors in the following program, and suggest one statement that, although legal, will not do what the programmer presumably expected.

```
CHARACTER*5 A, B
CHARACTER C*10, E*15, F*20

A = 'HARRY'
B = 'HELEN'
C = A // B
C = A // '=FANG'
E = B // D
B = '12345'
C = A // B
C = A // '12345'
C = A // 12345
A = A // '12345'
A = 'DANNY'
B = ' IS A '
F = A // B // ' GOOD BOY.'
PRINT *, F(1:11) // 'BAD' // F(16:20)

END
```

2. Modify the program of Figure 7.9 so that it reads identifiers for the unknowns, and then prints those identifiers with the solution. Each line of the output should consist of one identifier and the corresponding value for that unknown.

3. Add a test to the program of Figure 8.19 so that if an input name begins with a comma, the program does not blow up. Choose your own solution: an error message, printing the last name only, or whatever else seems reasonable.

4. Extend the program of Figure 8.19 so that it can process middle names as well as first and last. The input will consist of the last name, a comma, the first name, a blank, and the middle name. The output is to consist of the first name, one blank, the middle initial, a period, one blank, and the last name. For instance, if given the name

```
JONES,SAMUEL ALBERT
```

the program should print

```
SAMUEL A. JONES
```

This requires searching for the first blank after finding the comma. The first letter of the middle name will be in the character position after that.

Make your own decision on how much error checking to do. If you try to anticipate too many kinds of bad data, the program will become unwieldy for the purposes of an exercise. But do try to prevent the program from bombing over bad data, which might require at least checking that variables representing character positions have valid values.

5. How would the program of Figure 8.21 operate if the statement

```
J = J - 1
```

were omitted?

*6. What would the program of Figure 8.21 do if the input consisted of exactly 50 records?

7. Modify the program of Figure 8.21 so that it does not require the "buffer" of blanks at the beginning of the table.

8. Modify the program of Figure 8.21 so that it will handle an ID consisting of a nine-digit social security number, treated as alphanumeric data. After reading the data, pad it with one blank so that the **CSWAP** subroutine can be used unchanged.

9. Modify the program of Figure 8.21 so that it uses the technique sketched in Exercise 7 on page 176.

10. Make the name processing core of the program of Figure 8.19 into a subroutine, then modify the program of Figure 8.21 so that it calls this subroutine to print the names in the normal order.

SUBPROGRAMS, II

Introduction

The basic ideas of subprograms were presented in Chapter 4, along with the reasons they are so important and so widely used. We have seen them in various applications since then. In this chapter we shall consolidate this knowledge in a number of examples, and introduce the EXTERNAL, COMMON, EQUIVALENCE, DATA, and PARAMETER statements, along with the concept of adjustable array dimensions. Some of these topics have wider applicability than just in connection with subprograms, but have not been needed in earlier examples and may be collected here.

Several of the example programs use FUNCTION or SUBROUTINE subprograms provided by IMSL, Inc., of Houston. Full use of such libraries is strongly recommended, as noted several times in earlier chapters, because they save programming time and because they incorporate a level of mathematical and programming expertise that the occasional user cannot hope to match.

An IMSL SUBROUTINE to find roots of polynomials

Let us take one more look at the fourth-degree polynomial with which we began our study, to see how its roots could be found using a subroutine from the IMSL library.

The first step is to choose an appropriate routine. The manuals accompanying the IMSL library comprise four fairly thick volumes, describing a total of over 500 subroutines. The table of contents points us to the section on finding roots of functions, where we find a further choice of about half a dozen programs that might be used. A little further study reveals that the most appropriate is one called ZPOLR, which finds the roots of a polynomial having real coefficients. The documentation consists of:

- A brief statement of the purpose of the subroutine.

- A list of the arguments, giving their purpose, type, whether input or output, and other descriptive material as appropriate. (The programs in the rest of this chapter that use IMSL routines include that part of the documentation.)

- An indication of the computer configurations on which the subroutine will run. In practice this mostly means whether there are both single and double precision versions available, since some machines designed for scientific calculations have such long word lengths that double precision is not needed.

■ A brief description of the mathematical algorithm used, together with a bibliographic reference to a more complete description.

■ An example, including all the statements required to invoke the subroutine and the output that would be produced. (This is nice, because it gives you a sample with which to test that you are using the subroutine correctly.)

ZPOLR takes four arguments:

■ A one-dimensional real array holding the coefficients of the polynomial. If the degree of the polynomial is n, then there will be $n+1$ coefficients. The coefficient of x^n must be in element 1 of the array, the coefficient of x^{n-1} in element 2, etc., with the constant term in element $n+1$.

■ An integer giving the degree of the polynomial.

■ A complex array with n elements, in which the roots are placed. (Real roots simply have a zero imaginary part.)

■ An integer variable that contains an error code on return from the subroutine. A value of zero indicates that no problems were detected; values of 129, 130, or 131 indicate various errors, as detailed in the documentation and as shown in the program that follows.

It is permissible for the two arrays to be larger than the sizes implied by the degree of the polynomial. ZPOLR will know how many elements are used in any one invocation. The fact that there may be additional unused elements creates no problems. The program is, accordingly, set up to handle a polynomial of degree up to 20 so that we can try out a few additional examples.

At the time and place this work was done, there wasn't any way to invoke the IMSL subroutines from a Fortran 77 program. Therefore, I wrote the program in an older version of Fortran, one that implements the 1966 standard instead of the 1977 standard. For our purposes in this particular case, the differences are of no consequence. Comments have to be designated by a C in column 1 instead of an asterisk, and the indexing parameters in a DO statement or an implied DO in a READ or WRITE must be either integer constants or integer variables. Because of the latter, we must set up a variable to hold a number that is one greater than the degree of the polynomial. In Fortran 66 (or in WATFIV) we cannot write a statement such as:

```
READ (5, 200) (A(K), K = 1, NDEG+1)
```

The inconvenience in this case is strictly minor. Fortran 66's lack of the IF-THEN-ELSE, list-directed input and output, and character variables, to name three deficiencies, are much more serious matters—but they are not needed here.

Since compilers based on Fortran 66, or some level of extension of it, are still widely in use, it is just as well that an occasion has arisen to show some programs written in it. The reader may expect now and then to have to modify, or at least be able to read, such programs. This is not difficult, except to the extent that the absence of such features as the IF-THEN-ELSE logical construct tends to make them harder to understand.

The program shown in Figure 9.1 was, in fact, developed and run using the facilities of the City University of New York. It was entered using my Diablo 1620 terminal, which is a typewriterlike device, as distinguished from a TV-like

```
//DANCC  JOB  IPOLYGEN
//       EXEC  FORTGCG,ADDLIB='SYS2.IMSLS'
//FORT.SYSIN DD *
C A PROGRAM TO FIND THE ROOTS OF POLYNOMIALS WITH REAL COEFFICIENTS
C USING THE IMSL SUBROUTINE ZPOLR.
C
C THE SUBROUTINE IS OF THE FORM:
C     CALL ZPOLR (A, NDEG, Z, IER)
C     A        - REAL VECTOR OF LENGTH NDEG+1, CONTAINING THE
C                  COEFFICIENTS IN ORDER OF DECREASING
C                  POWERS OF THE VARIABLE. (INPUT)
C     NDEG     - INTEGER DEGREE OF THE POLYNOMIAL. (INPUT)
C                  NDEG MUST BE GREATER THAN 0 AND LESS
C                  THAN 101.
C     Z        - COMPLEX VECTOR OF LENGTH NDEG CONTAINING
C                  THE COMPUTED ROOTS OF THE POLYNOMIAL.
C                  (OUTPUT)
C     IER      - ERROR PARAMETER. (OUTPUT)
C                  TERMINAL ERROR
C                    IER = 129 INDICATES THAT THE DEGREE OF THE
C                        POLYNOMIAL IS GREATER THAN 100 OR LESS
C                        THAN 1.
C                    IER = 130 INDICATES THAT THE LEADING
C                        COEFICIENT IS ZERO. THIS RESULTS IN AT
C                        LEAST ONE ROOT, Z(NDEG), BEING SET TO
C                        POSITIVE MACHINE INFINITY.
C                    IER = 131 INDICATES THAT ZPOLR FOUND
C                        FEWER THAN NDEG ZEROS. IF ONLY M ZEROS
C                        ARE FOUND, Z(J), J = M+1,...,NDEG,
C                        ARE SET TO POSITIVE MACHINE INFINITY.
C                        ZPOLR WILL TERMINATE WITH THIS ERROR
C                        IF IT CANNOT FIND ANY ONE ROOT WITHIN
C                        200*NDEG ITERATIONS OR IF IT DETERMINES,
C                        WITHIN THOSE 200*NDEG ITERATIONS, THAT
C                        IT CANNOT FIND THE ROOT IN QUESTION.
C
C
C THIS PROGRAM IS SET UP TO HANDLE A POLYNOMIAL OF DEGREE
C UP TO 20.  THE ACTUAL DEGREE IS READ AS DATA AND PASSED
C TO THE SUBROUTINE. THE FACT THAT THE ARRAY A IS NOT FULL
C WHEN THE DEGREE IS LESS THAN 20 CAUSES ZPOLR NO PROBLEMS.
C
      INTEGER NDEG, NCOEFF, IER, K
      REAL A(21)
      COMPLEX Z(20)
C
      READ (5,100) NDEG
  100 FORMAT (I2)
      NCOEFF = NDEG + 1
      READ (5, 200) (A(K), K = 1, NCOEFF)
  200 FORMAT (F10.0)
      WRITE (6, 300) NDEG, (K, A(K), K = 1, NCOEFF)
  300 FORMAT ('1', 'NDEG = ', I4//(' ', 'A(', I2, ') = ', F12.7))
C
      CALL ZPOLR ( A, NDEG, Z, IER )
C
      WRITE (6, 400) IER, (K, Z(K), K = 1, NDEG)
  400 FORMAT ('0', 'IER = ', I4//(' ', 'Z(', I2, ') = ', 2F12.7))
C
      STOP
      END
```

Figure 9.1

```
//GO.SYSIN DD *
04
2.
-15.
-2.
120.
-130.
/*
```

Figure 9.1 (cont'd) A program based on the IMSL subroutine ZPOLR to find the roots of the fourth degree polynomial studied in prior chapters.

device. The connection to the CUNY system was by a commercial telephone line. The program was entered, edited, and run using the WYLBUR system developed at Stanford University and marketed by On-Line Business Systems, Inc., of San Francisco.

What is shown as "the program" is, this time, the entire file as submitted for execution. That is, it includes the Job Control Language (JCL) lines at the beginning and two other places, and it includes the data. The first JCL line is a JOB card; something of the sort must almost always appear at the beginning of a job submitted for batch execution (as distinguished from the interactive execution that has been used in most other programs so far). The second JCL line says to execute (EXEC) a set of programs cataloged under the name FORTGCG, and to use the single precision version of the IMSL library (IMSLS) that is a part of a collection named SYS2. (The name may be different at your installation.) With CUNY having leased the IMSL library and loaded it (from the magnetic tape on which it is supplied) into the computer disk storage as part of SYS2, that's all I have to do to get access to it with my Fortran programs.

The cataloged procedure named FORTGCG will compile the program (that's what the "C" means), using a compiler named Fortran G (that's the FORTG), and then "Go" (that's what the second G means). In other words, the program is compiled and then immediately run. This is what always happens automatically with WATFIV, but with other compilers we have choices. The procedure FORTGCLG, for example, permits subprograms not part of this source program to be linked in. There are many other variations.

The third JCL statement specifies that the source program follows immediately in the job stream. (It could have been elsewhere—in a magnetic disk file, for example.) The JCL line between the end of the program and the beginning of the data specifies that the data also follows in the job stream, and the last line signifies the end of the job.

Naturally, unless you are using an IBM operating system like this one, your JCL will be different. Many of the general concepts carry over, however. Any operating system, for example, will have some provision for designating where to search for subroutines in a library, and some way to indicate where the data is.

The body of the program itself contains no new concepts, except that the SUBROUTINE named in the CALL statement is not, in this case, part of the program. This (main) program will be compiled without reference to that fact. When the compilation is complete, the "Go" part of the procedure named FORTGCG searches for ZPOLR and sets up the communications between the main

```
NDEG =     4

A( 1) =       2.0000000
A( 2) =     -15.0000000
A( 3) =      -2.0000000
A( 4) =     120.0000000
A( 5) =    -130.0000000
IER =       0

Z( 1) =       6.4573193    0.0
Z( 2) =      -2.8713579    0.0
Z( 3) =       2.5264320    0.0
Z( 4) =       1.3876047    0.0
```

Figure 9.2 The output of the program of Figure 9.1, run with data defining the fourth degree polynomial.

program and the subprogram. In fact, ZPOLR itself calls quite a number of subsidiary subprograms from the IMSL library. All this also occurs automatically. As we begin to see, an operating system does a great deal of work for us behind the scenes. This kind of semi-automatic file management is a central part of what any operating system does, whether it is running on the largest machines or on a home computer.

The program is set up to echo the input, then call ZPOLR, and finally, print the results. The output from the program, when run with the data shown, appears in Figure 9.2. We see that the four roots have been found, each as a complex number with zero for the imaginary part. The error return is zero, indicating that no problems were encountered in finding the roots.

With the program running, let's try it with a different polynomial: $x^5 + x^4 + 2x^3 + 3x^2 + 4x + 5 = 0$. WYLBUR can now be used to modify the data at the end of the same program and resubmit it. The output is shown in Figure 9.3. We see that there is one real root and four complex, the latter occurring in conjugate pairs.

```
NDEG =     5

A( 1) =       1.0000000
A( 2) =       1.0000000
A( 3) =       2.0000000
A( 4) =       3.0000000
A( 5) =       4.0000000
A( 6) =       5.0000000
IER =       0

Z( 1) =      -1.2662582    0.0
Z( 2) =      -0.5813786   -1.2001343
Z( 3) =      -0.5813786    1.2001343
Z( 4) =       0.7145080   -1.3076324
Z( 5) =       0.7145080    1.3076324
```

Figure 9.3 The output of the program of Figure 9.1, when run with data defining a fifth degree polynomial having one real root and four complex roots.

```
NDEG =       8

A( 1) =      1.0000000
A( 2) =     -2.0000000
A( 3) =      2.0000000
A( 4) =     -1.0000000
A( 5) =      0.0
A( 6) =      0.0
A( 7) =     -6.0000000
A( 8) =     11.0000000
A( 9) =     -5.0000000
IER =        0

Z( 1) =     -1.2662582    0.0
Z( 2) =     -0.5813786   -1.2001343
Z( 3) =     -0.5813786    1.2001343
Z( 4) =      0.7145080   -1.3076324
Z( 5) =      0.7145080    1.3076324
Z( 6) =      1.0000029    0.0000052
Z( 7) =      1.0000029   -0.0000052
Z( 8) =      0.9999939    0.0
```

Figure 9.4 The output of the program
of Figure 9.1, when run with data
defining a polynomial that is the same
one in Figure 9.3, except multiplied by
$(x - 1)^3$.

Finally, let's define an eighth degree polynomial that is the same as the previous one, except multiplied by $(x - 1)^3$. The roots should therefore be the same five just found, plus a triple root at 1. Figure 9.4 shows the output. We see that some kind of problem with the finite precision of number representation and with roundoff error has caused the solution method to find a very small imaginary part in two of the triple roots at 1. This is no fault of the mathematics or of the writers of the IMSL subroutine. As we have seen many times by now, such quirks are, to a certain extent, unavoidable.

The EXTERNAL **statement**

ZPOLR is suitable for finding roots of polynomials. But suppose we wanted to find roots of some function that is not defined as a polynomial? In such a case, we need to use some subroutine that permits us to specify a FUNCTION as one of its arguments. That, in turn, requires us to have some way to specify that the argument is not any kind of expression, but rather is the name of a external subprogram. Such is the purpose of the EXTERNAL statement.

For a first look at this new feature, let us determine the value of π to 14 places, by finding the root of $\sin x = 0$ that lies between 3 and 4. Inspection of the IMSL documentation discloses a suitable routine, named ZBRENT. The method, named after its developer, R. P. Brent, requires that we know an interval in which the function changes sign, but that is hardly a problem with the sine function. The algorithm is summarized in the documentation as ". . . a combination of linear interpolation, inverse quadratic interpolation, and bisection. Convergence is usually superlinear, and is never much slower than it is for bisection." (Still want to invest a couple of months writing a good root-finder?

Plus a couple of years to learn enough mathematics to do it? When you can lease this whole library of about 500 subroutines for a year, for about the cost of a couple of months of programming time?)

The arguments of ZBRENT are as shown in the program of Figure 9.5. We see that the first argument must be the name of a function for which we wish

```
//DANCC JOB FINDPI
//       EXEC FORTGCG,ADDLIB='SYS2.IMSL'
//FORT.SYSIN DD *
C A PROGRAM TO FIND PI AS A ROOT OF SIN X = 0, USING THE
C DOUBLE PRECISION VERSION OF THE IMSL SUBROUTINE ZBRENT.
C
C THE SUBROUTINE IS OF THE FORM:
C     CALL ZBRENT (F, EPS, NSIG, A, B, MAXFN, IER)
C       F        - AN EXTERNAL FUNCTION SUBPROGRAM F(X)
C                    PROVIDED BY THE USER, WHICH COMPUTES F FOR
C                    ANY X IN THE INTERVAL (A, B). (INPUT)
C                    F MUST APPEAR IN AN EXTERNAL STATEMENT
C                    IN THE CALLING PROGRAM.
C       EPS      - FIRST CONVERGENCE CRITERION. (INPUT). A ROOT,
C                    B, IS ACCEPTED IF ABS(F(B)) IS LESS THAN OR
C                    EQUAL TO EPS. EPS MAY BE SET TO ZERO.
C       NSIG     - SECOND CONVERGENCE CRITERION. (INPUT). A ROOT,
C                    B, IS ACCEPTED IF THE CURRENT APPROXIMATION
C                    AGREES WITH THE TRUE SOLUTION TO NSIG
C                    SIGNIFICANT DIGITS.
C       A,B      - ON INPUT, THE USER MUST SUPPLY TWO POINTS, A
C                    AND B, SUCH THAT F(A) AND F(B) ARE OPPOSITE
C                    IN SIGN.
C                    ON OUTPUT, BOTH A AND B ARE ALTERED. B
C                    WILL CONTAIN THE BEST APPROXIMATION TO THE
C                    ROOT OF F.  SEE REMARK 1.
C       MAXFN    - ON INPUT, MAXFN SHOULD CONTAIN AN UPPER BOUND
C                    ON THE NUMBER OF FUNCTION EVALUATIONS
C                    REQUIRED FOR CONVERGENCE.  ON OUTPUT, MAXFN
C                    WILL CONTAIN THE ACTUAL NUMBER OF FUNCTION
C                    EVALUATIONS USED.
C       IER      - ERROR PARAMETER. (OUTPUT)
C                    TERMINAL ERROR
C                    IER = 129 INDICATES THAT THE ALGORITHM
C                       FAILED TO CONVERGE IN MAXFN EVALUATIONS.
C                    IER = 130 INDICATES THAT F(A) AND F(B)
C                       HAVE THE SAME SIGN.
C
C REMARK 1.  ON EXIT FROM ZBRENT, WHEN IER=0, A AND B SATISFY THE
C            FOLLOWING,
C            F(A)*F(B) .LE. 0,
C            ABS(F(B)) .LE. ABS(F(A)), AND
C            EITHER ABS(F(A)) .LE. EPS OR
C            ABS(A-B) .LE. MAX(ABS(B),0.1)*10.0**(-NSIG).
C            THE PRESENCE OF 0.1 IN THIS ERROR CRITERION CAUSES
C            LEADING ZEROES TO THE RIGHT OF THE DECIMAL POINT TO BE
C            COUNTED AS SIGNIFICANT DIGITS. SCALING MAY BE REQUIRED
C            IN ORDER TO ACCURATELY DETERMINE A ZERO OF SMALL
C            MAGNITUDE.
C
```

Figure 9.5

```
      DOUBLE PRECISION EPS, A, B, FD
      INTEGER NSIG, MAXFN, IER
      EXTERNAL FD
C
      EPS = 0.0D0
      NSIG = 13
      A = 3.0D0
      B = 4.0D0
      MAXFN = 50
C
      CALL ZBRENT(FD, EPS, NSIG, A, B, MAXFN, IER)
      WRITE (6, 100) EPS, NSIG, A, B, MAXFN, IER
  100 FORMAT ('0', 'EPS   = ', D23.14/
     $         ' ', 'NSIG  = ', I4/
     $         ' ', 'A     = ', D23.14/
     $         ' ', 'B     = ', D23.14/
     $         ' ', 'MAXFN = ', I4/
     $         ' ', 'IER   = ', I4)
      STOP
      END
C
      DOUBLE PRECISION FUNCTION FD(X)
      DOUBLE PRECISION X
      FD = DSIN(X)
      RETURN
      END
```

Figure 9.5 (*cont'd*) A program using the IMSL subroutine ZBRENT to find the value of π to 14 decimal places.

to find a root. (To be precise, we should speak of roots of equations and zeros of functions, but the terms are often used interchangeably when the meaning is clear.) There are two convergence criteria. The first accepts point B as a root if F(B) is less than EPS. The second convergence criterion asks that B agree with the true solution to NSIG decimal digits. If EPS is specified as zero, only the second convergence criterion is applied. (The notion of the "true solution" may sound a little odd: if the program knows the "true solution," why doesn't it just say what it is? The remark in the prologue explains what is meant by this terminology.)

A and B are our first examples, in the IMSL library, of arguments that are both input and output. On input they specify a range of the independent variable within which a root is known to lie. On output both have been changed, and B contains the best approximation to the root. MAXFN is also both input and output, as noted. IER this time has two possible error values. The first indicates very serious mathematical problems with the function for which a root is being sought or that the value of MAXFN is too small, and the second indicates bad data (A and B did not bracket a root).

To get π to 14 places we need to use double precision, so we specify the library IMSL on the JCL instead of IMSLS. (As noted above, the names may be different in your installation.) We also declare EPS, A, and B as being double precision rather than real.

The statement:

```
EXTERNAL FD
```

tells the compiler that, when it later sees FD as an argument in the CALL ZBRENT statement, that argument should be treated, not as an expression, but as the name of a function external to this program. In this case, the function DSIN, one supplied with the system, is used within FD to compute the function value.

In Fortran 77, when we wish to specify the name of a built-in function for use as a subprogram argument, we must utilize the INTRINSIC statement instead of the EXTERNAL statement. Other solutions to this problem are available, but discussion of them would take us too far afield.

Here is the output when this program was run:

```
EPS     =     0.0
NSIG    =     14
A       =     0.314154625558920D 01
B       =     0.314159265358980D 01
MAXFN   =     6
IER     =     0
```

The value printed for B is in fact correct to 14 places, as we shall see in the next chapter, where π is computed to 100 places.

A SUBROUTINE **that invokes two** EXTERNAL **functions**

In this example, we shall compute two numerical integrals, using a subroutine that invokes two EXTERNAL functions, one built-in (SIN) and one a FUNCTION subprogram that is part of the program.

The numerical integration is done by the simplest method of any practical value, the *trapezoidal rule*. This method approximates the area under a curve defined by the integrand, with trapezoids. In other words, it approximates a curve by a series of straight lines. In principle, assuming a well-behaved integrand, the approximation can be made as accurate as we please by taking more intervals. We shall see, however, for one last time, that this standard tenet of elementary calculus breaks down under the impact of finite representation and roundoff error.

The program in Figure 9.6 declares FS and RECIP as being EXTERNAL. The first uses the sine function, and the second is a simple FUNCTION, which appears at the end of the program, to compute the reciprocal of its argument.

The program begins by computing the integral of sin x from zero to π. It does this first with just two intervals, which amounts to approximating the sine function with a triangle. Then it does it again, this time with four intervals, then with eight, sixteen, etc., ending with 2^{18} intervals. This latter is a rather large number, and doing the job this way, to be frank, takes an inordinate amount of computer time. It is done here to make a point that will appear shortly.

After doing this task, the program invokes the integration routine one last time, specifying the FUNCTION RECIP and providing different values for the other arguments. This time, we are computing the integral of 1/x from 1 to 10, so the result should be an approximation to the natural logarithm of 10.

```
* A main program to demonstrate the trapezoidal rule
* The main program calls TRAPEZ, which approximates an integral
*    for which the integrand is given by an EXTERNAL FUNCTION
*
* The program first evaluates the integral of sin(x) from zero to pi.
* It does so repeatedly, for numbers of intervals that are powers of 2,
*    to demonstrate the effects of truncation error and roundoff error.
* It then evaluates integral of 1/x from 1 to 10.

      REAL A, B, INTGRL, LN10
      REAL FS, RECIP
      INTEGER N, I, NINTER

      EXTERNAL FS, RECIP

* Print headings
      PRINT 100
  100 FORMAT ('1', '              N       INT SIN(X)'/)

* Evaluate the integral for intervals that are powers of 2
      DO 10 I = 1, 18
         NINTER = 2**I
         CALL TRAPEZ (FS, 0, 3.14159265, NINTER, INTGRL)
         PRINT 200, NINTER, INTGRL
  200    FORMAT (' ', I10, F15.7)
   10 CONTINUE

* Compute the integral of 1/X from 1 to 10
      CALL TRAPEZ (RECIP, 1.0, 10.0, 9, LN10)
      PRINT 300, LN10
  300 FORMAT ('0', 'THE NATURAL LOG OF 10 APPROX = ', F15.7)

      END

* A SUBROUTINE to evalute an integral by the trapezoidal rule
*
* Parameters:
*    F: the name of a function that defines the integrand
*    A: lower limit of integration (input)
*    B: upper limit of integration (input)
*    N: number of intervals (input)
*    RESULT: the value of the integral (output)

      SUBROUTINE TRAPEZ (F, A, B, N, RESULT)

      REAL F, A, B, RESULT, H, SUM
      INTEGER N, I

* Compute H and find the sum of integrand, except endpoints
      H = (B - A) / N
      SUM = 0.0
      DO 10 I = 1, N - 1
         SUM = SUM + F(A + I*H)
   10 CONTINUE

* Compute trapezoidal rule approximation to integral
      RESULT = (H / 2) * ( F(A) + 2*SUM + F(B) )

      RETURN
      END
```

Figure 9.6

```
* A FUNCTION to supply the sine as an external function
      REAL FUNCTION FS(X)
      REAL X
      FS = SIN(X)
      RETURN
      END

* A FUNCTION to compute the reciprocal of its argument
      REAL FUNCTION RECIP(X)
      REAL X
      RECIP = 1.0/X
      RETURN
      END
```

Figure 9.6 (cont'd) The first version of a program that uses two **EXTERNAL** subprograms in computing two integrals by the trapezoidal rule.

The **SUBROUTINE** named **TRAPEZ** computes the approximation to the integral of the function named by its first argument. The lower limit of integration is the second argument, the upper limit is the third, the number of intervals is the fourth, and the value of the approximation is given to the fifth (output) argument.

This not being a book on numerical methods, we shall assume that you either already know how the trapezoidal rule works (it's covered in every elementary calculus text), or that you don't care. Just observe that, within the **DO** loop, **SUM** is continually being incremented by the value of the integrand as the independent variable moves along its axis.

The output is shown in Figure 9.7. The exact value of the integral of the sine from zero to π should be 2. Observe that the accuracy of the approximation improves up through 256 intervals, then gets worse! Why should that be?

The answer is in the statement:

```
SUM = SUM + F(A + I*H)
```

For the largest number of intervals used, the value of **SUM** is in the range of 160,000. When the small values of sin x that occur toward the end of the interval are added to such a number, most of the significance in the small numbers is irretrievably lost.

It is a simple matter to prove that this is the only problem here. A modified version of this program, not shown, involved exactly one change—the statement:

```
DOUBLE PRECISION SUM
```

was added to the **SUBROUTINE TRAPEZ**. This provided roughly nine more decimal digits of precision for storing the sum of the integrands, preventing the loss of significance when the small values at the end were added. *Nothing else was changed.* The same single precision **SIN** function was used, the computation of the approximation still involved single precision numbers except for **SUM**, etc.

N	INT SIN(X)
2	1.5707951
4	1.8961182
8	1.9742298
16	1.9935675
32	1.9983835
64	1.9995861
128	1.9998856
256	1.9999552
512	1.9998655
1024	1.9997683
2048	1.9997234
4096	1.9996901
8192	1.9979057
16384	1.9962511
32768	1.9954309
65536	1.9949083
131072	1.9665222
262144	1.9328661

THE NATURAL LOG OF 10 APPROX =   2.3789654

Figure 9.7 The output of the program of Figure 9.6. Observe that the accuracy of the approximation to the integral of the sine decreases after the number of intervals reaches 256, because of the effect of roundoff error.

Figure 9.8 shows the result, proving that the loss of significance in addition was the only problem here.

The natural logarithm of 10 is about 2.302585. This program doesn't come very close to that, proving only that nine line segments don't approximate $1/x$ very well over the interval from 1 to 10.

With this last excursion into the real world of finite precision ended, let us turn to another Fortran topic and see it in application with another IMSL subroutine.

The COMMON statement

All illustrative programs so far have set up the "communication" between calling and called programs through argument lists. Using techniques presented to date, there is no way to tell the operating system that T in the main program is the same variable as T in a subprogram, except to put T into the argument list.

For the most part, that is just fine. A strong case can be made for the notion that program segments *should not be able* to communicate except through the argument list. A wide variety of problems can be avoided thereby. This book is not the place to try to present both sides of this debate and settle the issue. Suffice to say that although most serious students of the craft of programming probably take the argument-list-only approach, there are cases where that approach is not sufficient. The example program of this section leads into one such situation.

The application involves finding the percentage of the total energy radiated by a hot object, that falls within the visible portion of the electromagnetic spectrum. For reference, and to provide a basis for the programs that follow, here are the equations that describe the physical situation.

An object that radiates all of the energy it absorbs is called a *blackbody* radiator. A blackbody radiates energy at a rate proportional to the fourth power of its absolute temperature, according to the Stefan-Boltzmann equation:

$$E = 36.9 \cdot 10^{-12} \, T^4$$

where E = emissive power, watts/cm^2 and T = temperature, °K.

We are interested in the fraction of this total energy contained in the visible portion of the spectrum, which is taken here to be 4×10^{-5} to 7×10^{-5} cm. We can get the visible part by integrating Planck's equation between these limits:

$$\int_{4\cdot10^{-5}}^{7\cdot10^{-5}} \frac{2.39 \cdot 10^{-11} \, dx}{x^5(e^{1.432/Tx} - 1)}$$

where x = wavelength, cm; E and T as before.

The *luminous efficiency* is defined as the ratio of the visible energy to the total energy. Our problem, then, is to evaluate the integral, divide by the expression giving the total energy, and multiply by 100 to convert to a percentage.

N	INT SIN(X)
2	1.5707951
4	1.8961182
8	1.9742308
16	1.9935694
32	1.9983931
64	1.9995985
128	1.9998999
256	1.9999743
512	1.9999933
1024	1.9999981
2048	1.9999990
4096	1.9999990
8192	2.0000000
16384	2.0000000
32768	2.0000000
65536	1.9999990
131072	2.0000000
262144	2.0000000

THE NATURAL LOG OF 10 APPROX = 2.3789673

Figure 9.8 The output of the program of Figure 9.6 when the statement DOUBLE PRECISION SUM was added to the subroutine TRAPEZ.

This time we shall use an IMSL subroutine to do the integration. DCADRE uses "cautious Romberg extrapolation," combined with automatic subdivision of the interval of integration, until it has satisfied the convergence criteria. Here again, as with ZBRENT, there are two: absolute and relative. If we know the approximate size of the integral, it may make sense to specify that the approximation be correct to within some absolute amount. More commonly, we will specify that the approximation agree with the true value to within some fraction of the true value. If the absolute error is specified as zero, only the relative error criterion is applied.

This should remind us, from previous chapters, that there is a limit on the accuracy that can reasonably be demanded. In a machine where real values are represented with seven decimal digits, it would be futile to ask for a relative error of 10^{-9}. But how should the IMSL documentation handle this problem, bearing in mind that the IMSL subroutines run on 18 different computers, with widely differing word sizes? Here is the wording used: "RERR should be large enough so that, when added to 100.0, the result is a number greater than 100.0."

(What do you suppose the man in the street would make of a sentence like that? How could it be that adding something to 100.0 would not give a number greater than 100.0? By now, you should be able to answer that question. As a matter of fact, you should be able to write a small program that would find an approximation to the smallest number that meets this test—without knowing, when you write the program, anything about the computer's word size.)

The program in Figure 9.9 shows the arguments for DCADRE, which involve no new concepts. Note, however, the requirement that the function F have only a single argument. This is reasonable, and the argument is the variable of integration. But look back at the formula for the integral: it involves both wavelength (the integration variable) and temperature. If we wish to compute the luminous efficiency for a fixed temperature, there is no problem: we can write in that fixed temperature. Let's pause and do just that, before proceeding to solve the communications problem that is developing.

For a bit of variety, the program this time is in WATFIV, which will be handy when we get to the second version. Observe that near the beginning of the main program, T is given the value 7000.0. The value of T is then used in the denominator of the expression that computes the efficiency. But that same expression uses the value returned by DCADRE (it's a FUNCTION this time rather than a SUBROUTINE), which also involves the temperature. The integrand is defined

```
//DANCC JOB INTEGRAT
//      EXEC WATFIV,ADDLIB='SYS2.WATIMSLS'
$JOB
C A PROGRAM USING THE IMSL FUNCTION DCADRE TO DO INTEGRATION.
C THE APPLICATION IS TO FIND THE LUMINOUS EFFICIENCY OF A
C BLACKBODY RADIATOR, I.E., THE PERCENTAGE OF THE TOTAL
C RADIATED ENERGY THAT IS VISIBLE.
C
C THE EFFICIENCY IS A FUNCTION OF THE TEMPERATURE,
C WHICH IN THIS VERSION IS ASSIGNED A VALUE WHEREVER IT IS NEEDED.
C
```

Figure 9.9

```
C THE FUNCTION OF IS THE FORM:
C      FUNCTION DCADRE(F, A, B, AERR, RERR, ERROR, IER)
C      DCADRE - ESTIMATE OF THE INTEGRAL OF F(X) FROM A TO B.
C               (OUTPUT)
C      F      - A SINGLE-ARGUMENT REAL FUNCTION SUBPROGRAM
C               SUPPLIED BY THE USER. (INPUT)
C               F MUST BE DECLARED EXTERNAL IN THE
C               CALLING PROGRAM.
C      A,B    - THE TWO ENDPOINTS OF THE INTERVAL OF
C               INTEGRATION. (INPUT)
C      AERR   - DESIRED ABSOLUTE ERROR IN THE ANSWER. (INPUT)
C      RERR   - DESIRED RELATIVE ERROR IN THE ANSWER. (INPUT)
C      ERROR  - ESTIMATED BOUND ON THE ABSOLUTE ERROR OF
C               THE OUTPUT NUMBER, DCADRE. (OUTPUT)
C      IER    - ERROR PARAMTER. (OUTPUT)
C               WARNING ERROR (WITH FIX)
C                 IER = 65 IMPLIES THAT ONE OR MORE
C                   SINGULARITIES WERE SUCCESSFULLY HANDLED.
C                 IER = 66 IMPLIES THAT, IN SOME
C                   SUBINTERVAL(S), THE ESTIMATE OF THE
C                   INTEGRAL WAS ACCEPTED MERELY BECAUSE THE
C                   ESTIMATED BEHAVIOR WAS SMALL, EVEN THOUGH NO
C                   REGULAR BEHAVIOR WAS RECOGNIZED.
C               TERMINAL ERROR
C                 IER = 131 INDICATES FAILURE DUE TO
C                   INSUFFICIENT INTERNAL WORKING STORAGE.
C                 IER = 132 INDICATES FAILURE DUE TO TOO
C                   MUCH NOISE IN THE FUNCTION (RELATIVE
C                   TO THE GIVEN ERROR REQUIREMENTS) OR
C                   DUE TO AN ILL-BEHAVED INTEGRAND.
C                 IER = 133 INDICATES THAT RERR IS GREATER
C                   THAN 0.1, OR RERR IS LESS THAN 0.0, OR
C                   RERR IS TOO SMALL FOR THE PRECISION OF
C                   THE MACHINE.
C
C
       REAL T, EFFIC, AERR, RERR, ERROR, ENERGY
       REAL DCADRE
       INTEGER IER
       EXTERNAL ENERGY
C
       AERR = 0.0
       RERR = 0.001
       T = 7000.0
C
       EFFIC = 100.0 * DCADRE(ENERGY, 4E-5, 7E-5, AERR, RERR, ERROR, IER)
     $      / (36.9E-12 * T**4)
C
       WRITE (6, 100) IER, EFFIC
 100   FORMAT ('0', 'IER = ', I3, '  LUMINOUS EFFICIENCY = ', F12.7)
       STOP
       END
C
       REAL FUNCTION ENERGY (X)
       REAL X, T
       T = 7000.0
       ENERGY = 2.39E-11/(X**5*(EXP(1.432/(T*X)) - 1.0))
       RETURN
       END
$ENTRY
```

Figure 9.9 (*cont'd*) The first version of a program to find the luminous efficiency of a blackbody radiator. This version sets the temperature of the radiator with separate assignment statements in the main program and in the FUNCTION that defines the integrand.

by a FUNCTION that we write, called ENERGY here, which needs to know the temperature. In this version of the program we simply insert another assignment statement in the FUNCTION definition, to give a value to the temperature.

An aside, in the nature of a review of basics. Suppose the statement:

```
T = 7000.0
```

had not been placed in the FUNCTION. Wouldn't the compiler know that, when it sees T in the formula for the integrand in ENERGY, we mean the variable of that name that was given a value in the main program? The answer: absolutely not. The compiler treats the variables of each program segment completely independently. Remember that the main program and the subprogram might well be compiled at entirely different times. We say that the variables of each program segment are *local* to it, meaning that they are unknown outside of that segment.

Here is the output when this program was run:

```
IER =    0  LUMINOUS EFFICIENCY =    39.3115900
```

In other words, at 7000° K ($\cong$ 6700° C) about 39% of the radiated energy is visible.

But back to the problem we set for ourselves. Suppose we now want a table giving the efficiency as a function of temperature. This means that the main program will have a loop running through a range of temperatures, and each temperature must somehow be communicated to the FUNCTION ENERGY. One obvious solution would be to set up a new version of ENERGY, this time with two arguments: wavelength and temperature. But we can't do that if we want to use DCADRE, because of the requirement that the FUNCTION defining the integrand have only a single argument! We might think of asking IMSL to provide a second version of DCADRE, this one having two arguments. But then a problem would pop up requiring three! If this line were followed, IMSL (and everybody else who provides such facilities) would have to supply many versions of each subprogram that calls other FUNCTIONs. And even then an occasional need would arise for something with even more parameters. This can't be the solution.

Enough suspense. The answer is the COMMON statement, with which we are able to tell the compiler that two variables in two independent program segments are in fact the same variable—without making them a subprogram argument. Such a variable is said to be *global*. The technique is simply to place the statement:

```
COMMON T
```

in both the main program and in the FUNCTION ENERGY. This causes the compiler to assign these two variables (and they still *are* two different variables, remember) to the same location in a special section of computer storage. Even though the two segments might still be compiled separately, the variable T in the main program will be assigned to the same location as the variable T in the FUNCTION, and that's all we need.

(As a matter of fact, the variables would not have to have the same name. What happens is that the *first* variable in *any* COMMON statement is assigned to the first location in the special section of storage.)

The modified program, not including the initial documentation, appears in Figure 9.10. The luminous efficiency is computed for temperatures from 1000° to 9000° in steps of 500°. These temperatures are obtained with a WATFIV WHILE loop.

```
//DANCC JOB INTEGRAT
//      EXEC WATFIV,ADDLIB='SYS2.WATIMSLS'
$JOB
C A PROGRAM USING THE IMSL FUNCTION DCADRE TO DO INTEGRATION.
C THE APPLICATION IS TO FIND THE LUMINOUS EFFICIENCY OF A
C BLACKBODY RADIATOR, I.E., THE PERCENTAGE OF THE TOTAL
C RADIATED ENERGY THAT IS VISIBLE.
C
C THE EFFICIENCY IS A FUNCTION OF THE TEMPERATURE,
C WHICH IN THIS VERSION IS HANDLED THROUGH COMMON
C
C THE FUNCTION OF IS THE FORM:
C       FUNCTION DCADRE(F, A, B, AERR, RERR, ERROR, IER)
C       DCADRE - ESTIMATE OF THE INTEGRAL OF F(X) FROM A TO B.
C                (OUTPUT)
C       F      - A SINGLE-ARGUMENT REAL FUNCTION SUBPROGRAM
C                SUPPLIED BY THE USER. (INPUT)
C                F MUST BE DECLARED EXTERNAL IN THE
C                CALLING PROGRAM.
C       A,B    - THE TWO ENDPOINTS OF THE INTERVAL OF
C                INTEGRATION. (INPUT)
C       AERR   - DESIRED ABSOLUTE ERROR IN THE ANSWER. (INPUT)
C       RERR   - DESIRED RELATIVE ERROR IN THE ANSWER. (INPUT)
C       ERROR  - ESTIMATED BOUND ON THE ABSOLUTE ERROR OF
C                THE OUTPUT NUMBER, DCADRE. (OUTPUT)
C       IER    - ERROR PARAMTER. (OUTPUT)
C                WARNING ERROR (WITH FIX)
C                   IER = 65 IMPLIES THAT ONE OR MORE
C                      SINGULARITIES WERE SUCCESSFULLY HANDLED.
C                   IER = 66 IMPLIES THAT, IN SOME
C                      SUBINTERVAL(S), THE ESTIMATE OF THE
C                      INTEGRAL WAS ACCEPTED MERELY BECAUSE THE
C                      ESTIMATED BEHAVIOR WAS SMALL, EVEN THOUGH NO
C                      REGULAR BEHAVIOR WAS RECOGNIZED.
C                TERMINAL ERROR
C                   IER = 131 INDICATES FAILURE DUE TO
C                      INSUFFICIENT INTERNAL WORKING STORAGE.
C                   IER = 132 INDICATES FAILURE DUE TO TOO
C                      MUCH NOISE IN THE FUNCTION (RELATIVE
C                      TO THE GIVEN ERROR REQUIREMENTS) OR
C                      DUE TO AN ILL-BEHAVED INTEGRAND.
C                   IER = 133 INDICATES THAT RERR IS GREATER
C                      THAN 0.1, OR RERR IS LESS THAN 0.0, OR
C                      RERR IS TOO SMALL FOR THE PRECISION OF
C                      THE MACHINE.
C
C
C THIS VERSION COMPUTES THE LUMINOUS EFFICIENCY FOR A
C RANGE OF TEMPERATURES.
C THE TEMPERATURE CANNOT BE COMMUNICATED TO THE FUNCTION THAT
C EVALUATES THE INTEGRAND AS AN ARGUMENT, BECAUSE DCADRE
C REQUIRES THAT FUNCTION TO HAVE ONLY ONE ARGUMENT.
C THE TEMPERATURE IS ACCORDINGLY PLACED IN COMMON.
C
```

Figure 9.10

```
      REAL T, EFFIC, AERR, RERR, ERROR, ENERGY
      REAL DCADRE
      COMMON T
      INTEGER IER
      EXTERNAL ENERGY
C
      AERR = 0.0
      RERR = 0.001
C
      WRITE (6, 100)
 100  FORMAT ('1', 'LUMINOUS EFFICIENCY OF A BLACKBODY RADIATOR'/
     $' ', 'AS A FUNCTION OF TEMPERATURE IN DEGREES KELVIN'//
     $'        T              % EFFICIENCY'//)
C
      T = 1000.0
      WHILE ( T .LE. 10000.0 ) DO
C
         EFFIC = 100.0 * DCADRE(ENERGY,4E-5,7E-5,AERR,RERR,ERROR,IER)
     $          / (36.9E-12 * T**4)
C
         WRITE (6, 200) T, EFFIC
 200     FORMAT (' ', 5X, F6.0, F16.1)
C
         T = T + 500.0
      END WHILE
C
      STOP
      END
C
      REAL FUNCTION ENERGY (X)
      REAL X, T
      COMMON T
      ENERGY = 2.39E-11/(X**5*(EXP(1.432/(T*X)) - 1.0))
      RETURN
      END
$ENTRY
```

Figure 9.10 (*cont'd*) A second version of the luminous efficiency program, modified to use a WHILE loop to generate a range of temperatures that are communicated to the FUNCTION through COMMON.

Figure 9.11 shows the output.

(A true aside, not on the subject at all. The melting point of tungsten, of which the filaments of incandescent light bulbs are made, is about 3300° C ($\cong$ 3600° K). At this temperature only about a sixth of the energy being radiated is visible. The rest goes into the room as infrared and ultraviolet light and ends up heating the walls, which is not a total loss in winter, but it's no help in reading the newspaper. The actual temperature obviously has to be somewhat below the melting point. Furthermore, even at temperatures below the melting point, a few atoms continuously boil off the surface; when they strike the glass, they produce the blackening that appears as a bulb ages. The hotter the filament, the more boiling and the shorter the life. Eventually, the filament must burn out because of the atoms lost in this boiling. The light bulb designer must therefore choose a burning temperature that compromises between a short hot life with good efficiency, which reduces electricity costs but raises bulb replacement costs, or a long dull life that uses a lot of electricity to produce very little light. Ponder these things the next time you hear someone say that the light bulb manufacturers deliberately design their bulbs to burn out. Well, in a way, it's true, but the reason lies in the laws of physics, not greed. Yes, I learned my trade working for the General Electric Company.)

One final example of the use of this integration routine. The IMSL documentation asserts that DCADRE can, in many cases, handle jump discontinuities. Now, in advanced calculus one learns of various methods for integrating functions that are thoroughly discontinuous, but it is nontrivial to set up a program to do it. Most programming students, assigned to write a program to do numerical integration, would probably not even think of the possibility.

As one small—and not very difficult—illustration of this capability, a program was set up to call DCADRE with this FUNCTION definition:

```
REAL FUNCTION  F(X)
REAL X
IF ( X .LE. 5.0 ) F = X
IF ( X .GE. 5.0) F = X - 5.0
RETURN
END
```

This defines what is usually called a *sawtooth* function, with a discontinuity at 5. The integral of such a function clearly has a meaning, given rather simple definitions of what is meant by the value of the function at the point of discontinuity. But an integration routine not designed to look for such possibilities would almost certainly give a wrong answer. DCADRE, when asked to integrate this function from zero to 10 with a relative error of 0.00001, promptly—one is tempted to say nonchalantly—returned the result 24.9998.

```
LUMINOUS EFFICIENCY OF A BLACKBODY RADIATOR
AS A FUNCTION OF TEMPERATURE IN DEGREES KELVIN
```

T	% EFFICIENCY
1000.	0.0
1500.	0.1
2000.	0.8
2500.	3.4
3000.	8.3
3500.	14.5
4000.	21.1
4500.	27.0
5000.	31.8
5500.	35.4
6000.	37.7
6500.	38.9
7000.	39.3
7500.	39.0
8000.	38.2
8500.	37.1
9000.	35.7
9500.	34.2
10000.	32.6

Figure 9.11 The output of the program of Figure 9.10.

The EQUIVALENCE statement

The COMMON statement lets us specify that two variables in different program segments should be assigned to the same storage location. The EQUIVALENCE statement, on the other hand, lets us specify that two variables in the *same* program segment should occupy the same storage location. Why would we ever want to do such a thing?

Probably the most important reason is to conserve storage space. Not for single variables; today's computers are never so short of storage as that. However, suppose we have two very large arrays that are processed by a single program, but not at the same time: it is never necessary to save the information in one while the other is processed. Suppose further that either one can fit in storage, but together the main storage could not hold them. All we have to do is name both of them in an EQUIVALENCE statement, and the compiler will allocate storage so that they both start in the same location.

You might wonder why a special language feature is needed to handle the problem. Why not just use the same array twice? But suppose one array is one-dimensional with 200,000 elements, whereas the other is two-dimensional with 100 rows and 3000 columns? The two arrays aren't even the same size, and using them requires one subscript in the first case and two in the second. Actually, we *could* set up a linear array with 300,000 elements, use it as is for the first array, and program our own formula to turn a row number and a column number into a location in the linear array for the second. But that kind of thing is what compilers are supposed to keep us from having to do!

Or suppose one array has 100 rows and 200 columns of integer values, but the second has 100 rows and 100 columns of complex values? In most computers one complex number takes the same space as two integers, so the same total amount of storage is involved. But the compiler is going to object violently to any attempt to process the same array as holding integers in one statement and as holding complex numbers later in the same program. In this kind of case, using the same storage for two different purposes is essentially impossible in Fortran without the EQUIVALENCE statement.

Any time a variable must be used with more than one type characteristic, integer and complex, for example, the EQUIVALENCE statement must be used. An example occurs in ZPOLY, an IMSL subroutine that uses a different method for finding the roots of a polynomial with real coefficients. The documentation states that the subroutine treats the linear array that holds the complex roots as a real array. In some Fortran systems, the linkage between calling and called programs is such that this causes no problems. In others, it is necessary to set up a separate real array and use EQUIVALENCE. For example, suppose we want the roots of a cubic. There are four real coeficients and three complex roots. Matching the three-element complex array requires a six-element real array. All of this can be done thus:

```
COMPLEX Z(3)
REAL RZ(6)
EQUIVALENCE (Z(1), RZ(1))
```

RZ and Z each require a total of six storage locations. The EQUIVALENCE statement says that they are to start in the same locations, so they will in fact occupy the same six locations. The CALL statement tells ZRPOLY about a real argument; the calling program can operate on an equivalent complex array without trouble.

When the COMMON and EQUIVALENCE statements are used in the same program, they interact in ways that can become exceedingly complicated. In fact, this presentation has not even covered all the complexities when the two statements are used separately. The general consensus at this writing is that global variables and the renaming or retyping of variables—which are the generic terms for what COMMON and EQUIVALENCE do—should be used sparingly if at all. We have shown, for each, an application that cannot be done without them.

Free use otherwise is to be discouraged. Inserting an EQUIVALENCE statement to correct a memory lapse under which a variable got two different names at different points in the program, simply cannot be defended. Neither can putting everything into COMMON, so that subprograms need not have arguments. Such tricks have an immediate superficial appeal, but long experience has, by now, proven that the long-term effect is pernicious.

It is possible that these strictures need modification when it comes to very large programs, those with many thousands of statements. There, where many programmers work on one project for months or years, the problems of communications between people loom much larger than the problems of communication between parts of the program. Used wisely, COMMON and EQUIVALENCE may have a role to play that is rather different from what has been suggested here.

We leave such considerations and move on, however, since the assumption is that the reader of this book does not anticipate working on such projects. If that assumption is not valid for some readers—if you do turn to work on large programming projects—you are going to have to learn a great deal more about many topics than could ever go into a book like this in any event.

A pipe temperature program with four SUBROUTINEs

Our final illustrative program for this chapter computes the temperature in the interior of a square pipe. (Square pipes are, admittedly, hard to find, but the computations are somewhat simpler to set up. We leave round pipes to the writers of packages, of which you will see examples in the following chapter.) Physically, the situation is that the temperature at each point on the interior and exterior surfaces of the pipe is known; we are to find the steady-state temperature within the metal of the pipe. Mathematically, the problem is to solve LaPlace's Equation:

$$\frac{\partial^2 T}{\partial x^2} + \frac{\partial^2 T}{\partial y^2} = 0$$

given the surface temperatures. The solution turns out to have a characteristic that is comprehensible even if you don't know much about partial differential equations: the temperature at each point in the metal is the average of the temperatures at the points immediately surrounding it. This formulation also, in effect, defines the computational task. We set up a rectangular grid covering the pipe. (This is where we appreciate square pipes.) Each grid intersection represents a point at which we want to know the temperature. We write an equation for each of the grid points, stating that its temperature is the average of the four surrounding points. We now have a system of many simultaneous equations—as many as there are grid points within the metal.

Figure 9.12 shows a schematic representation of the pipe. It is half immersed in ice water at 0° C; the top surface is held at 100° C; the temperature along the tops of the sides varies linearly from 0° to 100°; a fluid at 200° C is flowing through the pipe.

We shall cover the pipe with a grid having 61 rows and 61 columns in the first version of the program. Row 1 is the top surface of the pipe; row 19 is the top surface of the hole through the pipe; row 43 is the bottom of the hole; row 61 is the bottom (exterior) surface. Similar numbers apply to the 61 columns. We shall set up an array with 61 rows and 61 columns to represent this grid. That is 3721 elements, but there are not that many simultaneous equations. First, 529 grid points (23 × 23) correspond to locations in the fluid of the pipe, and we are not concerned with those. Second, the temperatures at the 240 grid points on the exterior surface and the 88 grid points on the interior surface are constants, representing the boundary conditions. We are left with 2864 grid points that lie within the metal (but not on either surface).

It would therefore seem that we must set up a system of 2864 simultaneous equations and solve it. That is, in effect, exactly what we do, but not by establishing an array of 2864 rows and 2864 columns, which would greatly exceed

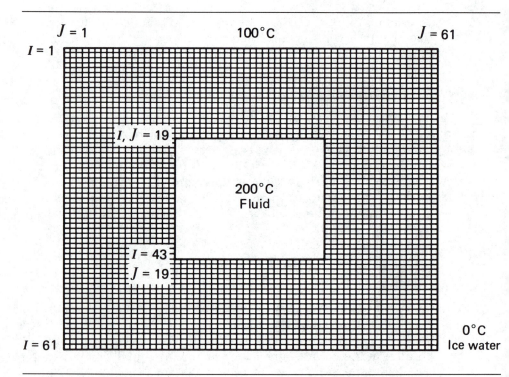

Figure 9.12 Schematic representation of the square pipe for which the temperature distribution is calculated by the program of Figure 9.13.

the internal storage capacity of almost all available computers. But even on those few computers that might have enough storage, doing the job this way would be terribly wasteful. Each and every equation in this system has exactly five nonzero coefficients, and they are -1, -1, 4, -1, and -1. That's all it takes to represent the fact that each temperature is the average of the four temperatures around it. Why set up an array with over eight million zeros in it, even if we had the space? Instead, we set up an array with 61^2 elements, one for each grid point, then use an iterative approach that boils down to the Gauss-Seidel iteration method presented in Chapter 7, with the coefficients appearing simply as constants in an assignment statement.

Let's try that again, walking through the method of solution that the program implements. We set up a 61×61 array. Each array element corresponding to a point on the interior or exterior surface of the pipe is initialized to its fixed temperature. All other points are initialized to zero. We then systematically take up each point in the interior of the metal, and compute its new temperature as the average of the temperatures at the four surrounding points. Doing this for each of the 2864 grid points within the metal constitutes one *sweep* of the system. Each time we compute a new temperature, we compute the difference between the new and old temperature at that point; this is called the *residual* at that point. This process is repeated until the largest residual found anywhere in a complete sweep is less than a convergence criterion that is read as input. At that point we stop the process and present the results.

And how shall that be done? Printing the 2864 temperatures would be almost useless, if intended to be helpful to a human being. Instead, we shall produce a graphical output, using a technique that will be described in connection with the program.

The first version of the program to do this task appears in Figure 9.13. After the usual prologue describing the program, its variables, and subprograms, we have the short main program and four subroutines. Most the work of the program is done by the subroutines, with the main program handling only the highest level logic. This is ordinarily considered to be a good general approach to program organization. It promotes understandability, it places boundaries on the effects of errors and changes, and it permits stepwise program development.

Based on the earlier description of the work of the program, reading it should not be difficult. The **SUBROUTINE READIN** contains one **READ** statement as its only executable statement. It is slightly unusual for a subroutine to do so little, but the advantages of program modularization argue for doing things this way. There are no important penalties in program space, execution speed, or programming time.

The **SUBROUTINE INITLZ** gives initial values to the elements of the array, in a fairly straightforward manner. If the problem specifications were to change, all program modifications having to do with establishing the boundary conditions would be localized in this one subroutine. Any change we might make to this subroutine could be guaranteed to have no effect on anything except the variables in its argument list, which is what is meant by localizing the effect of program modifications. This desirable localization would be destroyed by casual use of the **COMMON** statement, which is the major reason we avoid it except under carefully controlled circumstances.

```
********************************************************************
* A program to solve LaPlace's Equation for a square pipe
********************************************************************
*
* Bottom half of pipe is immersed in ice water; top is at 100 C;
*    pipe is carrying a fluid at constant temperature;
*    temperature of top half of sides varies linearly from 100 to 0 C
*
* The main program calls the following subroutines:
*
*    READIN: Gets various parameters
*    INITLZ: Initializes the pipe temperature array
*    ITRATE: Performs one iteration of the solution method
*    PLOT:   Produces the results as a plot of isothermal regions
*
*    Variables:
*
*    TEMPER: A 61x61 array that holds the temperature values
*    OMEGA:  Acceleration factor in the iterative solution
*    EPSLON: The convergence criterion
*    BIGRES: (Biggest residual) largest residual found on one sweep;
*            convergence is assumed to have been achieved when
*            BIGRES < EPSLON
*    INNERT: (Inner temperature) temperature of the fluid in the pipe
*    MAXIT:  Maximum number of iterations to be permitted
*    ITERNO: (Iteration number) a counter to check against MAXIT
*    I:      Row number
*    J:      Column number
*
********************************************************************

        REAL TEMPER(61, 61)
        REAL OMEGA, EPSLON, BIGRES, INNERT
        INTEGER MAXIT, ITERNO

        CALL READIN (INNERT, OMEGA, EPSLON, MAXIT)
        CALL INITLZ (TEMPER, INNERT)

* Use DO loop to check for nonconvergence in MAXIT iterations
        DO 10 ITERNO = 1, MAXIT
            CALL ITRATE (TEMPER, OMEGA, BIGRES)
            IF ( BIGRES .LT. EPSLON ) THEN
                PRINT *, ' WITH EPSILON = ', EPSLON,
     $                   ', CONVERGENCE ACHIEVED IN', ITERNO, ' ITERATIONS'
                CALL PLOT (TEMPER)
            STOP
            END IF
   10   CONTINUE

* If DO is satisfied, convergence was not achieved in MAXIT iterations
        PRINT *, 'NO CONVERGENCE IN', MAXIT, ' ITERATIONS'
        STOP
        END
```

Figure 9.13

```
*****************************************************************************
* READIN: The subroutine to read parameters
*****************************************************************************
      SUBROUTINE READIN (INNERT, OMEGA, EPSLON, MAXIT)
      REAL INNERT, OMEGA, EPSLON
      INTEGER MAXIT
      READ *, INNERT, OMEGA, EPSLON, MAXIT
      RETURN
      END


*****************************************************************************
* INITLZ: The subroutine to initialize the pipe temperature array
*****************************************************************************
      SUBROUTINE INITLZ (TEMPER, INNERT)
      REAL TEMPER(61, 61), INNERT
      REAL EDGET
      INTEGER I, J

* The outer boundaries of the pipe are at I and J = 1, and I and J = 61
* The inner boundaries are at 19 and 43

* Initialize entire array to zero
      DO 20 I = 1, 61
         DO 10 J = 1, 61
            TEMPER(I, J) = 0.0
  10     CONTINUE
  20  CONTINUE

* Initialize top surface
      DO 30 J = 1, 61
         TEMPER(1, J) = 100.0
  30  CONTINUE

* Initialize top half of sides
      DO 40 I = 1, 31
         EDGET = 100.0 - 100.0 * (I - 1) / 30.0
         TEMPER(I, 1) = EDGET
         TEMPER(I, 61) = EDGET
  40  CONTINUE

* Initialize inside of pipe
      DO 50 I = 19, 43
         TEMPER(I, 19) = INNERT
         TEMPER(I, 43) = INNERT
  50  CONTINUE
      DO 60 J = 19, 43
         TEMPER(19, J) = INNERT
         TEMPER(43, J) = INNERT
  60  CONTINUE

      RETURN
      END
```

Figure 9.13 (*cont'd*)

```
********************************************************************
* ITRATE: The subroutine to do one iteration of the solution of the
*         LaPlace Equation
* Note that the method is essentially the Gauss-Seidel iteration
*    technique, applied to a system of 2864 simultaneous equations--but
*    there are only five nonzero coefficients in each row,
*    and they are always the same (-1, -1, 4, -1, -1), so they can be
*    generated in the program rather than being stored in a huge array.
********************************************************************
      SUBROUTINE ITRATE (TEMPER, OMEGA, BIGRES)
      REAL TEMPER(61, 61), OMEGA, BIGRES, TEMPT
      INTEGER I, J

* Initialize BIGRES
      BIGRES = 0.0

* Compute new approximation to temperature at each nonboundary point
* Must not compute temperature for fluid portion

      DO 20 I = 2, 60
         DO 10 J = 2, 60
            IF (        (I .LT. 19)
     $           .OR. (I .GT. 43)
     $           .OR. (J .LT. 19)
     $           .OR. (J .GT. 43) ) THEN
               TEMPT = (OMEGA/4.0) * (   TEMPER(I+1,J)
     $                              + TEMPER(I-1,J)
     $                              + TEMPER(I,J+1)
     $                              + TEMPER(I,J-1) )
     $              + (1.0 - OMEGA) * TEMPER(I,J)
               BIGRES = AMAX1(BIGRES, ABS(TEMPT - TEMPER(I,J)))
               TEMPER(I,J) = TEMPT
            END IF
   10    CONTINUE
   20 CONTINUE

      RETURN
      END


********************************************************************
* PLOT: The subroutine to print the results in graphical form
********************************************************************
      SUBROUTINE PLOT (TEMPER)
      REAL TEMPER(61,61)
      CHARACTER LINE(61)
      INTEGER I, J, K
      CHARACTER SYMBOL(17), BLANK
      DATA (SYMBOL(K), K = 1, 17)/'A',' ','B',' ','C',' ','D',' ',
     $   'E',' ','F',' ','G',' ','H',' ','I'/, BLANK/' '/

* Move to top of new page
      WRITE (6, 100)
  100 FORMAT ('1')

* Run through all rows, from top (I = 1) down
      DO 20 I = 1, 61
```

Figure 9.13 *(cont'd)*

```
*  Convert each temperature to a letter or blank,
*  but do not print anything for fluid region or for out-of-range values
           DO 10 J = 1, 61
              IF (          (I .GT. 19)
     $              .AND. (I .LT. 43)
     $              .AND. (J .GT. 19)
     $              .AND. (J .LT. 43)
     $              .OR. (TEMPER(I, J) .LT. 0.0)
     $              .OR. (TEMPER(I, J) .GT. 200.0) ) THEN
                 LINE(J) = BLANK
              ELSE
                 K = TEMPER(I, J) / 11.765 + 1.0
                 LINE(J) = SYMBOL(K)
              END IF
  10          CONTINUE
              WRITE (6, 200) LINE
 200          FORMAT (' ', 61A1)
  20       CONTINUE

           RETURN
           END
```

Figure 9.13 (*cont'd*) A main program and four subprograms to compute the temperature distribution of the pipe sketched in Figure 9.12.

The subroutine to do the actual calculation is named ITRATE. Basically, it runs through all the points within the metal, computing a new value for the temperature at each. Avoiding the exterior surface is a matter of having the DO loops run from 2 to 60 rather that from 1 to 61. Avoiding the points on the interior surface of the pipe and the points in the fluid is handled here by an IF statement having four logical tests.

The computation of a new temperature is a little more elaborate than the earlier discussion suggested. It turns out that, if we merely replace each temperature by the average of the four surrounding temperatures, convergence is very slow. Instead, we take the point of view that each time we compute a change in the temperature, we are probably going in the right direction, but not as far as will eventually be needed. So why not extrapolate a bit: replace each temperature by its old value plus some factor times the difference between it and the average of its neighbors? This technique, which is called *over-relaxation* or *acceleration,* makes the Gauss-Seidel iteration method practical for this kind of problem; without it, convergence takes several times as many iterations. The rate of convergence is fairly sensitive to the value of the acceleration factor, which in this program is named OMEGA. Experimentation with this problem led to a value of 1.84.

When we calculate a new temperature by this method, we do not immediately place it back in its array element. Rather, we first compute the residual at that point, which is the absolute value of the difference between the new and old temperatures. This value is then compared, using the built-in function AMAX1, with the value of BIGRES. Since BIGRES was initialized to zero, at the end of a sweep it will contain the largest residual found anywhere in the array. This is returned as an output argument to the main program, which uses it to determine whether the iteration process has converged.

Presenting the results graphically is the task of the subroutine named PLOT. This subroutine, incidentally, was written last, after the rest of the program had been checked out. During that checkout, PLOT consisted only of a few WRITE statements to produce the temperatures at a few representative points, as numbers. This is an example of stepwise refinement. It is simple when a program is well modularized as a small main program and a collection of subprograms.

The idea in this graphical output is that we shall establish 17 temperature bands, each about 12° wide. Temperatures in the first band, from 0° to about 12°, will be represented by the letter A. Then a 12° band will be represented by blanks. The next band will be represented by B, then blank, then C, etc. The output will therefore show bands of blanks and letters, on the order of a contour map of the temperatures. (Glance ahead to Figure 9.14 to see what is meant.)

The conversion into a blank or letter of a temperature represented by a Fortran real quantity, is handled by two statements in the subroutine. Dividing a temperature in the range (0, 200) by 11.765 produces a quotient in the range (0, 16.999). Adding 1.0 produces a number in the range (1, 17.999); truncating this to an integer gives a number in the range (1, 17). This value, called K in the subroutine, is then used to select one element from an array of 17 one-character elements in an array named SYMBOL.

The initialization of this array brings in one new language element, the DATA statement. The general form is:

$$\text{DATA list}/d_1, d_2, \ldots, d_n/, \text{list}/d_1, d_2, k*d_3, \ldots d_m/ \ldots$$

In this symbolic description, a "list" contains the names of the variables to receive values, the ds are the values, and k, if present, is an integer constant. For example, the statement:

```
DATA /A, B, C/14.7, 62.1, 1.5E-20/
```

would assign the value 14.7 to A, 62.1 to B, and 1.5×10^{-20} to C.

This is done prior to the time of execution of the compiled object program. The DATA statement is not executable. The values assigned by the DATA statement are placed in storage when the object program is loaded into storage, and that is the end of the actions instituted by the DATA statement. It is permissible (although risky) to change the values of these variables, if that is useful. But, having done so, it is *not* possible to "re-execute" the DATA statement to put these variables back to their initial values.

In our program, the DATA statement uses an implied DO list, combined with a final assignment of a value to the sysmbol BLANK.

The conversion of numerical values to letters or blanks is embedded within two DO loops. The first works through the rows, developing a line of print corresponding to each horizontal line of the grid. The second works through the columns of the row selected by the first. The conversion is, finally, placed within an IF to prevent the printing of anything corresponding to the fluid within the pipe.

The output of the program is shown in Figure 9.14. The picture is not square because on the printer used here, there are not the same number of vertical lines per inch as there are characters per inch horizontally.

```
EEEEEEEEEEEEEEEEEEEEEEEEEEEEEEEEEEEEEEEEEEEEEEEEEEEEEEEEEEEEEEEE
EEEEEEEEEEEEEEEEEEEEEEEEEEEEEEEEEEEEEEEEEEEEEEEEEEEEEEEEEEEEEEEE
   EEEEEEEEEEEEEEE                                EEEEEEEEEEEEEEE
     EEEEEEEEE                                      EEEEEEEEE
      EEEEEEE            FFFFFFFFFFFFFF             EEEEEEE
       EEEEEE        FFFFFFFFFFFFFFFFFFFFFFF        EEEEEE
D        EEEEE      FFFFFFFFFFFFFFFFFFFFFFFFFFF      EEEEE        D
DD       EEEEE       FFFFFFF          FFFFFFF       EEEE        DD
DDD       EEE         FFFFF             FFFFF       EEE        DDD
 DDD      EEE        FFFF    GGGGGGGGGGGGGGG    FFFF       EEE       DDD
 DDD      EEE        FFF    GGGGGGGGGGGGGGGGGGGGG   FFF       EEE       DDD
  DDD     EEE     FFF    GGGGGG          GGGGGG    FFF     EEE       DDD
   DD     EEE    FF    GGGG                GGGG    FF     EEE       DD
C    DD    EE     FF   GGG     HHHHHHHHHHHH     GGG   FF    EEE      DD   C
C    DD    EE    FF    GG    HHHHHHHHHHHHHHHHHHHHH    GG    FF    EE     DD   C
CC    DD  EE    FF    G   HHHH                HHHH    G    FF    E     DD   CC
 C    DD    E    FF  GG  HH                      HH  GG   FF    E     DD    C
CC   DD    E    F    G  HH    IIIIIIIIIIIIIIIIIIII    HH  G    F    E     DD  CC
CC    D    E    F    G   H  IIIIIIIIIIIIIIIIIIIIIIIII  H  G    F    E     D    CC
  C    D    E    F  GG   H  I                      I  H  GG   F    E     D    C
B   CC   D    E    F  G   H  I                      I  H   G    F    E     D  CC  B
B    C   DD   E    F  G  H   I                      I   H  G   F    E     DD   C    B
B    C   DD   E    F  G  H  II                      II  H  G   F    E     DD   C    B
 B    C    D   E    F  G  H  II                      II  H  G   F    E     D    C   B
 B   CC    D   E    F  G  H  II                      II  H  G   F    E     D   CC   B
 B    C    D   EE   F  G  H  II                      II  H  G   F    EE   D    C    B
 BB   C    D   EE   F  G  H  II                      II  H  G   F    EE   D    C   BB
A    B    C    D    E  F  G  H  II                      II  H  G  F    E     D    C   B   A
A    B    C   DD    E  F  G  H  II                      II  H  G  F    E     DD   C   B   A
A    B    C    D    E  F  G  H  II                      II  H  G  F    E     D    C   B   A
A    B   CC    D    E  F  G  H  II                      II  H  G  F    E     D   CC   B   A
A   BB    C    D    E  F  G  H  II                      II  H  G  F    E     D    C   BB  A
AA   B    C    D    E  F  G  H  II                      II  H  G  F    E     D    C   B   AA
AA    B   C    D    E  F  G    H  I                      I  H  GG  F    E     D    C   B   AA
AA    B   C    D    E  FF   G  H  I                      I  H  G  F    E     D    C   B   AA
AA    B   C    D    EE  F  G  H  I                      I  H  G  F  E     D    C   B   AA
AA    B   C   DD    E  F  G  H  I                      I  H  G  F  E     DD   C   B   AA
AA    B   C    D    E  F  G  H  I                      I  H  G  F  E     D    C   B   AA
AA    B   CC   D    E  F  G  H  I                      I  H  G  F  E     D   CC   B   AA
AA    B    C   D    E  F  G  H  I                      I  H  G  F  E     D    C   B   AA
AA    B    C    D    E  F  GH  I                      I  HG  F  E     D    C   B   AA
AA   BB   C   DD    E  F  G    I                      I     G  F  E     DD   C   BB  AA
AA   BB   C    D    E  F  G  HIIIIIIIIIIIIIIIIIIIIIIIIIIIIIIH  G  F    E     D    C   BB  AA
AA   BB  CC   D    E  F  G  H                      H  G  F    E     D   CC   BB  AA
AA    B    C    D  E    F  G  HHHHHHHHHHHHHHHHHHHHHH   G  F  EE   D    C    B    AA
AA    B   CC  DD   E   F  GGG                      GGG  F   E    DD  CC    B    AA
AA   BB  CC   D    E   FF   GGGGGGGGGGGGGGGGGGGGG   FF   E    D   CC  BB    AA
AA    BB   CC   DD   EE   FFF                 FFF   EE    DD  CC    BB    AA
AAA    BB  CC   DD    EE    FFFFFFFFFFFFFFFFFFF    EE   DD   CC   BB    AAA
AAA    BB   CC   DDD   EEE              EEE   DDD   CC    BB    AAA
AAA   BBB   CC    DDD    EEEEEEEEEEEEEEEEEEE    DDD    CC   BBB    AAA
AAA    BB   CCC    DDD                 DDDD    CCC    BB    AAA
AAAA   BBB    CCC    DDDDDDDD     DDDDDDDD    CCC    BBB    AAAA
AAAA   BBBB   CCCC     DDDDDDDD     CCCCC    BBBB    AAAA
AAAAA    BBBB   CCCCCCC             CCCCCCC    BBBB    AAAAA
AAAAA     BBBBB     CCCCCCCCCCCCCCCC     BBBBB    AAAAA
AAAAAAA      BBBBBBBB                BBBBBBBB      AAAAAAA
AAAAAAAAA       BBBBBBBBBBBBBBBBBBBBBBBBBBBB       AAAAAAAAA
AAAAAAAAAAAAAA                                AAAAAAAAAAAAAA
AAAAAAAAAAAAAAAAAAAAAAAAAAAAAAAAAAAAAAAAAAAAAAAAAAAAAAAAAAAAAAAA
AAAAAAAAAAAAAAAAAAAAAAAAAAAAAAAAAAAAAAAAAAAAAAAAAAAAAAAAAAAAAAAA
```

Figure 9.14 The output of the program of Figure 9.13, when run with data specifying that the temperature of the internal fluid is 200°C.

The program was run four more times, with input specifying fluid temperatures of 150°, 100°, 50°, and 0°. The results, reduced considerably, are shown in Figure 9.15.

From the mathematics of the situation, one would expect the results to be symmetric around a vertical axis. Close inspection of the output reveals that they are not. The explanation is that a convergence criterion of 0.5° was used, and some pairs of points in symmetric positions can, therefore, differ by a fraction of a degree. In a few cases, this put the two points in different temperature bands. Using a smaller convergence criterion would reduce this effect, of course, but it would also increase the number of iterations required. Even with the relatively coarse criterion of 0.5°, 39 iterations were required for the first plot, and that's a lot of computation when each iteration involves computing a temperature at each of 2864 points.

Program optimization

If this program were to be run a great many times, we would rather quickly look for some way to speed up program execution. One approach would be to experiment with the value of the acceleration factor. In fact, in this case, convergence can be achieved in slightly fewer iterations by fine-tuning that value. In some applications that might not be nearly good enough. It is not unusual, for example, for a computer to be embedded in some kind of process where it runs continuously. Computers in automobile engine controls, for example, obviously run all the time the car is running. Or digital signal processing computers in television sets must perform relatively simple processing at very high rates anytime the set is on. In these and many other applications, it is impossible to take the rather casual approach to program execution speed that has characterized our work so far. Sometimes the job can't be done at all without careful attention to speed. In others—the automotive application comes to mind—the issue is cost. If you are buying a million computer chips, it is worth a lot of extra programming effort to see if a $3 chip can be stretched to do the job instead of a $4 chip.

As our one excursion into this topic, let us see what a modest investment in programming effort can do to speed up the execution of this program.

The first thing we do is focus all attention on the subroutine ITRATE. Roughly 99% of the total program time is spent executing the one IF statement in it. Until we have drastically reduced the time there, it is pointless to worry about the initialization or the plotting. This will tend to be true of most programs: some small fraction of the statements in a program will represent a large fraction of the total execution time. Sometimes it is not easy to locate the crucial program sections. In such cases, it is possible to use *performance evaluation* software to identify them.

The simplest thing to try first, in our situation, is to see if the compiler itself can do a better job, given orders to work at it. This program was compiled with the IBM VS Fortran compiler, slightly modified by systems programmers at National CSS, Inc, to make it work with their operating system. The VS Fortran compiler permits the user to specify three *levels of optimization*. With level zero, which is the default when we say nothing about optimization, the compiler produces object code that correctly implements the source program, but with

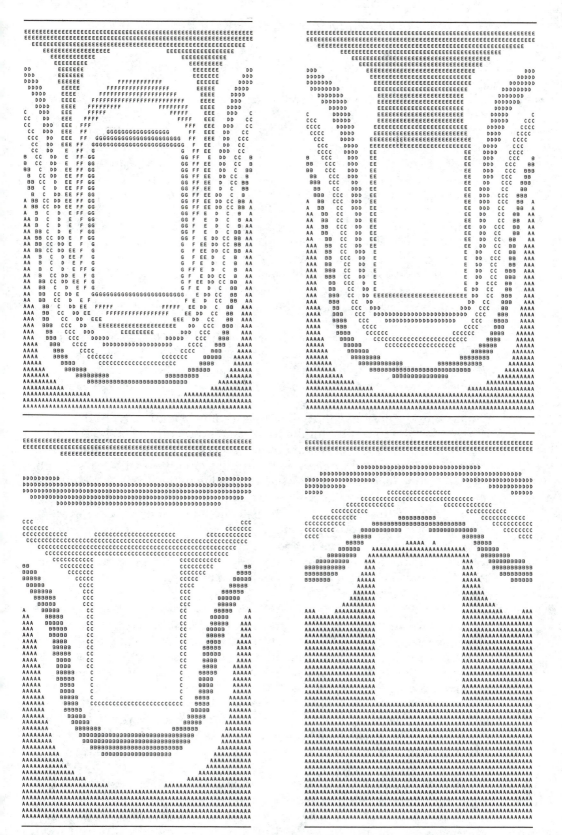

Figure 9.15 The output of the program of Figure 9.13, when run with data specifying internal temperatures of 150°, 100°, 50°, and 0°

no extra effort either to save space in the object program or to speed its execution. If we specify level 1, the compilation takes somewhat longer, as the compiler seeks to reduce storage and time requirements. With level 2, the process takes still longer, to produce a program that is almost as good as a skilled programmer could do, working directly in the language of the machine.

To demonstrate these effects, the program was recompiled with the option OPT 1 added to the command to compile. The compilation took about 15% longer, but the object program execution time was 42% less. Doing the job again at optimization level 2 produced a compile time 50% greater than that for level zero, but an execution time 70% less. The compiled object code for the ITRATE subroutine was also smaller for the optimized versions, by about a third.

At this point, we need to ask how "smart" our compiler is. Let us scrutinize that IF statement that is using most of the total time. What does the object program do if it finds that the first logical test is true? There is then no need to carry out the other three tests, because once any one condition in a set combined with .OR. has been found to be true, the whole expression is true and the others need not be considered. Some optimizing compilers take advantage of this fact, and some don't. Or how about the terms (OMEGA/4.0) and (1.0 - OMEGA)? The value of OMEGA does not change during the execution of the nested DO statements, so the division by 4.0 and the subtraction from 1.0 could be done before entering the DO loop, rather than doing the same arithmetic 2864 times.

Sometimes the answers to such questions can be found in a reference manual that describes the compiler, but the information did not seem to be available in this case. The compilation was therefore repeated, this time asking that the output listing contain the object code. Getting any information from such a program listing requires considerable knowledge of how the particular computer works. Developing that knowledge, with the background assumed for a typical reader of this book, would normally involve taking a course in what is called assembly language programming. Such a course usually appears early in the sequence taken by computer science majors, and is a good idea for anyone who will be doing design work involving microcomputers, since the course usually teaches not only the particular machine but considerable about how computers in general operate.

In any event, the answer for the VS Fortran compiler was, no, the compiler does not short-cut the OR testing and it does not move computations involving unchanging variables outside loops. To get an idea of the effect of these changes, the source program was modified to precompute the two arithmetic factors and to use four separate IF statements. With compilation at optimization level 2, the execution time was reduced about 20% over the previous best time.

Seeking further improvement, the program was rewritten to use four separate DO loops, one for each of four regions that together covered all of the 2864 points. None of these DO loops required an IF statement at all, and the execution time dropped about another 25%. Naturally, the object program was larger, in the classic trade of time for space.

Summarizing, recompiling with the best available optimization reduced execution time by about 70%, compared with the unoptimized version, and changes at the source level reduced the execution time to under 20% of what it was for the original unoptimized version. Clearly, if this program had to run several hours a day for many months, such improvements would be essentially mandatory.

In such a case, we would also need to inspect the object code to see if there is more to gain. As it turns out, in this case, the object code can be improved slightly. This compiler is quite good, at optimization level 2, but the loop can be shortened by a few instructions, leading to a further time reduction on the order of 10%.

If it is necessary to recode in assembly language to gain the last measure of speed, we can usually convert only the portion that represents the major time factor. In this example, it would not be difficult to recode only the ITRATE subroutine. Setting up the communication between a Fortran calling program and an assembly language subroutine is not difficult once one has learned the required techniques in an assembly language programming course.

Sometimes such considerations are crucial; often they hardly matter at all. If a program that takes a day or two to write is going to be executed only a few times and takes only a few minutes to execute, it's a waste of your time even to consider these issues. In other cases, the entire feasibility of a project may depend on getting adequate speed at lowest computing cost. In these latter situations, you will either need to know quite a bit more than you should expect to gain in an introductory course, or you will need to get help from an expert.

The goal in this book has been to help you gain enough background to know the difference, and to be able to talk intelligently with the experts when that course is indicated.

The PARAMETER **statement and adjustable dimensions**

The pipe temperature program clearly does its job, but it has one less than desirable characteristic: changing the size of the temperature array would involve finding something over a dozen places in the program where the dimensions are embedded. This would include all dimensioning statements where 61 occurs, all DO loops involving the numbers 19, 31, 43, or 60, and all the IF statement references to the size of the hole. Clearly, such a task would be slow and error-prone. If we can reasonably anticipate that such a change might be necessary, it would be much better to set up the program so that all such information is defined in one place—especially if writing the program that way is actually not much extra trouble. The PARAMETER statement, combined with adjustable dimensions, provides the capability we need.

The PARAMETER statement is used to give a symbolic name to a constant. If, for example, at the beginning of a program we have these statements:

```
INTEGER N
REAL PI
PARAMETER (N=100)
PARAMETER (PI=3.141593)
```

then anywhere later in the same program segment N will have the value 100 and PI will have an approximation to π. These names becomes symbolic representations of these constants; they are not variables. Any attempt to redefine them with assignment statements will be flagged as an error by the compiler. They may be used only where constants are permitted.

One such place, fortunately, is in defining arrays. With these parameter defintions established, a subsequent statement such as:

```
REAL TEMPER (N, N)
```

would be legal, establishing TEMPER as an array with 100 rows and 100 columns. If we later wished to change the array to have 80 rows and 80 columns, only the PARAMETER statement would need be changed.

What we shall do, in our pipe temperature program, is to make PARAMETERs of the numbers defining the temperature array, including the location of the hole. We must then arrange to pass this information to any subroutines that need it, which means that such subroutines must have additional arguments. Let us look at the modified program, in Figure 9.16, to see how this works out.

We see, at the beginning of the main program, a PARAMETER statement that gives values to the six numbers that define the pipe geometry. Since parameters follow the usual default typing rules if not named in type statements, these symbols are named in an INTEGER statement that begins the program. Whether to use one continued PARAMETER statement or to use several PARAMETER statements, is a matter of personal preference.

The subroutine READIN does not involve the array size, so it is unchanged. All of the others are modified to accept six additional arguments. We now need to be very clear on the distinction between dummy and actual arguments. In the definitions of these three arrays, the dummy arguments NROWS, NCOLS, etc., are *only* dummy arguments. To make the program easier to follow, we use the same names as those of the parameters in the main program—but the compiler takes no information from that correspondence. In processing the definitions, the dummy arguments simply tell, as always, what to do with the actual arguments that are supplied at execution time.

There is nothing new here, except for one crucial factor: within the subroutine definitions, dummy arguments appear in the array dimensioning information. This is how *adjustable dimensions* are established. When dummy arguments are used to give the dimensions of arrays in subprograms, the compiler sets up the subprogram so that it can handle any array size that may later appear when the subprogram is invoked.

Let us now review the ways in which the symbols NROWS and NCOLS are used. In the PARAMETER statement, they are used to give symbolic names to two constants. In the CALL statements for INITLZ, ITRATE, and PLOT, they are used as actual arguments that refer to constants. In the definitions of the three subroutines, they are used as dummy arguments. Finally, in the statement:

```
REAL TEMPER (NROWS, NCOLS)
```

that appears in each of the three subroutines, they are used to specify adjustable dimensions.

When this program was compiled and run, it produced the same results as the earlier version.

```
***********************************************************************
* A program to solve LaPlace's Equation for a square pipe
*
* Second version: uses PARAMETER and adjustable dimensions
*   to make program much more easily adaptable to differing arrays sizes.
* All dimensioning information for all program components is contained
*   in the PARAMETER statement.
*
***********************************************************************
*
* Bottom half of pipe is immersed in ice water; top is at 100 C;
*    pipe is carrying a fluid at constant temperature;
*    temperature of top half of sides varies linearly from 100 to 0 C
*
* The main program calls the following subroutines:
*
*    READIN: Gets various parameters
*    INITLZ: Initializes the pipe temperature array
*    ITRATE: Performs one iteration of the solution method
*    PLOT:   Produces the results as a plot of isothermal regions
*
*    Variables:
*
*    TEMPER: An array that holds the temperature values;
*             its size in this version is given by NROWS and NCOLS.
*    OMEGA:  Acceleration factor in the iterative solution
*    EPSLON: The convergence criterion
*    BIGRES: (Biggest residual) largest residual found on one sweep;
*             convergence is assumed to have been achieved when
*             BIGRES < EPSLON
*    INNERT: (Inner temperature) temperature of the fluid in the pipe
*    MAXIT:  Maximum number of iterations to be permitted
*    ITERNO: (Iteration number) a counter to check against MAXIT
*    I:      Row number
*    J:      Column number
*
*    NROWS:  Number of rows, = maximum value of I
*    NCOLS:  Number of columns, = maximum value of J
*    TOPI:   Number of row in pipe just above interior
*    BOTTMI: Number of row in pipe just below interior
*    LEFTJ:  Number of column just to left of interior
*    RIGHTJ: Number of column just to right of interior
*
***********************************************************************
      INTEGER NROWS, NCOLS, TOPI, BOTTMI, LEFTJ, RIGHTJ
      PARAMETER (NROWS = 61,
     $           NCOLS = 61,
     $           TOPI = 19,
     $           BOTTMI = 43,
     $           LEFTJ = 19,
     $           RIGHTJ = 43)
      REAL TEMPER(NROWS, NCOLS)
      REAL OMEGA, EPSLON, BIGRES, INNERT
      INTEGER MAXIT, ITERNO

      CALL READIN (INNERT, OMEGA, EPSLON, MAXIT)
      CALL INITLZ (TEMPER, INNERT,
     $             NROWS, NCOLS, TOPI, BOTTMI, LEFTJ, RIGHTJ)
```

Figure 9.16

```
* Use DO Loop to check for nonconvergence in MAXIT iterations
      DO 10 ITERNO = 1, MAXIT
         CALL ITRATE (TEMPER, OMEGA, BIGRES,
     $                    NROWS, NCOLS, TOPI, BOTTMI, LEFTJ, RIGHTJ)
         IF ( BIGRES .LT. EPSLON ) THEN
            PRINT *, ' WITH EPSILON = ', EPSLON,
     $                    ', CONVERGENCE ACHIEVED IN', ITERNO, ' ITERATIONS'
            CALL PLOT (TEMPER,
     $                    NROWS, NCOLS, TOPI, BOTTMI, LEFTJ, RIGHTJ)
            STOP
         END IF
   10 CONTINUE

* If DO is satisfied, convergence was not achieved in MAXIT iterations
      PRINT *, 'NO CONVERGENCE IN', MAXIT, ' ITERATIONS'
      STOP
      END


**************************************************************************
* READIN: The subroutine to read parameters
**************************************************************************

      SUBROUTINE READIN (INNERT, OMEGA, EPSLON, MAXIT)
      REAL INNERT, OMEGA, EPSLON
      INTEGER MAXIT
      READ *, INNERT, OMEGA, EPSLON, MAXIT
      RETURN
      END


**************************************************************************
* INITLZ: The subroutine to initialize the pipe temperature array
**************************************************************************

      SUBROUTINE INITLZ (TEMPER, INNERT,
     $                    NROWS, NCOLS, TOPI, BOTTMI, LEFTJ, RIGHTJ)
      INTEGER NROWS, NCOLS, TOPI, BOTTMI, LEFTJ, RIGHTJ
      REAL TEMPER(NROWS, NCOLS), INNERT
      REAL EDGET
      INTEGER I, J

* The outer boundaries of the pipe are at I and J = 1,
*    and I = NROWS and J = NCOLS
* The inner boundaries are at I = TOPI and BOTTMI, J = LEFTJ and RIGHTJ

* Initialize entire array to zero
      DO 20 I = 1, NROWS
         DO 10 J = 1, NCOLS
            TEMPER(I, J) = 0.0
   10    CONTINUE
   20 CONTINUE

* Initialize top surface
      DO 30 J = 1, NCOLS
         TEMPER(1, J) = 100.0
   30 CONTINUE
```

Figure 9.16 *(cont'd)*

```
* Initialize top half of sides
      DO 40 I = 1, (NROWS - 1) / 2
         EDGET = 100.0 - 100.0 * (I - 1) / ((NROWS - 1) / 2.0)
         TEMPER(I, 1) = EDGET
         TEMPER(I, NCOLS) = EDGET
   40 CONTINUE

* Initialize inside of pipe
      DO 50 I = TOPI, BOTTMI
         TEMPER(I, LEFTJ) = INNERT
         TEMPER(I, RIGHTJ) = INNERT
   50 CONTINUE
      DO 60 J = LEFTJ, RIGHTJ
         TEMPER(TOPI, J) = INNERT
         TEMPER(BOTTMI, J) = INNERT
   60 CONTINUE

      RETURN
      END

***********************************************************************
* ITRATE: The subroutine to do one iteration of the solution of the
*         LaPlace Equation
* Note that the method is essentially the Gauss-Seidel iteration
*    technique, applied to a system of 2864 simultaneous equations--but
*    there are only five nonzero coefficients in each row,
*    and they are always the same (-1, -1, 4, -1, -1), so they can be
*    generated in the program rather than being stored in a huge array.
***********************************************************************

      SUBROUTINE ITRATE (TEMPER, OMEGA, BIGRES,
     $                   NROWS, NCOLS, TOPI, BOTTMI, LEFTJ, RIGHTJ)
      INTEGER NROWS, NCOLS, TOPI, BOTTMI, LEFTJ, RIGHTJ
      REAL TEMPER(NROWS, NCOLS), OMEGA, BIGRES, TEMPT
      INTEGER I, J

* Initialize BIGRES
      BIGRES = 0.0

* Compute new approximation to temperature at each nonboundary point
* Must not compute temperature for fluid portion

      DO 20 I = 2, NROWS - 1
         DO 10 J = 2, NCOLS - 1
            IF (       (I .LT. TOPI  )
     $          .OR. (I .GT. BOTTMI)
     $          .OR. (J .LT. LEFTJ )
     $          .OR. (J .GT. RIGHTJ) ) THEN
               TEMPT = (OMEGA/4.0) * (    TEMPER(I+1,J)
     $                                  + TEMPER(I-1,J)
     $                                  + TEMPER(I,J+1)
     $                                  + TEMPER(I,J-1) )
     $                   + (1.0 - OMEGA) * TEMPER(I,J)
               BIGRES = AMAX1(BIGRES, ABS(TEMPT - TEMPER(I,J)))
               TEMPER(I,J) = TEMPT
            END IF
   10    CONTINUE
   20 CONTINUE

      RETURN
      END
```

Figure 9.16 *(cont'd)*

```
************************************************************************
* PLOT: The subroutine to print the results in graphical form
************************************************************************
      SUBROUTINE PLOT (TEMPER,
     $                 NROWS, NCOLS, TOPI, BOTTMI, LEFTJ, RIGHTJ)
      INTEGER NROWS, NCOLS, TOPI, BOTTMI, LEFTJ, RIGHTJ
      REAL TEMPER(NROWS, NCOLS)
      CHARACTER LINE(132)
      INTEGER I, J, K
      CHARACTER SYMBOL(17), BLANK
      DATA (SYMBOL(K), K = 1, 17)/'A',' ','B',' ','C',' ','D',' ',
     $   'E',' ','F',' ','G',' ','H',' ','I'/, BLANK/' '/

* Move to top of new page
      WRITE (6, 100)
  100 FORMAT ('1')

* Run through all rows, from top (I = 1) down
      DO 20 I = 1, NROWS

* Convert each temperature to a letter or blank,
* but do not print anything for fluid region or for out-of-range values

      DO 10 J = 1, NCOLS
         IF (        (I .GT. TOPI  )
     $        .AND. (I .LT. BOTTMI)
     $        .AND. (J .GT. LEFTJ )
     $        .AND. (J .LT. RIGHTJ)
     $        .OR. (TEMPER(I, J) .LT. 0.0)
     $        .OR. (TEMPER(I, J) .GT. 200.0) ) THEN
            LINE(J) = BLANK
         ELSE
            K = TEMPER(I, J) / 11.765 + 1.0
            LINE(J) = SYMBOL(K)
         END IF
   10    CONTINUE
         WRITE (6, 200) (LINE(K), K = 1, NCOLS)
  200    FORMAT (' ', 132A1)
   20 CONTINUE

      RETURN
      END
```

Figure 9.16 A modified version of the pipe temperature program, using the PARAMETER statement and adjustable dimensions to promote program modifiability.

Let us now make several changes, to see how much effort is involved. Suppose we make the array 91 × 81 and enlarge the hole so that its defining parameters are 19 and 63. This makes the bottom part thicker than the top. Next, let's imagine that the pipe is tipped in the ice water, so that the bottom two-thirds of the left side is submerged but only the bottom one-third of the right side is submerged. Finally, let's change the plotting routine so that there are 41 temperature bands of 5° each, with only every fourth band assigned a nonblank character. This will make the printed characters give a better feel of being contour lines.

The modified sections of the program are shown in Figure 9.17. Changing the shape of the pipe required only modifying the PARAMETER statement in the main program. All of the changes involved in redefining the boundary conditions are localized in INITLZ, and all the changes in producing the output appear in PLOT.

Making these changes and rerunning the program was done rather hurriedly, so it may be imagined that mistakes were made in doing it. When first compiled, there were two errors: I changed the DATA statement to contain more symbols, but forgot to change the definition of SYMBOL correspondingly, and I omitted the final slash in the DATA statement. On recompilation there were no diagnostic messages, but when the program was run the entire right side of the pipe was at zero degrees, instead of varying for two-thirds of its length. This symptom pointed rather strongly at the initialization routine, where it was quickly noted that what was intended as NROWS had been entered as NCROWS. Since Fortran does not force variables to be typed, this non-existent (from my viewpoint) variable had been used, and apparently it had been initialized by the system to zero. With this error corrected, the program ran uneventfully, producing the output shown in Figure 9.18. The entire process of making the changes, correcting the errors made in doing so, compiling three times, and producing the output, took about 25 minutes.

A good case can be made for *never* using numeric constants in programs, with the possible exception of numbers like 0 and 1 when used to initialize variables or increment counters. The point is that any program that is used over a period of time has to be modified, and program modification can be both difficult and dangerous. The difficulty is presumably obvious, in terms of this example. Finding and correctly changing all the places where the pipe geometry

```
      INTEGER NROWS, NCOLS, TOPI, BOTTMI, LEFTJ, RIGHTJ
      PARAMETER (NROWS = 91,
     $           NCOLS = 81,
     $           TOPI = 19,
     $           BOTTMI = 63,
     $           LEFTJ = 19,
     $           RIGHTJ = 63)


* Initialize top third of left side
      DO 40 I = 1, (NROWS - 1) / 3
          EDGET = 100.0 - 100.0 * (I - 1) / ((NROWS - 1) / 3.0)
          TEMPER(I, 1) = EDGET
   40 CONTINUE

* Initialize top two thirds of right side
      DO 45 I = 1, 2 * (NROWS - 1 ) / 3
          EDGET = 100.0 - 100.0 * (I - 1) / (2*(NROWS - 1) / 3.0)
          TEMPER(I, NCOLS) = EDGET
   45 CONTINUE
```

Figure 9.17

```
*****************************************************************************
* PLOT: The subroutine to print the results in graphical form
*****************************************************************************
      SUBROUTINE PLOT (TEMPER,
     $                 NROWS, NCOLS, TOPI, BOTTMI, LEFTJ, RIGHTJ)
      INTEGER NROWS, NCOLS, TOPI, BOTTMI, LEFTJ, RIGHTJ
      REAL TEMPER(NROWS, NCOLS)
      CHARACTER LINE(132)
      INTEGER I, J, K
      CHARACTER SYMBOL(41), BLANK
      DATA (SYMBOL(K), K = 1, 41)/'A',' ',' ',' ','B',' ',' ',' ','C',
     $   ' ',' ',' ','D',' ',' ',' ','E',' ',' ',' ','F',' ',' ',' ',
     $   'G',' ',' ',' ','H',' ',' ',' ','I',' ',' ',' ','J',
     $   ' ',' ',' ','K'/, BLANK/' '/

* Move to top of new page
      WRITE (6, 100)
  100 FORMAT ('1')

* Run through all rows, from top (I = 1) down
      DO 20 I = 1, NROWS

* Convert each temperature to a letter or blank,
* but do not print anything for fluid region or for out-of-range values

         DO 10 J = 1, NCOLS
            IF (         (I .GT. TOPI  )
     $           .AND. (I .LT. BOTTMI)
     $           .AND. (J .GT. LEFTJ )
     $           .AND. (J .LT. RIGHTJ)
     $           .OR. (TEMPER(I, J) .LT. 0.0)
     $           .OR. (TEMPER(I, J) .GT. 200.0) ) THEN
               LINE(J) = BLANK
            ELSE
               K = TEMPER(I, J) / 5.0 + 1.0
               LINE(J) = SYMBOL(K)
            END IF
   10    CONTINUE
         WRITE (6, 200) (LINE(K), K = 1, NCOLS)
  200    FORMAT (' ', 132A1)
   20 CONTINUE

      RETURN
      END
```

Figure 9.17 *(cont'd)* The sections of the program of Figure 9.16 that had to be modified to compute the temperature distribution in a somewhat different pipe, with different boundary conditions, and using different symbols for the temperature distribution.

was embedded in the first version of this program would be burdensome, bearing in mind that these numbers often appear incremented or decremented by 1 when used in DO loops. And this is a very small program, compared with much of the routine work of computing!

The danger is that, because of mistakes in making the changes, the modified program will be syntactically correct but logically wrong, and the error will not show up as obviously as a completely wrong temperature on a plot. Such changes are commonly made under serious time pressure, and there is an almost overwhelming temptation to accept the changes as correct so long as the compiler

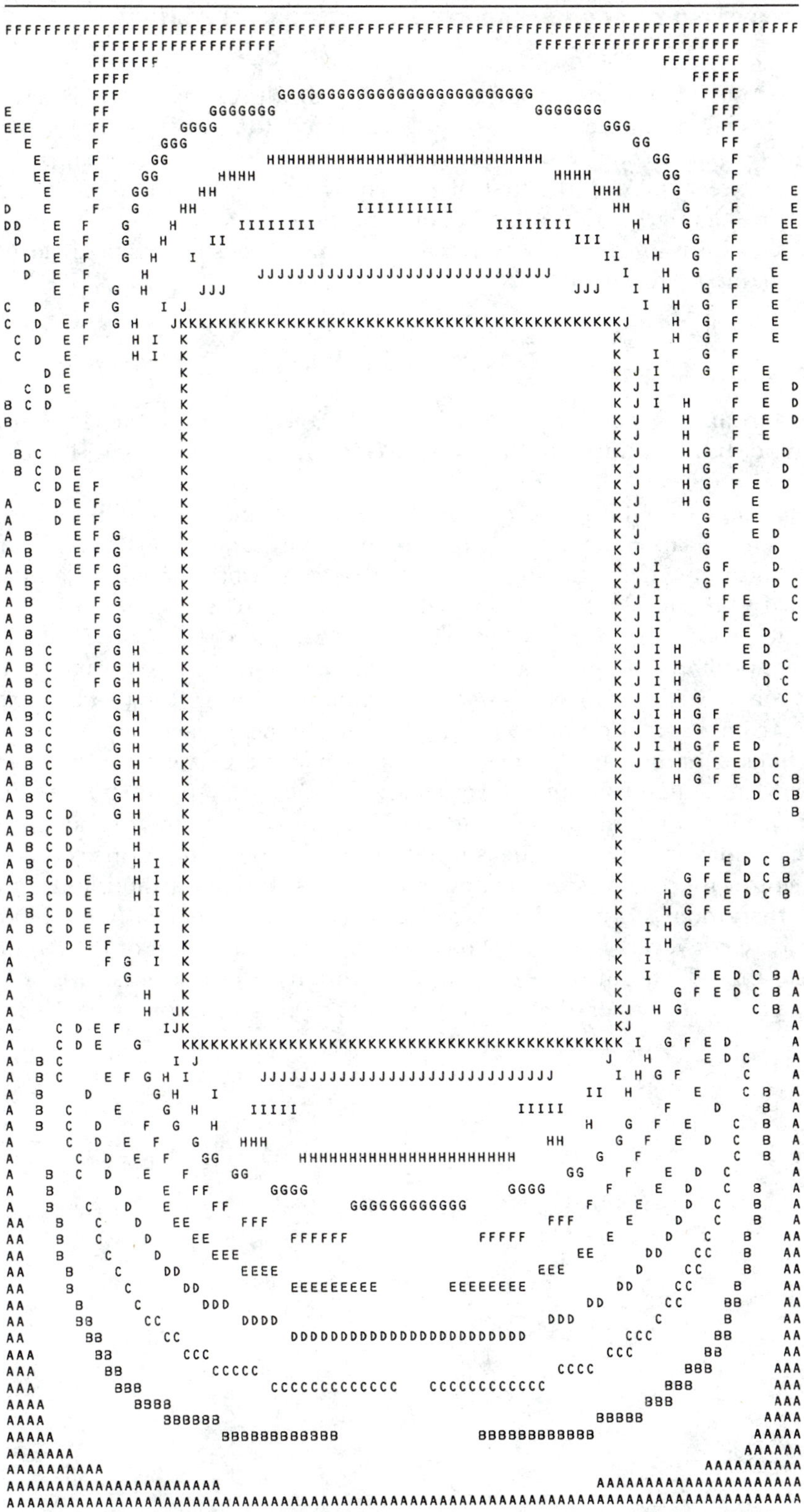

Figure 9.18 The output of the program for which the modified sections were shown in Figure 9.17.

produces no diagnostics. A subtle error that produces disastrous results only under special circumstances, may go undetected for *years*. Or a mistake that produces *slightly* wrong results may *never* be detected. All this can happen when programs are written in the first place, too, of course, but the danger is much greater in making hurried program modifications.

The process is much simpler and safer if as many constants as possible appear only as named parameters.

Conclusion: The Cover

This program was modifed as needed to make it run on an Apple IIe computer, with the output produced in the form of color graphics on a home TV set. The cover reproduces the TV picture.

It would be nice if it had been possible to present more than this on the topic of computer graphics, which is currently enjoying rapid growth in capabilities and applications. Unfortunately, the subject is somewhat specialized and rather awkward to discuss in an elementary book without color.

In the process of finding a suitable size, distribution of colors, etc., I ran this program perhaps a hundred times, using three programming languages (Pascal and Basic as well as Fortran), and used various software for producing color output from a small computer on an ordinary home TV set.

Perhaps a middle-aged college professor may be permitted a brief glance backward. The first computer I worked on was the IBM 701 in 1953. In its time it was an important and pioneering machine, on which many of the leaders in the field today cut their teeth. But its arithmetic speed was approximately 50 times slower than today's simple home computers, and many thousands of times slower than today's fastest machines. The pace of advancement in the field may perhaps be conveyed with the following comparison. The cost of the Apple, the TV set, and all the software I used, if converted to 1953 prices, would barely have paid for the *rental* of enough time to solve this equation *once* on the IBM 701.

TEN

NONPROCEDURAL APPROACHES TO APPLICATION DEVELOPMENT

Introduction

Up until now, the entire presentation has been based on programming in Fortran. You have been learning programming because it is a useful skill in its own right and because there seems to be no other way to gain a good understanding of both the powers and the limitations of computers. You have been learning Fortran because it is the most widely used programming language for applications in engineering and science.

But, as emphasized in the Preface, in the first chapter, and occasionally since, you will probably not do much actual programming in your work in engineering or science. The tasks that are within your capability are better handled by using subroutines from a library, as illustrated in the previous chapter, or by turning to any of a wide variety of *nonprocedural* approaches, to a sampling of which this final chapter is devoted.

Procedural vs. nonprocedural approaches

Fortran is an example of a procedural application development method. To "solve a problem" in Fortran, we do not just tell Fortran what the problem is and wait for an answer. Rather, we give it a carefully devised set of commands: "get some data," "look at the first number and stop if it is negative," "call the subroutine named such-and-such," "print the values of the following variables," "go back to get more data," etc. If we have correctly devised the set of commands, which together are called a procedure, then when the computer carries them out the numbers printed will be "the answer." We never told the computer what our problem was; we only instructed it to carry out a set of commands. In the nonprocedural approaches to be sampled in this chapter, we move much closer to the goal of being able to say to a computer: "Here's my problem; what's the answer?" In other words, we come closer to telling the computer *what to do*, rather than *how to do it*.

263

The six examples of this approach are as follows:

■ TWODEPEP, a Fortran software package for solving partial differential equations.

　■ ISPICE, a package for circuit analysis.

　■ muMATH, a package that does symbolic mathematics on small computers.

　■ DIALOG, a service for retrieving information from any of about 200 commercial databases.

　■ NOMAD, a largely nonprocedural language for specifying computer applications.

　■ Script, a text formatting program.

The intention in this chapter is to give you a glimpse of the world beyond Fortran, so to speak. Fortran is important; you have not wasted your time learning it. But you should know that there are many other ways to get work done with a computer. That statement is true even today; it will become increasingly so in the years of your career.

Naturally, we can do no more than sample the offerings. There are literally thousands of products of the kinds illustrated here, and many of them require detailed knowledge of a specific application area for even minimal understanding. Since many readers of this book will be in their first or second year of college, making impossible a detailed study of packages requiring knowledge of specific applications (or of more advanced mathematics), we shall have to be content with a hint of what is available.

In short: At least browse through this chapter, and make a mental note ever to be alert for ways to avoid programming altogether!

The pipe temperature problem solved with a software package

TWODEPEP is a Fortran software package for solving partial differential equations. It is a product of IMSL, Inc., whose subroutine library we sampled in the previous chapter. Since your mathematical background is not assumed to include much experience with partial differential equations, let us simply say that they arise in many engineering and scientific applications. Common examples include problems in elasticity, diffusion, heat conduction, potential energy, and fluid mechanics. That list, as it happens, covers a significant fraction of applied mathematics, so a software package that lets us deal with partial differential equations provides a powerful tool indeed.

Examples showing the full scope of this package and others similar to it being beyond the scope of this book, we must be content with one simple example. The pipe temperature program of Chapter 9 solved a partial differential equation that describes a problem in diffusion. Using the methods of procedural programming in Fortran, we had to set up a method of solution and describe it in complete detail. Using TWODEPEP, on the other hand, we need only provide a description of the temperature at each point on the boundaries, along with a description of the "grid" of points at which the temperature is to be approximated. TWODEPEP takes over, solves the problem, and—in this case— produces the output as a graphical plot.

Figure 10.1 shows the complete definition of the problem, and Figure 10.2 shows the plot that was produced. You should not expect to be able to understand the problem definition, which would require more mathematics and an understanding of the TWODEPEP format for problem definition. On the output plot, the x and y axes represent the horizontal and vertical dimension of the pipe, and the Z axis gives the temperature. Thus, we not only have a simpler method of specifying the problem, but we get an easily visualized presentation of the results.

```
    1   18    2
**** 18 FOURTH DEGREE TRIANGULAR FINITE ELEMENTS USED
QUARTICS
**** TWODEPEP SOLVES D(OXX)/DX+D(OXY)/DY+F1=0, WHERE OXX,OXY,F1 ARE
**** (FORMALLY) ARBITRARY FUNCTIONS OF X,Y,U,UX(=DU/DX),UY(=DU/DY).
**** HERE OXX=UX,OXY=UY,F1=0
OXX         UX
OXY         UY
**** BOUNDARY CONDITIONS
**** ICE WATER
ARC=-1
FB1         0.0
**** TOP OF RIGHT SIDE
ARC=-2
FB1         100.*(Y-30.)/30.
**** TOP OF LEFT SIDE
ARC=-3
FB1         100.*(Y-30.)/30.
**** TOP OF PIPE
ARC=-4
FB1         100.
**** HOLE
ARC=-5
FB1         200.
**** DATA TO DEFINE INITIAL TRIANGULATION OF PIPE
VXY        0 0, 60 0, 60 60, 0 60, 30 9, 51 30, 30 51, 9 30
VXY        18 18, 42 18, 42 42, 18 42, 60 30, 0 30
IABC       1 2 5, 2 10 5, 10 9 5, 9 1 5, 2 13 6, 13 3 6, 3 11 6, 11 10 6
IABC       10 2 6, 3 4 7, 4 12 7, 12 11 7, 11 3 7, 4 14 8, 14 1 8, 1 9 8
IABC       9 12 8, 12 4 8
I          -1 0 -5 0  -1 -2 0 -5 0  -4 0 -5 0  -3 -1 0 -5 0
**** OUTPUT RESULTS ON 30 BY 30 GRID
NX         30
NY         30
END.
```

Figure 10.1 The TWODEPEP coding to specify the pipe temperature problem studied in Chapter 9.

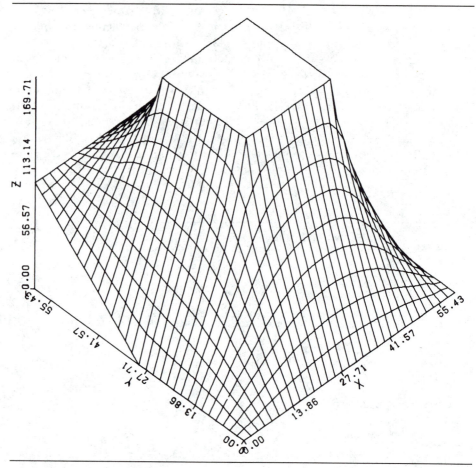

Figure 10.2 The graphical output from TWODEPEP, showing the pipe temperature distribution. The x axis is the horizontal dimension of the pipe, the y axis is the vertical, and the temperature scale is given on the z axis.

A circuit analysis package

TWODEPEP is a package that deals with a broad class of mathematical problems. There are also many packages that deal with a wide variety of problems within a particular applications area. The application to be illustrated briefly here is the analysis of electronic circuits.

ISPICE is a National CSS software package based on work done at the Electronics Research Laboratory, University of California at Berkeley. Using it, you can analyze circuits containing resistors, capacitors, inductors and mutual inductors (transformers), independent voltage and current sources, diodes, a variety of types of transistors, and operational amplifiers. The size of the circuit that ISPICE can simulate is limited only by the size of computer memory available; a circuit containing a thousand elements may be simulated in a single operation.

The basic types of analysis that can be performed in a circuit simulation include:

■ Nonlinear DC analysis: operating point, input, and transfer characteristics.

■ AC small-signal sinusoidal steady-state analysis: operating points are automatically calculated from the DC analysis for driving point and transfer response.

■ Nonlinear transient analysis: for determination of time response to arbitrary inputs.

■ Sensitivity analysis: both AC and DC small-signal sensitivities of voltages and currents with respect to element values and device parameters.

■ Worst case analysis: maximum or minimum circuit voltages and currents for either an AC or a DC steady-state operating point.

■ Monte Carlo analysis: statistical representation of circuit voltages or currents when component or parameter values are distributed over some range.

■ Fault analysis: the effects a faulted device (opened or shorted component or device junction) has on the operating point of the circuit.

■ Noise analysis: the effects of thermal, shot, or flicker noise at a single frequency or for a bandwidth.

The results can be selectively printed as reports or plotted as graphs for any node and/or device in the circuit.

Clearly, since some readers may know only elementary physics, we cannot hope to demonstrate the full capability of such a package. Perhaps, however, some of the "flavor" of the subject may be communicated through one example.

Figure 10.3 is a circuit diagram of a Schmidt trigger, a device that has two stable states and, when triggered, switches from one state to the other. Such devices find frequent application in computer hardware, among many other places. Those of you who know about such things will see five resistors, a capacitor, two transistors, a voltage supply, and an input voltage. The circled numbers represent *nodes,* points in the circuit where devices are connected to each other. The output voltage, which would be used to control some other device or circuit, appears at node 2.

The numbering of the nodes is completely arbitrary. Having numbered them, we describe the circuit in terms of the nodes to which each device is connected and in terms of the characteristics of the devices. Figure 10.4 shows the complete description of this circuit. (This is a listing of a file entered into the National CSS system using separate operations not shown.) The first line is a comment. The second line says that the input voltage is connected between node 7 and node 0, the latter by default being ground. It also says that the input is a pulse, having characteristics given in parentheses. The third line says that the supply voltage is connected between nodes 1 and 0, and is 5 V. The next six lines describe the connections and the values of the resistors and the capacitor. Q1 and Q2 are the two transistors, with the node connections given in the order of collector, base, and emitter. The OFF with Q1 says that it is is initially not conducting. Without this, ISPICE would report inability to analyze the circuit, since the circuit has just two stable states and the package has to know its initial condition. "QM1" promises that there will later be a description of the charac-

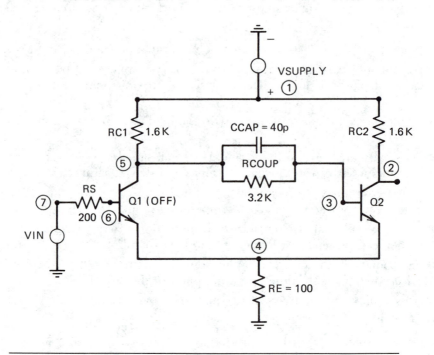

Figure 10.3 Schematic diagram of a Schmidt trigger circuit.

teristics of these transistors, which follows immediately. We'll not try to explain how transistors are specified!

Figure 10.5 shows the DC analysis of this circuit, together with the commands necessary to tell ISPICE what we want to do. The ISPICE command "sim" means simulate; "strigger" (for Schmidt trigger) is the name I gave to the file describing the circuit; the "dc" says we want a simple direct current analysis. The "CORE 576K" means that ISPICE automatically obtained more storage space as needed to complete the analysis. The command "probe op *" means, in effect, "tell me

```
* SCHMIDT TRIGGER CIRCUIT DESCRIPTION
VIN 7 0 PULSE (.15 3.6 0 250N 250N 100N 800N)
VSUPPLY 1 0 5V
RS 7 6 200
RE 4 0 100
RC1 1 5 1.6K
RC2 1 2 1.6K
RCOUP 5 3 3.2K
CCAP 5 3 40P
Q1 5 6 4 QM1 OFF
Q2 2 3 4 QM1
MODEL QM1 NPN(BF=100 BR=1 RB=50 TF=.22N TR=15N CJE = 7.5P CJC=3.8P IS=1E-14 VA=50
```

Figure 10.4 The coding of the Schmidt trigger circuit in a form suitable for use by the circuit analysis package ISPICE.

everything you can about the operation of this circuit." (Bearing in mind that we asked only for a DC analysis.) An electrical engineer would presumably find everything of interest about the stable-state operation of the circuit in this listing.

Suppose, now, that we wanted a description of the circuit. Figure 10.6 provides two different descriptions. The command "display node *" means to show, for each node, every element connected to it. The command "display element *" says to show all circuit elements in terms of the nodes between which they are connected and the element values.

Given your assumed background, this will have to suffice as a "teaser," indicating in a general way the kinds of things that can be done with packages of this type. You have seen possibly 2% of the power of ISPICE.

There are hundreds of packages of this general sort, each devoted to an applications area. In each area, there may be several packages, either competitive products or products dealing in greater depth with a specialized facet of the field. Some of the fields covered include:

■ Fluid flow in networks of piping. The fluid may be compressible or incompressible; the networks may be very complex; the effects of surges may be computed (these are the "knocking" heard in water pipes); the analysis may be steady-state or transient.

```
ISPICE:  >sim strigger;dc
CORE   576K

ISPICE:  DC

     SINGLE POINT DC    SIMULATION OF CIRCUIT: STRIGGER COMPLETED
ISPICE: >probe op *

**** RESISTORS

NAME              VOLTAGE       CURRENT       POWER

RS               -7.554D-10   -3.777D-12    2.853D-21
RE                3.680D-01    3.680D-03    1.354D-03
RC1               1.301D+00    8.130D-04    1.058D-03
RC2               4.587D+00    2.867D-03    1.315D-02
RCOUP             2.602D+00    8.130D-04    2.115D-03

**** CAPACITORS

NAME              VOLTAGE       CURRENT       POWER

CCAP              2.602D+00    0.0           0.0

**** VOLTAGE SOURCES

NAME              VOLTAGE       CURRENT       POWER

VIN               1.500D-01    3.777D-12     5.666D-13
VSUPPLY           5.000D+00   -3.680D-03    -1.840D-02

**** BIPOLAR JUNCTION TRANSISTORS

NAME       MODEL          VBE        VBC       VCE        IC           IB

Q1         QM1          -0.218     -3.549    3.331    3.559D-12   -3.777D-12
Q2         QM1           0.729      0.684    0.045    2.867D-03    8.130D-04
```

Figure 10.5 The DC analysis of the Schmidt trigger produced by ISPICE.

```
ISPICE: >display node *

**** NODES

NODE         ELEMENTS CONNECTED

    0        RE           VIN          VSUPPLY
    1        RC1          RC2          VSUPPLY
    2        RC2          Q2
    3        RCOUP        CCAP         Q2
    4        RE           Q1           Q2
    5        RC1          RCOUP        CCAP         Q1
    6        RS           Q1
    7        RS           VIN

ISPICE: >display element *

**** RESISTORS

NAME                 NODES        VALUE

RS              7       6        2.000D+02
RE              4       0        1.000D+02
RC1             1       5        1.600D+03
RC2             1       2        1.600D+03
RCOUP           5       3        3.200D+03

**** CAPACITORS

NAME                 NODES        VALUE        IN CURR    IN VOLT

CCAP            5       3        4.000D-11    0.0       *NOT SET*

**** VOLTAGE SOURCES

NAME             +       -        DC/TR VAL    AC VAL     AC PHS

VIN             7       0        1.500D-01    0.0       0.0
VSUPPLY         1       0        5.000D+00    0.0       0.0

**** BIPOLAR JUNCTION TRANSISTORS

NAME             C       B        E          MODEL

Q1              5       6        4          QM1              OFF
NODE    5                AREA =  1.000D+00
Q2              2       3        4          QM1
NODE    2                AREA =  1.000D+00
```

Figure 10.6 The ISPICE description of the Schmidt trigger network, together with descriptive information about the location and circuit values of all components.

■ Stress analysis in individual elements or in trusses.

■ Optical lens design. The package may simply analyze a proposed design, or it may be asked to optimize performance given a set of objectives and constraints.

■ Project management. Various packages assist in planning and scheduling large projects such as building construction, the introduction of a new banking service, or the development of large computer programs.

■ Simulation. Several packages permit study of the interrelated events in a system, such as how a proposed building elevator controller would respond to overloads at rush hour, or how a complex combination of orders would flow through a factory.

■ Statistics. This is one of the earliest areas in which it became obvious that a package would be useful, and several have been available for many years, with continual enchancement.

There are many others. In fact, the availability of packages is so great that the user often needs some way to know just what *is* available. International Computer Programs (ICP), of Indianapolis, provides catalogs of packages available in a number of broad applications areas, including commercial applications.

At the time of writing, software packages were estimated to be a billion dollar industry, and growing in excess of 30% per year.

Symbolic mathematics with muMATH

Back in Chapter 1 we said that a computer cannot follow the order "solve this equation," but then quickly added the qualification "at least not in Fortran." Now we shall see that, once a computer has been given appropriate instructions, it can do exactly that.

The subject of symbolic mathematics is a major area of specialization in computer science, with hundreds of workers. The first research was done in the early 1950s, making this one of the most mature parts of the field. Systems with names like FORMAC, MACSYMA, SCRATCHPAD, and REDUCE have been the target of many person-decades of work.

The subject is a complex one, which we cannot hope to more than sketch here.

Doing symbolic mathematics on a computer involves writing a program to tell the computer how to process the symbols of mathematics, and also informing the computer—through that program—how to simplify equations, do differentiation and integration, and do whatever other capabilities are built into the program.

Such a program could in fact be written in Fortran, and some of them are, using the character operations that we saw in Chapter 8. Most symbolic math programs, however, are written in another language called LISP, or something like it, which is better suited to such work.

We will make no attempt to demonstrate the inner workings of such programs, but we will show one of the smaller ones in action. The system to be exhibited is called muMATH, which runs on the Apple, the IBM Personal Computer, the Radio Shack TRS-80, and any other system that has the CP/M operating system. It is a product of The Soft Warehouse in Honolulu and is also marketed by the Microsoft Corporation, Bellevue, Washington. Since it is able to run on a computer with a tiny fraction of the capabilities of those on which systems like MACSYMA run, it naturally has more limited powers. Nevertheless, it can do most of the manipulations of a college freshman calculus course, which I find astonishing.

To provide some idea of what the system and others like it can do, here is a sample session when I ran muMATH on my Apple computer. The paragraph numbers are keyed to the circled numbers on the accompanying printout.

1. muMATH does rational arithmetic only, that is, quantities are represented as the quotient of integers. (We can get results printed as decimal fractions, as we shall see, but that's only for output.) When muMATH is ready for input, it types a question mark; its results are preceded by the at-sign (@). Each command must be terminated with a semicolon.

2. The exponentiation operator is an upward arrow on some printers, but on my terminal it is the circumflex. (This result is the number of bytes addressable by a 24-bit computer address.)

3. Roots are taken only if the result is an integer. This result is merely a rearrangement of the input.

4. The fourth root of 16 is an integer, so we get it.

5. The system knows about factorials.

6. BIG factorials!

7. On my system, no number may have more than about 600 decimal digits. On other computers, with more internal storage, numbers may have tens of thousands of digits.

8. I specify that I want results printed with 30 places after the decimal point, and try an expression that combines various operators.

9. A negative value for POINT means to go back to quotient-of-integers output. (This is not the same as setting POINT to zero, which would truncate rational results to an integer.)

10. In writing an equation that I want solved, I have to use two equal signs. (One equal sign means to test two expressions for equality, which returns **TRUE** or **FALSE**.) The brackets around the two solutions mean that they are elements of an array; muMATH knows all about matrix algebra, too, although there are no examples of it here.

11. muMATH knows all about complex numbers, of course. The combination #I means the imaginary unit.

12. Quick! Do you know what the 10 tenth roots of unity are??? I would have had to do some reviewing, myself. (#E is, of course, the symbol for e and #PI is the symbol for π.)

13. Symbolic differentiation was where symbolic mathematics began, with the work of Harry Kahramanian in the early 1950s. muMATH knows all the standard forms and can plow through expressions where I would have about a 99% likelihood of making algebraic mistakes. The X after the comma indicates that the differentiation is with respect to x. In the fourth example, the function is:

$$\frac{e^{-x^2} \sec x}{1 + 1/x}$$

(1)
```
? 1/2 + 2/3 - 7/23;
@: 119 / 138
```

(2)
```
? 2^24;
@: 16777216
```

(3)
```
? 27^(1/2);
@: 3 ^ (3/2)
```

(4)
```
? 16^(1/4);
@: 2
```

(5)
```
? 5!;
@: 120
```

(6)
```
? 60!;
@:
8320987112741390144276341183223364380754172606361245952449277
69640960000000000000000
```

(7)
```
? 250!;
@:
3232856260909107732320814552024368470994843717673780666747942
4271128237475551112094888179153710281994509285073531894329226
7309317128089908227910302790712819216765272401892647332180411
8626100683292536513367893908956993571353017504051317876007724
7933065402339006164825555224881943657258605739922264125483298
2204849137721776650641276858807153128978777672951913990844377
4787025891729732551502832417873206581884820624785826598084884
8255488000000000000000000000000000000000000000000000000000000
0000000000
```

(8)
```
? POINT: 30;
@: 30

? 20/10 - 1/239 + 14/6 - 1/7! + 2^3;
@: 12.3289508202165105930796307365334
```

(9)
```
? POINT: -1;
@: -1
```

(10)
```
? SOLVE (X^2 - 4 == 0, X);
@: [X == -2,
    X == 2]
```

(11)
```
? SOLVE (X^2 + 4 == 0, X);
@: [X == -2*#I,
    X == 2*#I]
```

(12)
```
? SOLVE (X^10 == 1, X);
@: [X == -#E^(4*#I*#PI/5),
    X == -#E^(3*#I*#PI/5),
    X == -#E^(2*#I*#PI/5),
    X == -#E^(#I*#PI/5),
    X == -1,
    X == #E^(4*#I*#PI/5),
    X == #E^(3*#I*#PI/5),
    X == #E^(2*#I*#PI/5),
    X == #E^(#I*#PI/5),
    X == 1]
```

(13)
```
? DIF (X^4, X);
@: 4 * X^3

? DIF (SEC(X), X);
@: TAN(X) * SEC(X)

? DIF (SIN(X) / (1 + COS(X)), X);
@: SIN(X)^2/(1+COS(X))^2 + COS(X)/(1+COS(X))

? DIF (#E^-X^2 * SEC(X) / (1 + 1/X), X);
@: -2*#E^X^2*X*SEC(X)/(#E^X^2/X+#E^X^2)^2 + #E^X^2*SEC(X)/(X
^2*(#E^X^2/X+#E^X^2)^2) - 2*#E^X^2*SEC(X)/(#E^X^2/X+#E^X^2)^
2 + TAN(X)*SEC(X)/(#E^X^2/X+#E^X^2)
```

14. Operations can be nested to any reasonable depth. This gets the second derivative.

15. Indefinite integration, of course.

16. muMATH knows the basics!

17. Let's cheat: Let's give it an integrand that is known not to be integrable in closed form. Turns out muMATH is aware of the error function.

18. Well, but what if I really wanted to calculate that integral? OK, let's integrate its Taylor series expansion around zero, out through terms in x^{10}.

19. muMATH knows about summations. This is:

$$\sum_{i=1}^{20} i$$

20. And it can do the same thing in symbolic form, too. This is:

$$\sum_{i=1}^{N} i$$

21. Well, maybe, but that's not the form I'm used to seeing for the sum of the first N integers. Let's give that result a name, so we can play with it a bit.

22. We have great flexibility in the degree of factoring or combining of terms. EXPD ("expand") is a muMATH function that performs a particular set of recombinations, the exact nature of which is controllable, but these are details beyond our scope here. This is closer to the form I like.

23. But with FCTR ("factor") I get what I really wanted!

24. Let's pretend I can't remember the Taylor series for the sine. Ah, yes:

$$\sin x = x - \frac{x^3}{3!} + \frac{x^5}{5!} - \frac{x^7}{7!} + \cdots$$

25. I could ask for the Taylor series for the sine to be evaluated for any rational value of the argument, but it will be rather faster to use the summation operator to evaluate the series directly, now that we've been reminded what it is. That series can be represented this way:

$$\sin x = \sum_{N=1}^{\infty} \frac{(-1)^{N+1} x^{2N-1}}{(2N-1)!}$$

I ask for this series to be evaluated for a value of 3/2. I give it a name, SIN1P5 ("sine of 1 point 5"), so that I can manipulate it later.

(14) ? DIF (DIF(X^5, X), X);
    @: 20 * X^3

(15) ? INT (1/X, X);
    @: LN (X)

    ? INT ((A + B*X)^N, X);
    @: (A+B*X)^(1+N) / (B*(1+N))

(16) ? INT (DIF (X^5, X), X);
    @: X ^ 5

(17) ? INT (#E^-X^2, X);
    @: #PI^(1/2)*ERF(X) / 2

(18) ? INT (TAYLOR (#E^-X^2, X, 0, 10), X);
    @: X - X^3/3 + X^5/10 - X^7/42 + X^9/216 - X^11/1320

(19) ? SIGMA (I, I, 1, 20);
    @: 210

(20) ? SIGMA (I, I, 1, N);
    @: (-1-N+(1+N)^2) / 2

(21) ? SUMN: SIGMA (I, I, 1, N);
    @: (-1-N+(1+N)^2) / 2

(22) ? EXPD (SUMN);
    @: (N+N^2) / 2

(23) ? FCTR (EXPD (SUMN));
    @: N*(1+N) / 2

(24) ? TAYLOR (SIN(X), X, 0, 9);
    @: X - X^3/6 + X^5/120 - X^7/5040 + X^9/362880

(25) ? SIN1P5: SIGMA ((-1)^(N+1)*(3/2)^(2*N-1)/(2*N-1)!, N, 1, 10);
    @: 1385183094856004427 / 1388661711044608000

26. I say that 10 decimal places should be printed and ask to see the value of SIN1P5. It is printed to 10 decimal places, which is more than I can check with my Texas Instruments calculator. So, let's test how accurate the representation might be. The way the summation formula was written, the last term computed was $x^{19}/19!$. How big is that? Evidently it is less than 10^{-10}; therefore the value for the sine of 1.5 radians must be accurate to 10 places. (Except for a possible error of 1 in the last place; muMATH doesn't round.) After all this, I say to go back to printing results as rational fractions.

27. Let's suppose I've forgotten the Taylor series for the arctangent. Well, the derivative of the tangent is $1/(1 + x^2)$, so why not integrate the Taylor series expansion of that? Ah, yes, now I remember:

$$\tan^{-1} x = x - \frac{x^3}{3} + \frac{x^5}{5} - \frac{x^7}{7} + \cdots$$

This can be written in summation form this way:

$$\tan^{-1} x = \sum_{N=1}^{\infty} \frac{(-1)^{N+1} x^{2N-1}}{2N - 1}$$

28. Searching through the bookshelf, I find my copy of *Problems for Computer Solution,* by Fred Gruenberger and George Jaffrey (Wiley, 1965) which contains on pages 228-231, fascinating material on the computation of approximations to the value of π. I ask muMATH to compute an approximation to π from Machin's formula:

$$\frac{\pi}{4} = 4 \tan^{-1} \tfrac{1}{5} - \tan^{-1} \tfrac{1}{239}$$

The numbers of terms were chosen to provide an answer accurate to at least 100 decimal places.

The fraction shown is exact, for the number of terms specified. All arithmetic has been *rational* arithmetic, that is, it has involved only fractions of integers. Now we all understand that π is an irrational number, and we understand that if I had specified only three terms of each series, the approximation would not have been very close. But the arithmetic that I *did* specify was done *exactly.* There was no round-off error, because only integers were involved.

29. After specifying 100 decimal places, I ask for this fraction to be printed. All muMATH has to do in this case is divide one 282-digit integer by another out through 100 decimal places!

30. Well, is the result really accurate to 100 places? To answer this, I specify 110 places, then check the size of the last terms retained in each of the arctangent series. Both terms are less than 10^{-100}; therefore the value of π shown is indeed accurate to 100 decimals.

(26) ```
? POINT: 10;
@: 10

? SIN1P5;
@: 0.9974949866

? (3/2)^19/19!;
@: 0.0000000000

? POINT: -1;
@: -1
```

(27) ```
? INT (TAYLOR (1/(1+X^2), X, 0, 11), X);
@: X - X^3/3 + X^5/5 - X^7/7 + X^9/9 - X^11/11
```

(28) ```
? PI: 4 * (4 * SIGMA ((-1)^(N+1)*(1/5)^(2*N-1)/(2*N-1), N, 1, 75)
  - SIGMA ((-1)^(N+1)*(1/239)^(2*N-1)/(2*N-1), N, 1, 25));
@:
47517551440523782578892875407761413565992936236838112342738 0
9371463995522210247791280439451903380595741520471202011324 74
1776150185508010643286557837657196106968451529733623259594 96
8526001193554761715557224490721415780305770428719300090111 32
773911725210356572144919581033807397588884 /
1512530639076554370206045588403972249792772684920836905100 45
2736145858362143760166401524080412207052089928546753181606 06
991679496490214347112214534689948862098832445740430858903 129
413624533844543835402362508528483709281400069485410196076 834
513090325629036669852212071418762207031 25
```

(29) ```
? POINT: 100;
@: 100

? PI;
@: 3.14159265358979323846264338327950288419716939937510582 09
7494459230781640628620899862803482534211706 79
```

(30) ```
? POINT: 110;
@: 110

? (1/5)^149/149;
@: 0.0000000000000000000000000000000000000000000000000000000 000
00000000000000000000000000000000000000000000000000004789

? (1/239)^49/49;
@: 0.00000000000000000000000000000000000000000000000000000000 000
000000000000000000000000000000000000000000000000000000000000
```
```

Conclusion

You have seen perhaps a third of the capabilities of muMATH. Others not illustrated: equation and expression simplification, array and matrix operations (including inverses), simplification of logarithmic and trigonometric expressions, limits of functions, definite integrals, and closed-form products. Furthermore, muMATH is extendable (by the user) using the LISP-like language in which it is written, muSIMP, which is supplied as part of the package.

And this is a *small* symbolic math system!

Symbolic mathematics systems of this type find practical use in doing elaborate operations that are tedious and error-prone, such as differentiation of complicated expressions and the use of rational arithmetic to check other calculations, among many others. In the nature of the way they are written and the way they do their work, they are inherently considerably slower—when there is a choice—than doing the same operations in a language like Fortran. Symbolic mathematics and rational arithmetic have their place; so does Fortran. The computer user who is fortunate enough to know both uses each where appropriate.

It is interesting to speculate what impact such systems may have on the teaching of mathematics. It seems only a few years ago that debates raged whether to permit hand calculators in engineering course examination rooms. How long will it be before similar debates focus on the use of hand-held muMATH calculators in calculus exams? It is not hard to imagine that, within a decade or two, the teaching of mathematics will have been seriously impacted—mostly for the better, we trust—by such developments.

Information retrieval from on-line databases

Two of the most prominent features of computers are the ability to store large amounts of data and the ability to search for data at high speeds. It should not be surprising, therefore, that retrieval of information from computer databases was investigated from almost the beginning of the computer era, in the early 1950s. In those early days, however, the only practical storage medium for the purpose was magnetic tape, which must be searched sequentially, and computers were very much more expensive than they are today. The first experiments, accordingly, showed more promise than practicality.

Since that time, the hardware has become much faster, less expensive, and more reliable. Magnetic disks, which do not have to be searched sequentially, have become widely available at reasonable cost. Furthermore, a great deal of intensive research has led to highly efficient methods for retrieving small subsets of the information in a very large database. This is an entire speciality within computer science and library science, called *information retrieval*.

Commercially available databases cover an astonishing variety of subject matter, from biographical data to welding research, from chemical formulas to labor law, from business statistics to abstracts of papers in philosophical journals. One database contains abstracts from 5000 journals that publish articles about coffee. There are databases devoted to electric power, aquaculture, child abuse and neglect, foundation grants, mental health, textile technology, and aluminum. There are seven separate databases devoted to information derived from the

Yellow Pages of U.S. telephone companies, and at least seven devoted to patent claims and patent law. One estimate is that there are more than 600 commercially available databases in the United States.

"Commercially available" means that any individual, business, school, or governmental agency can make arrangements to access the database with a computer terminal, through a telephone connection. Charges for the services are usually based on some combination of time used, number of searches requested, the amount of printing requested, and any physical retrieval of paper documents that is requested.

The type of information in a database varies. Some provide the complete data that a user might wish, such as financial data about a company, the officers of trade associations and professional societies, or citations of recent publications by the (U.S.) Government Printing Office. Others provide statistical summaries, such as labor statistics, the consumer price index, or economic time series data for all the nations of the world. A third group provides titles and (usually) abstracts of papers and articles in science, engineering, medicine, and general interest publications. A fourth general class contains frequently updated information such as stock market data, the news wire services, and indexes to the major newspapers.

The examples that follow were developed using the facilities of DIALOG® Information Services, of Palo Alto, California. DIALOG provides a sort of "supermarket" of database services. That is, the company obtains regular copies of about 200 databases from the organizations that produce them, then makes access to any of those databases available for a fee. Access is through any of several worldwide communications networks, or through a call directly to the DIALOG computer facilities. At the time of writing, the DIALOG storage system contained about 120 gigabytes of information ($= 1.2 \times 10^{11}$ bytes), and the disk storage units covered about one-fifth of an acre of floor space!

As with the preceding section on symbolic mathematics, the remainder of this section is keyed to the numbered lines on the right-hand pages.

1. My first inquiry is a simple one: What was the publication date of Joseph Weizenbaum's book, *Computer Power and Human Reason?* In a step not shown, I specified that I wanted to access File 470, which I had found in a handbook was the number of the DIALOG file containing information from the publication *Books in Print.* (I am not showing this step, which is quite simple, only because it first produces a cost summary for the use of the *previous* file. This would be confusing, because I did not do these searches in the order in which they are shown here. And it would be misleading, because—not being an experienced searcher—my costs were higher than a regular user would have incurred.) The system gives the name of the file and a brief description of it, then prints column headings for the results of the searches that follow.

2. All databases have one or (usually) more *indexes* that speed the search process. In many cases it is also possible to ask the system to search for the occurrence of terms anywhere in an abstract, for example, but when an index is available we may as well use it. For this file there are 16 indexes, covering the obvious items of title and author, but also year of publication, the language, the Library of Congress card number, etc. As I start the search I am not certain of the exact title, but I am certain of the spelling of the author's name, since Professor Weizenbaum is an old friend. I accordingly ask DIALOG to search the Books in Print file for all records in which the author's name is as shown. The response is that there is one such record, meaning that Professor Weizenbaum has written exactly one book. (Or, more precisely, that the Books in Print file knows about only one.) The result of this search is given the *set number* 1, by which we identify the set in any further operations.

3. The three numbers in a "type" command are the *set number,* the *format* for printing, and the *range* of records to be printed. There is only one set, with the number 1, which I specify. The word "format" is used here in a quite different sense from what we are familiar with in Fortran. Here, it refers to a choice among seeing only the title and indexing information, or perhaps the abstract and title, or maybe only the bibliographic citation, or the full record. There are several choices, which sometimes vary slightly from one database to another. Here, I specify just the bibliographic citation. The "range" is needed when there are many items in a set and we do not wish to see them all. In this case, there is only one item in the range and I simply ask to see it. The answer I was looking for is that the book was published in 1976.

4. Now I decide to see what's new in the manufacture of tantalum capacitors. Leafing through the handbook, and based on prior experimentation, I decide to interrogate a file named INSPEC, which stands for Information Services in Physics, Electrotechnology, Computers and Control. INSPEC is produced in England by the Institution of Electrical Engineers (IEE) and marketed in the U.S. by the Institute of Electrical and Electronic Engineers (IEEE). The "77-83" in the first line indicates the years covered by the file; another INSPEC file covers the years 1969-1976.

(1) File470:Books in Print - 1490-1983/Mar
(Copr. R. R. Bowker Company 1983)
        Set Items Description
        --- ----- -----------

(2) ? select au=weizenbaum, joseph
          1   1 AU=WEIZENBAUM, JOSEPH

(3) ? type 1/3/1
1/3/1
  0862283   653773XXX
   Computer Power & Human Reason: From Judgment to Calculation
   Weizenbaum, Joseph
   300p.
   W H Freeman 1976
   Trade $14.00x o.p.; pap. text ed. $10.95x
   ISBN: 0-7167-0464-1; 0-7167-0463-3
   Status: Active entry

(4) File13:INSPEC - 77-83/Iss08
(Copr. IEE)
See File 12(1969 thru 1976)
        Set Items Description
        --- ----- -----------

5. The "ss" means "select steps." It differs from the ordinary select, in that that separate set numbers are assigned to each of the intermediate steps in a search. This is often useful although I do not take advantage of it in this example.

My request this time is to know about records that meet the following criteria:

The words "tantalum" and "capacitor" appear anywhere in the title, abstract, or subject indexes of the record.

Along with the words "tantalum" and "capacitor," *either* the word "fabrication" must appear, *or* any word that begins with the letters "manufactur" must appear. This lets me specify "manufacturing," "manufacturer," "manufacturers," and any other words that may have this stem, all with one phrase. This is the function of the question mark. The "and" and "or" are called *Boolean operators,* and function like the Fortran .AND. and .OR. operators. A "not" operator is also available, but finds less use.

The response is that there are 2203 records in this file in which the word "tantalum" appears, 4155 in which the word "capacitor" appears, 6878 in which "fabrication" appears, and 18,671 in which there is some word that begins with the letters "manufactur." Each of these sets is given a number. Set 5 is the set produced by applying the Boolean operators. This says that the file has 19 records with some reference to the fabrication or manufacture of tantalum capacitors.

6. I ask to see items 1 and 2 of set 5, printed in format 6, which gives only the title and certain internal identifying data. I basically want to see if the search got me what I really wanted, and indeed it did.

7. I decide that the second record looks interesting and choose to print it in format 5, which gives the full record. The additional information includes the author, the publication data, some descriptive information, and an abstract.

8. I decide that these 19 abstracts would make interesting reading, but I don't want to pay approximately $1 a minute for connect time while they are typed at my terminal. Instead, I use the "print" command, asking for all items in set 4 to be printed in format 5. The printing is done on a high-speed printer at the DIALOG facilities and mailed to me. The system estimates the cost of this operation and gives me the option to cancel if I decide not to spend that much. I let the command stand. The printouts arrive a few days later.

(5) ? ss tantalum and capacitor and (fabrication or manufactur?)
           1   2203 TANTALUM
           2   4155 CAPACITOR
           3   6878 FABRICATION
           4 18671 MANUFACTUR?
           5   19 1 AND 2 AND (3 OR 4)

(6) ? type 5/6/1-2
     5/6/1
         996133   A83015290, B83010210
         MINIATURE CAPACITOR-TYPE STIMULATING ELECTRODES

     5/6/2
         993200   B83006943
         MINIATURE DECOUPLING CAPACITORS FOR HF APPLICATIONS

(7) ? type 5/5/2
     5/5/2
         993200   B83006943
         MINIATURE DECOUPLING CAPACITORS FOR HF APPLICATIONS
         SUNDA, J.A.
         SEMIKRON AUSTRALIA PTY, SPRINGVALE, AUSTRALIA
         AUST.   ELECTRON.   ENG. (AUSTRALIA)   VOL.15, NO.7   56-64   JULY 1982
     Coden: AUEEB5   ISSN: 0004-9042
         Treatment: APPLIC; PRACTICAL
         Document Type: JOURNAL PAPER
         Languages: ENGLISH
         THERE IS A WIDE CHOICE OF CAPACITORS FOR HF DECOUPLING APPLICATIONS.
     PLASTIC FILM AND MULTI-LAYER CERAMICS ARE SUITABLE FOR AT LEAST 80PERCENT
     OF REQUIREMENTS AND THEY ARE INTERCHANGEABLE IN MOST INSTANCES.   HOWEVER,
     THERE MAY BE SMALL VARIANCES IN SIZE, PARTICULARLY IN HEIGHT, AND SO THIS
     SHOULD BE TAKEN INTO ACCOUNT IF THE USER WANTS THE OPTION OF USING BOTH
     TYPES.   DISC CERAMICS MAY BE USED IN SOME CIRCUITS, AND THE HYBRID THICK
     FILM DESIGNER SHOULD REMEMBER THAT THERE ARE A NUMBER OF ROBUST AND
     ECONOMIC MULTILAYER CERAMIC AND TANTALUM CHIPS AVAILABLE FOR DECOUPLING
     PURPOSES.   IN MANY CIRCUITS,   IT IS OF CRITICAL IMPORTANCE TO MINIMISE   THE
     INDUCTANCE INTRODUCED INTO THE CIRCUIT BY THE CAPACITOR BY KEEPING THE LEAD
     LENGTH AS SHORT AS POSSIBLE.   FINALLY,   IF THERE IS ANY DOUBT WHEN
     INTERPRETING CAPACITOR DATA SHEETS, ONE SHOULD CONSULT THE MANUFACTURER
         Descriptors: CAPACITORS
         Identifiers: MINIATURE DECOUPLING CAPACITORS; HF APPLICATIONS; PLASTIC
     FILM; MULTI-LAYER CERAMICS; HYBRID THICK FILM; INDUCTANCE; LEAD LENGTH
         Class Codes: B2130

(8) ? print 5/5/1-19
     Printed5/5/1-19 Estimated Cost: $6.65 (To cancel, enter PR-)

9. Putting together my college background in chemistry (among other things) and my mild hypertension, I decide to seek some data about the anti-hypertensive medication I am taking. To get the basic chemical facts, I turn to CHEM-NAME®, one of many databases produced by Chemical Abstracts Service.

10. Both drugs are readily located, specifying simply their brand names. (If I had known how to ask, I could also have searched by chemical name, molecular formula, or several other ways.) I am fascinated by the list of synonyms for the diuretic I am taking, and charmed to learn that my wife and I are taking the same drug under two different names. Some day I'll have to ask my doctor if there is any significance in this.

11. Tenormin belongs to a class of drugs called *beta-blockers*, which sometimes produce fatigue as a side effect (they certainly did in me, at first). Let's see what I can find about this topic in the biomedical database named MEDLINE, a product of the U.S. National Library of Medicine, which includes indexing to articles from approximately 3000 journals published in over 70 countries.

12. I specify the drug by its Chemical Abstracts *registry number* (RN), which will tend to pull out far more references than the brand name. In fact, the response is given in terms of the generic name for the drug, as we can see by looking back up at the synonyms.

13. Fatigue seems to be a popular subject!

14. But there are only four papers on fatigue associated with this drug.

15. Let's see what the titles look like.

16. I'm getting the feeling I should leave this to my doctor, but for 60 cents I'll satisfy my curiosity!

```
 9  File301:CHEMNAME(tm) 1967-Jun82 1,279,331 subs
    (Copr. DIALOG Inf.Ser.Inc.1983)
              Set Items Description
              --- ----- -----------
10  ? select tenormin
              1       1 TENORMIN
    ? type 1/2/1
    1/2/1
        CAS REGISTRY NUMBER: 29122-68-7
        FORMULA: C14H22N2O3
        ANALYSIS OF RINGS: C6
        NUMBER OF RINGS: 1
        SIZE OF RINGS: 6
        CA NAME(S):
        HP=Acetamide (8CI), SB=2-(p-(2-hydroxy-3-(isopropylamino)propoxy)pheny-
    1)-
        HP=Benzeneacetamide (9CI), SB=4-(2-hydroxy-3-((1-methylethyl)amino)pro-
    poxy)-
        SYNONYMS: ICI 66082; Atenolol; Tenormin

    ? select hydro-diuril
                2       1 HYDRO-DIURIL
    ? type 2/2/1
    2/2/1
        CAS REGISTRY NUMBER: 58-93-5
        FORMULA: C7H8ClN3O4S2
        ANALYSIS OF RINGS: C3N2S-C6
        NUMBER OF RINGS: 2
        SIZE OF RINGS: 6,6
        FORMULA OF RINGS: NSC2NC
        REPLACED CAS REGISTRY NUMBER(S): 8049-49-8
        CA NAME(S):
        HP=2H-1,2,4-Benzothiadiazine-7-sulfonamide (8CI 9CI), SB=6-chloro-3,4--
    dihydro-, NM=1,1-dioxide
        SYNONYMS: Su 5879; Aquarills; Aquarius; Chlorosulthiadil;  Dichlotride;
    Diclotride;  Dihydrochlorothiazide;   3,4-Dihydrochlorothiazide;    Esidrex;
    Esidrix; HCTZ; HCZ; Hydril; Hydrochlorothiazide; Hydro-Diuril; Hydrosaluric
    ; Hypothiazid; Hypothiazide; Oretic; Vetidrex;  Hypothiazide(C7H8ClN3O4S2);
    Nefrix;  Hydrochlorthiazide;  Hydrochlorothiazid;   Dihydrochlorothiazidum;
    Dichlotiaziд; Disalunil; Hidril; Hidrochlortiazid; Idrotiazide; Drenol;
    Dihydrochlorothiazid; Megadiuril; Hydro-Aquil; Hidrotiazida; Hydrodiuretic

11  File154:MEDLINE - 80-83/JUN
              Set Items Description
              --- ----- -----------
12  ? ss rn=29122-68-7
              1     349 RN=29122-68-7   (ATENOLOL)
13  ? ss fatigue
              2    1225 FATIGUE
14  ? ss sl and s2
              3       4 1 AND 2
15  ? type 3/6/1-4
    3/6/1
    0817574    82262574
      [Evaluation by exercise test of effects of a single oral dose of atenolol
    in patients with stable angina (author's transl)]
      Gli  effetti  di una  dose singola  orale di  atenololo valutati con test da
    sforzo in pazienti con angina pectoris stabile.

    3/6/2
    0721534    82166534
      Running  performance  as  a function of the  dose-response relationship to
    beta-adrenoceptor blockade.

    3/6/3
    0426927    81156927
      Long-term  benefit  of  cardioselective  beta  blockade  with  once-daily
    atenolol therapy in angina pectoris.

    3/6/4
    0156216    80156216
      Atenolol in hypertension: a cardioselective drug.

16  ? print 3/5/1-4
    Printed3/5/1-4  Estimated Cost: $0.60 (To cancel, enter PR-)
```

17. Let's try something easier, although the best way to get the answer quickly is not entirely self-evident. (I adapted this one from the DIALOG instruction manual!) Who won the 1982 World Series? As it happens, I was teaching a class on the night of the last game, and offered to bet on the outcome (no takers), but my memory has let me down. I go to a file that covers over 400 general interest magazines in North America. The "ss" command this time says to make a set consisting of the identifiers of all records that meet all of these criteria:

They contain the words "world" and "series" together.

They were published in 1982.

They were published in *Sports Illustrated*.

It turns out that there were six such articles, and the title of the second one gives the answer. Ah, yes, I remember it well.

18. There are many databases that provide various kinds of financial data about business organizations. DISCLOSURE II is a DIALOG database that provides information extracted from reports filed with the U.S. Securities and Exchange Commission (SEC) about approximately 9000 publicly owned companies. Let's get the five-year short financial summaries on Apple and IBM. The "select" specifies that "apple" and "computer" must appear together in the company name. The full record includes full balance sheet and profit and loss statement data, plus officers and their compensation, and various other data, totaling over two pages of printing.

The summary shows that Apple Computer grew from sales of less than $8 million in 1978 (two years after its founding, as shown in another database), to sales of $583 million in the fiscal year ending September 24, 1982. ("EPS" stand for "earnings per share.") (At the time of writing, the annual sales rate is over $1 billion.) This is said to be the fastest growth in the history of American industry.

IBM's growth rate, by comparison, is positively leisurely! On the other hand, IBM's net income is several times as large as Apple's gross sales! Either way, it's a dynamic industry.

(17) File47:Magazine Index - 59-83/Mar
(Copr. IAC)
        Set Items Description
        --- ----- -----------
? ss world(w)series and py=1982 and jn=sports illustrated
        1   408 WORLD(W)SERIES
       2131994 PY=1982
      3 11478 JN=SPORTS ILLUSTRATED
      4   6 1 AND 2 AND 3
? type 4/6/1-2
4/6/1
1527470   DATABASE: MI File 47
  A hopping good series: Milwaukee rallied at home to take a 3-2 lead over
St. Louis in a veritable Oktoberfest of a World Series.

4/6/2
1527467   DATABASE: MI File 47
  For all you do, this hug's for you. (St. Louis wins World Series in 7)

(18) File100:DISCLOSURE II Mar 23 83
(Copr. DISCLOSURE INC.)
REPORTS:(800)638-8241
        Set Items Description
        --- ----- -----------
? s apple(w)computer/co
        1   1 APPLE(W)COMPUTER/CO
? type 1/8/1
1/8/1
0000536
APPLE COMPUTER INC
DISCLOSURE CO NO: A713500000
CROSS REFERENCE: NA

AUDITOR CHANGE: NA
AUDITOR: ARTHUR YOUNG
AUDITOR'S REPORT
FISCAL YEAR

FIVE YEAR SUMMARY

YEAR	SALES (000S)	NET INCOME	EPS
1982	583,061	61,306	1.06
1981	334,783	39,420	0.70
1980	117,126	11,698	0.24
1979	47,867	5,073	0.12
1978	7,856	793	0.03

INTERNATIONAL BUSINESS MACHINES CORP
DISCLOSURE CO NO: I510600000
CROSS REFERENCE: NA

AUDITOR CHANGE: NA
AUDITOR: PRICE WATERHOUSE
AUDITOR'S REPORT: UNQUALIFIED%SEGMENT   DATA   _(12/31/81)   SALES
(000S)   OP INCOME^DATA PROCESSING   24,073,000
5,832,000^OFFICE PRODUCTS   _ 4,219,000 _   263,000^FEDE‾
RAL   SYSTEMS   719,0‾0 _   56,‾000^OTHER BUSINESS
_   59,000 _   2,000   _
FISCAL   ‾NDING   12/31/81   12/31/80

FIVE YEAR SUMMARY

YEAR	SALES (000S)	NET INCOME	EPS
1981	29,070,000	3,308,000	5.63
1980	26,213,000	3,562,000	6.10
1979	22,863,000	3,011,000	5.16
1978	21,076,000	3,111,000	5.32
1977	18,133,000	2,719,000	4.58

19. Hmmm. One ventures to guess that a dynamic innovative industry would attract lawyers. How many patent attorneys do you suppose there are in San Francisco, Silcon Valley, and Boston? Well, that precise question probably can't be answered from any public database, but we can get an order-of-magnitude figure from a database that provides Yellow Page information on professionals in insurance, real estate, medicine, law, engineering, and accounting.

This database contains nearly two million records. To access it in reasonable times, at reasonable costs, there must be a variety of indexes. This file has 16, including, among others, company name, city, Standard Industrial Classification (SIC) Code, telephone number including area code, and ZIP code. The "ss" command in this case asks that the SIC be 8111A, which specifies patent attorneys, and that the telephone area code be either 415 (San Francisco), 408 (San Jose area), or 617 (Boston).

It turns out that there are 4096 patent attorneys listed in the Yellow Pages of the U.S. (as of 1982 or so), that there are 45,809 professional listings with telphones numbers in area 415, 14,139 in area 408, and 45,639 in area 617. Of the 4096 patent attorneys, 315 were in one of these three areas. If I wished, I could have these records printed, or search set 5 to see if there is an attorney named Zimmerman I vaguely remember from college, or whatever.

Naturally, there are presumably patent attorneys in, say, area 408, who work for IBM, Intel, or Lockheed (of which DIALOG is a subsidary). One presumes that these attorneys are not listed in the Yellow Pages; therefore we have only a general indication of the total. It is, nonetheless, an impressive number.

20. For a change of pace, let's see if we can find out what Herb Grosch wrote his doctoral dissertation on. Dr. Grosch was my manager for several years at General Electric and preceded me as president of the Association for Computing Machinery; so I have known for many years that his dissertation was in astronomy, but I don't know the exact subject.

There is a database called Comprehensive Dissertation Abstracts, developed by University Microfilms International, Ann Arbor, Michigan, and marketed by the Xerox Corporation. This database has brief data on all U.S. and many Canadian doctoral dissertations dating from 1861 (sic), and selected master's theses from 1962, about three-quarters of a million in all.

Utilizing the author index, I ask for records where the author's last name is Grosch and the first name begins with the letter "h." As it happens I know his full name, but I'm not sure how the database may have it. If there should happen to be two or more entries that meet this selection criterion, I can refine it later. But there is only one, and we see what the subject was.

(19) File502:Electronic Yellow Pages- Professional Dir.
   (Copr. Mrkt. Data Ret. Inc)
            Set Items Description
            --- ----- -----------
   ? ss sc=8111a and (te=415? or te=408? or te=617?)
            1 4096 SC=8111A (PATENT ATTORNEYS)
            2 45809 TE=415?
            3 14139 TE=408?
            4 45639 TE=617?
            5   315 1 AND (2 OR 3 OR 4)

(20) File35:Comprehensive Dissertation Abstracts 1861 to Apr 83
   (Copr. XEROX Corp)
            Set Items Description
            --- ----- -----------
   ? select au=grosch, h?
            1   1 AU=GROSCH, H?
   ? type 1/2/1
   1/2/1
   459015 ORDER NO: NOT AVAILABLE FROM UNIVERSITY MICROFILMS INT'L
     INTEGRATION ORBIT AND MEAN ELEMENTS OF JUPITER'S EIGHTH SATELLITE
     GROSCH, HERBERT REUBEN JOHN (PH.D. 1942 THE UNIVERSITY OF MICHIGAN).
     PAGE 6 IN VOLUME W1942
     ASTRONOMY
     DESCRIPTOR CODES: 0299
     INSTITUTION CODE: 0127

21. Now that we're here, let's explore a bit. Suppose we ask DIALOG to set up a file with the record identifiers of all records for which the Descriptor Code (DC) is 0984, which designates computer science, and for which the degree is either the Ph.D. or the Sc.D. We see that 3867 such degrees have been granted in the U.S. and Canada. The identifiers of those 3867 records are in set 5. Why not see what the trend of the number of such degrees has been over the years? (One of the indexes to this file is, as one might expect, publication year.) Let's get the data for a 10-year history. Since set 5 already has a set number, it is not necessary to repeat that search. All we have to do is specify the search criterion as being membership in set 5 and publication year (PY) = 1973. It turns out that there were 322 doctorates granted in computer science in 1973. OK, let's do the others. With such a dynamic field, growing as rapidly as the daily newspapers will tell you and as some of the earlier examples here have already indicated, and with computer science enrollments sky-rocketing, surely the number of doctorates has shown a steady increase.

As we proceed, messages appear about the number of "items to go." This relates to a limit of one million items stored on magnetic disk in the search of any one database, and the number shown is the amount of space we have left. Since it is possible to complete this inquiry without running out of space, we ignore these messages.

And what is the answer? Well, we should probably throw out 1982; it seems unlikely that all the doctorates from 1982 were in the database at the time this search was made. But even then, the number of computer science doctorates appears to be pretty flat. How can that be, when student enrollments have approximately tripled in the same period?

The answer, as best we understand it, is that the demand for people with computer science education is currently so intense that many graduate students are not finishing their doctorates, but, rather, are taking positions in government or industry. Those of us involved in education are happy for them, but we wonder where the next generation of professors is going to come from. And we puzzle, sometimes, that representatives of the very companies that hire away the graduate students can, in the next breath, complain about the quality of education that undergraduates receive.

But I digress.

(21) ? ss dc=0984 and (dg=ph.d. or dg=sc.d.)
```
                2   4158 DC=0984
                3695562 DG=PH.D.
                4   1682 DG=SC.D.
                5   3867   2 AND (3 OR 4)
```
? type 5/6/1-3
5/6/1
914814   ORDER NO: AAD83-06085
  DESIGN   STUDY   OF   A   FAULT-TOLERANT COMPUTER SYSTEM TO EXECUTE N-VERSION
SOFTWARE   168 PAGES.

5/6/2
914796   ORDER NO: AAD83-06041
  ON   PROGRAM   DECOMPOSITION   AND   PARTITIONING   IN   DATA-FLOW SYSTEMS   239
PAGES.

5/6/3
914792   ORDER NO: AAD83-06033
  CONTRIBUTIONS TO AUTOMATIC CODE GENERATION   208 PAGES.

? ss s5 and py=1973
```
                6 36796 PY=1973
                7    322   5 AND 6
```
? ss s5 and py=1974
```
                8 37709 PY=1974
                9    297   5 AND 8
```
? ss s5 and py=1975
```
               10 38302 PY=1975
               11    318   5 AND 10
```
? ss s5 and py=1976
```
               12 37985 PY=1976
               13    324   5 AND 12
```
? ss s5 and py=1977
```
               14 35342 PY=1977
               15    273   5 AND 14
```
? ss s5 and py=1978
```
               16 34464 PY=1978
               17    320   5 AND 16
```
? ss s5 and py=1979
```
               18 35713 PY=1979
                    (202,404 items to go)
               19    320   5 AND 18
                    (202,084 items to go)
```
? ss s5 and py=1980
```
               20 36141 PY=1980
                    (165,943 items to go)
               21    373   5 AND 20
                    (165,570 items to go)
```
? ss s5 and py=1981
```
               22 34906 PY=1981
                    (130,664 items to go)
               23    314   5 AND 22
                    (130,350 items to go)
```
? ss s5 and py=1982
```
               24 24459 PY=1982
                    (105,891 items to go)
               25    241   5 AND 24
                    (105,650 items to go)
```

22. To end on a positive note, looking toward the future, we return to the INSPEC database.

The Fifth Generation Computer Systems Project is a Japanese initiative to develop computer systems vastly more powerful than today's machines, over a 10-year period. "Fifth generation" refers to both hardware and software. Computer generations are defined approximately as follows:

First generation: vacuum tubes and germanium diodes in hardware, actual machine language in software. Approximately 1948-1955.

Second generation: transistors and assembly language. Approximately 1955-1963, but the boundaries are blurred.

Third generation: small-scale integrated circuits and procedural languages (Fortran, Cobol, etc.). Approximately 1963-1978, but the boundaries are *very* blurred.

Fourth generation: Very Large Scale Integration (VLSI) on the hardware side, meaning tens of thousands of circuit elements per chip, and nonprocedural languages on the software side. (An example of a nonprocedural language, NOMAD, appears in the next section.) Approximately 1978-1990? That remains to be seen, of course, quite apart from the arbitrariness of these definitions.

Fifth generation: extremely large scale integration (millions of circuit elements per chip) and entirely different computer architectures (highly parallel machines, for example) on the hardware side, and such things as natural language processing, inferential reasoning, and possibly large-vocabulary speech recognition on the software side.

It is conceivable that the notion of a fifth generation is used in fields other than computers and computing. Let's phrase the search request to restrict things to computers. Let's also allow for both "fifth" and "5th." The INSPEC file, at the time of this search, contained 51 records satisfying these criteria.

The first record in the set is always the most recent, as the DIALOG accession numbers work, and it turns out to be a survey article that gives a general picture of the goals of the project. (If you run the same search, you will not get this article, because between the day I ran the search and the day you read this, there will be a great many more papers published on this subject.)

(22) File13:INSPEC - 77-83/Iss08
(Copr. IEE)
See File 12(1969 thru 1976)
        Set Items Description
        --- ----- -----------
? ss (5th(w)generation or fifth(w)generation) and comput?
            1   5 5TH(W)GENERATION
            2   52 FIFTH(W)GENERATION
        3145371 COMPUT?
            4   51 (1 OR 2) AND 3
? type 4/5/1
4/5/1
 1023533   C83013849
 EXPERT SYSTEMS: AN INTRODUCTION FOR THE LAYMAN
 WINFIELD, M.J.
 NORTH STAFFORDSHIRE POLYTECH., STAFFORD, ENGLAND
 COMPUT. BULL. (GB)   SER. 2, NO.34   6-7, 18   DEC. 1982   Coden: COBUAH
 ISSN: 0010-4531
   Treatment: GENERAL,REVIEW
   Document Type: JOURNAL PAPER
   Languages: ENGLISH
   (9 Refs)
   IT HAS BEEN SAID THAT NON-EXPERTS OUTNUMBER EXPERTS IN ANY PARTICULAR
 FIELD BY THE RATIO 100:1. IF THIS IS THE CASE THEN THE NEXT GENERATION OF
 COMPUTERS NEEDS TO BE BUILT NOT JUST FOR THE EXPERT USER,   BUT ALSO FOR THE
 NAIVE USER.   THE JAPANESE HAVE RECENTLY PRODUCED A STUDY REPORT ON FIFTH
 GENERATION COMPUTERS WHICH WILL INVOLVE THE USE OF EXPERT SYSTEMS.   THE
 REPORT LISTS SEVERAL IMPORTANT ISSUES WHICH THE JAPANESE FACE IN THE FUTURE
 AND, IN FACT, WHICH MOST OF THE DEVELOPED COUNTRIES WILL ALSO FACE.   THIS
 LIST INCLUDES THE NEED TO IMPROVE PRODUCTIVITY IN LOW PRODUCTIVITY AREAS,
 COMPETE INTERNATIONALLY AND CONTRIBUTE TOWARD INTERNATIONAL CO-OPERATION,
 SAVE RESOURCES AND ENERGY,   AND COPE WITH AN AGED SOCIETY.   WITH THE AMOUNT
 OF MONEY THE JAPANESE WILL BE COMMITTING TO THE FIFTH GENERATION COMPUTER
 PROJECT,   THE WHOLE OF THE WESTERN WORLD'S COMPUTER FRATERNITY HAS BEEN
 AWAKENED TO THE NEED FOR DEVELOPMENT IN EXPERT SYSTEMS
   Descriptors: ARTIFICIAL INTELLIGENCE; SOFTWARE ENGINEERING
   Identifiers: AI; SOFTWARE ENGINEERING; LAYMAN; FIFTH GENERATION COMPUTERS
   EXPERT SYSTEMS
   Class Codes: C6110; C1230

23. One of the key elements in the Fifth Generation Project will be an approach called Expert Systems. Here, the knowlege of the experts in a field is entered into a computer, along with rules of inference and a person-machine dialog interface that permits the human being to understand how the computer arrived at its conclusions. On the basis of this information, the human being decides how much reliance to place on the computer's conclusions.

"Expert systems" triggers the name Ed Feigenbaum in my mind, since in his years at Stanford he has been a pioneer in the field. Let's see what INSPEC knows about his recent writings on this subject. The answer: He presented a paper at the conference in which the Fifth Generation Project was first widely publicized.

I ask to have all these records printed. When they arrive, I discover a few "false hits": a paper on human respiration, in which the author computed the total cross-sectional areas of the first through fifth generations of branching in the bronchial airways. This is, of course, an altogether different meaning of the concept "fifth generation" from the one I had in mind.

As of this writing, the first theoretical considerations of computerized information retrieval go back about 30 years, the first serious experiments about 20 and successful commericial application little more than 10. Today it is nearly a billion dollar industry.

What an astonishing field!

NOMAD: a fourth generation language

The examples so far in this chapter have dealt either with a broad applications area (partial differential equations, circuit analysis, or symbolic mathematics), or, in the case of DIALOG, involved the retrieval of information rather than any kind of application. Now we turn to a brief consideration of a language for the development of applications that are not limited to any particular field.

NOMAD®, a product of National CSS, Inc., is a system for the development of computer applications that includes:

■ A database management system, which handles the organization, storage, and manipulation of data.

■ A report writing language, with which information can be extracted from the database in a flexible, easy-to-learn, and easy-to-understand way.

■ Facilities for the entry and modification of data.

■ A statistics package.

■ A graphics capability.

The study of database management systems is an entire subfield of computer science. Let us avoid generalties and simply see a sample of what such a system can do.

(23) ? ss s4 and au=feigenbaum, e?
              5   11 AU=FEIGENBAUM, E?
              6   1 4 AND 5
      ? type 6/5/1
      6/5/1
        991156   C83005933
        INNOVATION AND SYMBOL MANIPULATION IN FIFTH GENERATION COMPUTER SYSTEMS
        FEIGENBAUM, E.A.
        COMPUTER SCI. DEPT., STANFORD UNIV., STANFORD, CA, USA
        MOTO-OKA, T. (Editors)
        FIFTH   GENERATION   COMPUTER   SYSTEMS.   PROCEEDINGS   OF THE INTERNATIONAL
      CONFERENCE   223-6   1982
        19-22 OCT. 1981   TOKYO, JAPAN
        Publ: NORTH-HOLLAND,   AMSTERDAM, NETHERLANDS
        VIII+287 pp.   ISBN 0 444 86440 7
        Treatment: PRACTICAL
        Document Type: CONFERENCE PAPER
        Languages: ENGLISH
        (6 Refs)
        FIFTH GENERATION COMPUTER SYSTEMS,   IF   SUCCESSFULLY   DEVELOPED   WILL   BE
      EXCELLENT   VEHICLES   FOR   EXPERT SYSTEMS APPLICATIONS.   FUNDAMENTAL TO THIS
      DEVELOPMENT IS A   SOFTWARE   METHODOLOGY   KNOWN   AS   KNOWLEDGE   ENGINEERING.
      KNOWLEDGE,   NOT   INFERENCE,   IS   THE KEY TO HIGH LEVELS OF PERFORMANCE OF
      EXPERT SYSTEMS. A CONSIDERABLE VARIETY OF EXPERT SYSTEMS APPLICATIONS, MOST
      HAVING GREAT POTENTIAL PAYOFF,   ARE ALREADY BEING   WORKED   ON.   SCIENTIFIC
      INNOVATIONS IN KNOWLEDGE ACQUISITION WILL BE REQUIRED.   INNOVATIONS LEADING
      TO AN EFFICIENT INDUSTRIAL TECHNOLOGY FOR KNOWLEDGE ENGINEERING   WILL   ALSO
      BE   NECESSARY.   MANAGERS MUST EXHIBIT IMAGINATION,   AND THE WILLINGNESS TO
      TAKE TECHNOLOGICAL RISKS TO REALIZE THE   ENORMOUS   POTENTIAL   BENEFIT   FROM
      FIFTH GENERATION SYSTEMS
        Descriptors: ARTIFICIAL INTELLIGENCE; PROGRAMMING
        Identifiers: FIFTH GENERATION COMPUTER SYSTEMS;   KNOWLEDGE ENGINEERING;
      EXPERT SYSTEMS APPLICATIONS
        Class Codes: C6100; C1230

      ? print 4/5/1-51
      Printed4/5/1-51 Estimated Cost: $17.85 (To cancel, enter PR-)

1. There are actually two versions of NOMAD. With the CSS command "nomad2," I say that I want to begin work with the latest version. The response tells me the release date of the version I am working with.

2. In a separate operation, not shown, I have already entered a description of the database into the computer. In the terminology of the database field, this description is called a *schema*. With the NOMAD command "slist" ("schema list"), I ask for this description to be printed.

The application for this example involves my gradebook for a computer course, here called Computer Science 123. The grades are for real people in a real class that I taught not long ago, but the names are fictitious and the social security numbers were produced by a random number generator program that I ran in Basic on my Apple.

The first line of the schema gives a name (**GRADE**) to this description of the data, along with a filename and filetype, and specifies that the database management system is to keep them in sequence on social security number (**SSN**). The second line says that **SSN** has 11 alphanumeric characters (that's the **A11**), and that the data entered for this field must be in the form of three numeric digits, then a hyphen, then two numeric digits, then another hyphen, and finally four more digits. The third line says that **STUDENT** is a *name*, and that it is to be printed with 18 characters although internally it is stored with 30. (This was to keep the reports of manageable width for the book.) The fourth line says that **DROPDATE** is a date, and the fifth that **CLASS** has four alphanumeric characters. The next line says that **HW** (for "homework") is an array with seven elements, that only numeric values are permitted, that two digits are to be printed on reports, that the *index* (subscript to us) is one digit, and that the acceptable values for the homework grades are zero to 15. The last two lines give the format for printing the two exam grades and the limits on them.

This schema provides the information that type declarations and array declarations do in Fortran but, as we already begin to see, considerably more. Without writing any procedural statements whatever, we have already provided for such things as checking the validity of data values. And we have two quite different data types, **NAME** and **DATE**, which permit powerful operations that will be illustrated below.

3. To make certain subsequent operations a little more meaningful, I specify that array contents are to be *verified*, which really just means to print them.

4. The "prompt" command tells the system to ask me for the data value that I name in the remainder of the command. I name all the fields in the record for each student, in the order in which they appear in my gradebook. (May as well make it convenient for myself.) This is why I say to ask for the first three homework grades, then the midterm, then the last four homeworks, then the final. The "do 4" means to prompt me for the data about four students. I enter the data for the first student in my gradebook. Names are entered in the form of last name followed by a comma, then first name, then middle initial(s) if any. This student did not drop the course, and so I simply pressed carriage return when asked for **DROPDATE**.

5. After I have entered all the data for this student, the system prints the values so that I can check them if I wish. There are options that condense this printing or eliminate it. The system proceeds to prompt me for the next student.

① 12.05.31 >nomad2
       NOMAD2 version 112Y 07MAR83

② >database csc123
       >slist

       MASTER GRADE FILENAME='CSC123' FILETYPE='GRADE' INSERT=KEYED(SSN,A);
           ITEM   SSN AS A11 MASK='999-99-9999';
           ITEM   STUDENT AS NAME'18' INT=A30;
           ITEM   DROPDATE AS DATE;
           ITEM   CLASS AS A4;
           ITEM   HW(7) AS 99 INT=I4 INDEX AS FORMAT'9' LIMITS=(0:15);
           ITEM   MIDTERM AS 999 INT=I4 LIMITS=(0:100);
           ITEM   FINAL AS 999 INT=I4 LIMITS=(0:100);

③ >verify array on

④ >prompt student ssn class dropdate hw(1:3) midterm hw(4:7) final do 4
       STUDENT=>alba,aurora
       SSN=>485-85-6156
       CLASS=>1107
       DROPDATE=>
       HW(1)=>10 10 10
       MIDTERM=>96
       HW(4)=>15 9 10 10
       FINAL=>95
⑤ SSN=485-85-6156 STUDENT=ALBA,AURORA DROPDATE=N/A   CLASS=1107

            1   2   3   4   5   6   7
           -- -- -- -- -- -- --
       HW=10 10 10 15 9 10 10
       MIDTERM=96 FINAL=95
       STUDENT=>alhassan,khalid
       SSN=>291-85-1623
       CLASS=>1107
       DROPDATE=>
       HW(1)=>10 7 10
       MIDTERM=>94

6. Getting in a hurry, I enter a wrong homework grade, which results in an error message and a re-prompt for that one item. You learned how to program this kind of testing in Fortran; here, we get it automatically as the result of entering one phrase about limits in the schema.

7. I said that hyphens had to be in the social security number, and so the system flags their omission.

8. My mind wandering, I enter a nonexistent date. NOMAD *knows* about dates! It knows a *lot* more about dates than there is any way to illustrate in this session, but at a very minimum, it will reject the 31st of November. (Yes, it knows absolutely everything about leap years.)

This kind of checking can be programmed in Fortran, but at the cost of vastly more effort than simply specifying that the data item is of the type "date."

9. This time I remember the hyphens, but I get them in the wrong places. A few lines later, I hit a wrong key after entering a homework grade. Since I had said it was to be numeric only, it is rejected. A mistyped exam grade is likewise rejected.

The entry of the remainder of the data is not shown.

10. Now I'd like to see all the data I have entered, in alphabetical order by student name. (Recall that within the database, the records are in sequence by social security number.) I use the "list" command, in which the "by student" tells NOMAD to sort the records into ascending alphabetical sequence on the students' names. The rest of the command gives the items I want printed, in the order I want them printed. Note that this is not the order in which I entered them, nor is it the order in which their descriptions appear in the schema.

We see that a complete report is produced as a result of this simple command, including page numbering and column headings. The column headings are simply the names of the data items as entered into the schema; we shall see shortly how simple it is to modify the column headings. Alternatively, column headings could also have been placed in the schema.

Any time I didn't enter data when prompted, the report contains N/A or N/, the latter when there are only two columns in the field. N/A mean "not available." This means that nothing was entered. It is *not* the same as either a blank or a zero. When no data has been entered for a data value, the system places a special mark in storage for that item. This permits us to distinguish between zero and nothing, or between blank and nothing. This capability—which is often highly useful, as we shall see—can require extraordinary effort to match in most procedural languages. (WATFIV has it, but standard Fortran doesn't.)

```
 6   HW(4)=>11 10 10 81
     PRO219: HW not within LIMITs.
     HW(7)=>8
     FINAL=>81
     SSN=291-85-1623 STUDENT=ALHASSAN,KHALID DROPDATE=N/A        CLASS=1107

         1  2  3  4  5  6  7
         -- -- -- -- -- -- --
     HW=10  7 10 11 10 10  8
     MIDTERM=94 FINAL=81
     STUDENT=>breuer,howard mark
     SSN=>180544710
 7   PRO220: SSN - MASK error.
     SSN=>180-54-4710
     CLASS=>1506

 8   DROPDATE=>11/31/83
     PRO425: Internal-format conversion failed for DROPDATE.
     DROPDATE=>11/14/83
     HW(1)=>10
     MIDTERM=>38
     HW(4)=>
     FINAL=>
     SSN=180-54-4710 STUDENT=BREUER,HOWARD MARK DROPDATE=11/14/83 CLASS=1506

         1  2  3  4  5  6  7
         -- -- -- -- -- -- --
     HW=10 N/ N/ N/ N/ N/ N/
     MIDTERM=38 FINAL=N/A
     STUDENT=>desai,rohit
 9   SSN=>32-226-4647
     PRO220: SSN - MASK error.
     SSN=>322-26-4647
     CLASS=>299-
     DROPDATE=>
     HW(1)=>10 10 10-
     PRO219: HW not within LIMITs.
     HW(3)=>10
     MIDTERM=>930
     PRO219: MIDTERM not within LIMITs.
     MIDTERM=>93
     HW(4)=>15 13 10 10
     FINAL=>87
     SSN=322-26-4647 STUDENT=DESAI,ROHIT DROPDATE=N/A       CLASS=299-

         1  2  3  4  5  6  7
         -- -- -- -- -- -- --
     HW=10 10 10 15 13 10 10
     MIDTERM=93 FINAL=87
10   >list by student ssn class dropdate midterm final across all hw


     PAGE   1
```

						HW						
						1	2	3	4	5	6	7
STUDENT	SSN	CLASS	DROPDATE	MIDTERM	FINAL	--	--	--	--	--	--	--
ALBA, AURORA	485-85-6156	1107	N/A	96	95	10	10	10	15	9	10	10
ALHASSAN,KHALID	291-85-1623	1107	N/A	94	81	10	7	10	11	10	10	8
BREUER, HOWARD MARK	180-54-4710	1506	11/14/83	38	N/A	10	N/	N/	N/	N/	N/	N/
DESAI, ROHIT	322-26-4647	299-	N/A	93	87	10	10	10	15	13	10	10
EDELMAN, DEBRA	418-96-9846	1107	N/A	98	80	10	7	8	10	N/	10	10
ELLIS, NORTON O	468-32-5114	1504	10/28/83	44	N/A	N/	N/	N/	N/	N/	N/	N/
GARCIA, RAUL	338-53-2919	1107	N/A	81	79	10	7	10	15	10	9	8
HOMER, ALBERT	103-35-9637	1107	N/A	N/A	N/A	N/	N/	N/	N/	N/	N/	N/
JASON, HAROLD	858-92-9431	1208	N/A	71	79	10	7	8	8	9	10	8
LEE, WILSON H	357-14-7308	1101	N/A	77	75	10	7	10	7	N/	8	N/
LEFCOURT, MARIE	018-33-7283	1107	N/A	99	83	10	10	10	10	13	10	8
LEVINE, DANIEL G	280-42-0534	1507	N/A	61	64	10	9	N/	6	7	4	7
LOONG, KUO-CHANG	647-39-4588	1107	N/A	93	94	10	10	10	10	10	10	10
MOSES, AARON	994-04-8781	1301	N/A	69	99	10	9	8	10	9	10	10
O'REILLY, JOHN W	062-13-8885	1107	N/A	98	96	10	10	10	14	10	10	10
STEVENS, KAY L	092-59-0625	299-	N/A	98	99	10	10	10	15	13	10	8
TATUM, PATRICK R	479-55-3749	1803	N/A	88	44	10	N/	N/	8	7	10	
THOMAS, KATHY V	884-23-8813	1107	N/A	63	72	10	7	7	9	7	N/	N/
YEUNG, TSAI	994-04-8741	1803	N/A	89	67	10	10	8	10	13	10	10
YOUNG, ALFRED G	629-60-7839	1101	N/A	91	79	10	7	10	N/	9	10	8

11. I printed this report mostly so I could proofread my data entry. I see two errors. The first is that I failed to enter a drop date for one student. The command "tlocate" means to search the database, starting at the top (that's the "t"), and "locate" the record in which the student name is as shown. NOMAD does so.

12. The "change" command lets me modify the contents of a record. Quotes are required around the date.

13. Now I see a student for whom I entered the wrong final exam grade. This time I observe that the social security number is greater than the one for the student I was just working with, so I don't have to say to start from the top if I am willing to search on social security number.

14. The "change" command corrects the grade. Quotes are not required around the numeric value.

15. Now I would like a report in social security number order, with the only other data being the student name, drop date, the two exams, and the sum of the homeworks. The command is almost an English equivalent of that last sentence. Observe that the first student who dropped did so before the midterm and turned in no homework; therefore the sum of the homeworks is N/A. The second student who dropped turned in one homework and dropped after the midterm. The third student to drop took the midterm although he had turned in no homework; such students *do* tend to drop!

16. It seems the Registrar wants a confirmation of the drop list. Let's get a report showing only the students who dropped the course, in name order, showing the social security number and the drop date. The selection of only those students who have dropped is handled with the "where dropdate ne &nav" clause. &NAV is a symbol that lets us test a data item to see if it is "not available," i.e., if nothing was ever entered for it. The logic is that if the drop date is still "not available," then the student has *not* dropped. We want the ones for which a drop date *has* been entered.

```
(11) >tlocate student='homer,albert'
     SSN=103-35-9637 STUDENT=HOMER,ALBERT DROPDATE=N/A        CLASS=1107

          1  2  3  4  5  6  7
         -- -- -- -- -- -- --
     HW=N/ N/ N/ N/ N/ N/ N/
     MIDTERM=N/A FINAL=N/A
(12) >change dropdate='11/02/83'
     SSN=103-35-9637 STUDENT=HOMER,ALBERT DROPDATE=11/02/83 CLASS=1107

          1  2  3  4  5  6  7
         -- -- -- -- -- -- --
     HW=N/ N/ N/ N/ N/ N/ N/
     MIDTERM=N/A FINAL=N/A
(13) >locate ssn='280-42-0534'
     SSN=280-42-0534 STUDENT=LEVINE,DANIEL G DROPDATE=N/A        CLASS=1507

          1  2  3  4  5  6  7
         -- -- -- -- -- -- --
     HW=10  9 N/  6  7  4  7
     MIDTERM=61 FINAL=64
(14) >change final=46
     SSN=280-42-0534 STUDENT=LEVINE,DANIEL G DROPDATE=N/A        CLASS=1507

          1  2  3  4  5  6  7
         -- -- -- -- -- -- --
     HW=10  9 N/  6  7  4  7
     MIDTERM=61 FINAL=46
(15) >list by ssn student dropdate midterm final sum(hw)


     PAGE    1

                                                          SUM
     SSN          STUDENT            DROPDATE MIDTERM FINAL HW
     -----------  ------------------ -------- ------- ----- ---
     018-33-7283 LEFCOURT,MARIE      N/A           99    83  71
     062-13-8885 O'REILLY,JOHN W     N/A           98    96  74
     092-59-0625 STEVENS,KAY L       N/A           98    99  76
     103-35-9637 HOMER,ALBERT        11/02/83     N/A   N/A N/A
     180-54-4710 BREUER,HOWARD MARK  11/14/83      38   N/A  10
     280-42-0534 LEVINE,DANIEL G     N/A           61    46  43
     291-85-1623 ALHASSAN,KHALID     N/A           94    81  66
     322-26-4647 DESAI,ROHIT         N/A           93    87  78
     338-53-2919 GARCIA,RAUL         N/A           81    79  69
     357-14-7308 LEE,WILSON H        N/A           77    75  42
     418-96-9846 EDELMAN,DEBRA       N/A           98    80  55
     468-32-5114 ELLIS,NORTON O      10/28/83      44   N/A N/A
     479-55-3749 TATUM,PATRICK R     N/A           88    44  45
     485-85-6156 ALBA,AURORA         N/A           96    95  74
     629-60-7839 YOUNG,ALFRED G      N/A           91    79  54
     647-39-4588 LOONG,KUO-CHANG     N/A           93    94  70
     858-92-9431 JASON,HAROLD        N/A           71    79  60
     884-23-8813 THOMAS,KATHY V      N/A           63    72  40
     994-04-8741 YEUNG,TSAI          N/A           89    67  71
     994-04-8781 MOSES,AARON         N/A           69    99  66


(16) >list by student ssn dropdate where dropdate ne &nav


     PAGE    1

     STUDENT             SSN         DROPDATE
     ------------------- ----------- --------
     BREUER,HOWARD MARK 180-54-4710 11/14/83
     ELLIS,NORTON O      468-32-5114 10/28/83
     HOMER,ALBERT        103-35-9637 11/02/83
```

17. Well, now, that's not very pretty, is it? Let's print the names in conventional order, with a period after the middle initial (if any). In the "list" command, the 18 says 18 characters, the "f" means first name, the "mi" means middle initial (whether the person has a full middle name or only an initial), the "." means to place a period after the middle initial, and "l" means last name. Now let's put a more meaningful heading over the social security number column, which the "heading" clause does. The colons designate points where the heading should be broken into separate lines. Now let's get a little prettier format for the date, too. The "mon" means to print the month as a three-letter abbreviation, the "bd" means to print the day with leading zeros suppressed, and the "yyyy" means to print the year as a four-digit number. (Dozens of variations are available.) The screening is as before. Finally, let's put a title on the report and suppress the numbering of pages. (The numbering of pages is suppressed on all following reports.)

Observe the handling of the three middle names. Breuer's middle name was cut down to an initial, Ellis' middle initial was printed properly, and NOMAD had enough sense not to print a period for a non-existent middle initial.

The exercise for this chapter is to write Fortran programs to do all these reports!

18. How many student records are there in the database?

19. OK, but from now on I'm going to want to exclude the three drops, and it gets tiresome adding the clause "where dropdate ne &nav" to every command. Let's just tell NOMAD that until informed otherwise, it should ignore students that have a drop date in their record, which is the same as saying to consider only those for which the drop date is N/A. The "select" command does this. Now, the same "list" command as before shows only 17 students.

20. The "count" was an example of a *report function,* as was the "sum" earlier. There are various other report functions, including some that obtain elementary statistics about the records in a database. Or, more precisely, about those records in the database that pass the selection criteria currently in effect: the students who dropped are not included in these statistics.

21. Hmmm. There were two graduate students in that course, and they both did very well. I wonder if their grades distorted the statistics much?

The way the "class" code works at my school, a 9 in the second position of the class code indicates a graduate student. I'd like to screen out those two students in rerunning the previous report. But how do I get at one character of an alphanumeric (character) item? Simple: With the "substr" ("substring") operator, which performs just as in Fortran but with a different syntax.

It seems that the two graduate students didn't pull up the average *that* much.

(17)
```
>list by student as name'18f mi. 1' -
>    ssn heading 'social:security:number' -
>    dropdate as date'mon bd, yyyy' -
>    where dropdate ne &nav -
>    title 'students who dropped computer science 123' -
>    nopageno
```

```
   STUDENTS WHO DROPPED COMPUTER SCIENCE 123

                          SOCIAL
                          SECURITY
   STUDENT                NUMBER      DROPDATE
   ------------------     ----------  ------------
   HOWARD M. BREUER       180-54-4710 NOV 14, 1983
   NORTON O. ELLIS        468-32-5114 OCT 28, 1983
   ALBERT HOMER           103-35-9637 NOV 2, 1983
```

(18)
```
>list count(student) nopageno
```

```
        COUNT
        STUDENT
       ----------
           20
```

(19)
```
>select dropdate = &nav
>list count(student) nopageno
```

```
        COUNT
        STUDENT
       ----------
           17
```

(20)
```
>list min(final) max(final) avg(final) median(final) stddev(final) nopageno
```

```
   MINIMUM MAXIMUM AVG   MEDIAN          STD DEV
    FINAL   FINAL  FINAL FINAL            FINAL
   ------- ------- ----- ------  ----------------
       44      99    80     80        16.142973
```

(21)
```
>list min(final) max(final) avg(final) median(final) stddev(final) -
>    where substr(class,2,1) ne '9' nopageno
```

```
   MINIMUM MAXIMUM AVG   MEDIAN          STD DEV
    FINAL   FINAL  FINAL FINAL            FINAL
   ------- ------- ----- ------  ----------------
       44      99    78     79        16.250128
```

22. Now I'm ready to get down to the serious business: grading. What I'd like now is a listing of the class, in descending order of the course averages. The grading system, announced at the beginning of the course, was that the midterm and the homeworks were worth 30% each and the final was worth 40%. To get this report, I define to NOMAD a new variable called "courseav." The 999 means that it should be printed with three digits. The "expr" stands for "expression," after which I give the formula. Since the midterm and the final were graded on the basis of zero to 100 points, they can be weighted with 0.3 and 0.4, respectively. Unfortunately—for the sake of simplicity here—I graded the homeworks on the basis of zero to 10, with extra credit possible in several cases. The homework average is the sum of the homeworks, divided by 7; 10 times that puts it on a zero to 100 basis; and 0.3 is the weighting factor. Obviously, the three constants here could have been combined, at the expense of understandability.

With NOMAD having been given this definition of a new variable, we can ask for a report using it. In fact, we can ask that a report be in descending sequence on it, even though the values in question do not occur in the database (and are not placed there by this operation).

23. Gee, that was fun. Let's do some more of that, this time to test one of my convictions (or prejudices, depending on your point of view). That is the conviction that you can't learn programming except by doing it, any more than you can learn to play the piano from a lecture course. Let's define a variable that records, for each student, the number of homeworks submitted. Fundamental to the operation of this definition, and to the report based on it, is the fact that N/A values are *not* included by the "count" report function. Now I produce a report in sequence on the number of homeworks submitted (not counting the drops, remember), and for each one ask that average of the course averages be printed. That is, for those students who submitted five homeworks, what was the average of their course averages, etc.

24. Well, that report makes my conviction/prejudice look pretty good. But it does, after all, involve a certain amount of self-correlation, since the number of homeworks enters into the computation of the course average. OK, let's set up a variable to compute the average of the two exams and see how that correlates with the number of homeworks. While we're at it, let's print the number of students in each group as well. The correlation is still persuasive, to my mind.

25. With statistics on my mind, I decide to make an informal test of a theory about what I would do in the exceedingly unlikely event that I should lose my gradebook just before the final. Surely, this has never happened in the history of education, but just suppose? One theory says that the best students get the best grades on everything, and that you wouldn't be committing too many sins by giving marks based on the final alone. I first define a variable that computes a "provisional grade" based on the midterm and the homeworks alone.

26. Now I tell NOMAD to turn on the statistics package.

27. To do statistics, I need to define two variables that make arrays of the data sets to be used by the statistics package. Let's ignore the syntax of these.

28. Now, for reasons that are of no interest to us here, I need for the database to be "positioned" on an "instance" of the records in question.

(22) >define courseav as 999 expr = 0.3*midterm + 0.4*final + 0.3*10*sum(hw)/7
     >list by courseav desc heading 'course:average' midterm final nopageno

COURSE AVERAGE	MIDTERM	FINAL
102	98	99
100	98	96
99	96	95
96	93	87
	93	94
93	99	83
89	94	81
	69	99
85	81	79
	98	80
84	89	67
82	91	79
79	71	79
71	77	75
65	63	72
63	88	44
55	61	46

(23) >define howmanyhw as 9 expr = count(hw)
     >list by howmanyhw heading 'how many:homeworks' -
     >   avg(courseav) heading 'course:average' nopageno

HOW MANY HOMEWORKS	COURSE AVERAGE
5	66
6	74
7	92

(24) >define examav as 999 expr = (midterm + final)/2
     >list by howmanyhw heading 'how many:homeworks' -
     >   avg(examav) heading 'exam:average' -
     >   count(student) heading 'number:in each:group' -
     >   title 'you learn to program by doing it!' -
     >   nopageno

YOU LEARN TO PROGRAM BY DOING IT!

HOW MANY HOMEWORKS	EXAM AVERAGE	NUMBER IN EACH GROUP
5	70	3
6	76	3
7	88	11

(25)
(26) >define provav as 999 expr = (midterm + 10*sum(hw)/7)/2
     >stat on
(27) >define provarray(0) as 999 expr = instance(provav)
     >define finalarray(0) as 999 expr = instance(final)
     >first grade
(28) SSN=018-33-7283 STUDENT=LEFCOURT,MARIE DROPDATE=N/A   CLASS=1107

```
       1   2   3   4   5   6   7
      --  --  --  --  --  --  --
HW=10  10  10  10  13  10   8
MIDTERM=99 FINAL=83
```

29. The command "bstat" stands for "basic statistics," and produces the report shown. For my immediate purposes, the interesting number is the reasonably high positive correlation between the provisional grade and the final grade.

30. Yes, but statistics inherently tend to blur the edges of the exceptional cases. How about a graph? Fine; NOMAD has a graphics capability. (This plot is greatly reduced from actual size.) And an exceptional case sticks out: a student who did so-so on the midterm but got a 99 on the final. (I remember her well. She messed up trying to do something creative where the wording of the midterm would have permitted her to copy from her notes, then did a superb job on the final. I was delighted and relieved to be able to give her an A.)

Observe that the "plot" command said nothing about where to start the axes; the graphics package figured that out from the data. Since I didn't specify otherwise, the data points (represented by the letter A) were connected by asterisks in the best approximations to straight lines. (Writing a Fortran program to produce such plots will be your final exam!)

Naturally, the graphics capabilities of a typewriter-based terminal are rather limited. NOMAD knows about high-resolution terminals, color, bar graphs, pie charts, multiple plots on one graph, etc. Likewise, there is considerably more to the statistics package than illustrated here.

NOMAD, and other fourth generation languages, are rapidly gaining acceptance. For the kinds of applications where they are appropriate, they permit reports to be obtained in a small fraction of the time that would be required using procedural languages like Fortran. They make it possible for people without computing or data processing background to do some of their own work, and they also make the expert more productive. Furthermore, they make it possible to respond much more gracefully to changing requirements than programs written in a procedural language do.

This demonstration involved 11 reports of varying complexity, one small statistical analysis, and one graph. Not counting data entry, I ran the whole thing in a little over an hour. Doing the same thing in Fortran would be a matter of at least several days of excruciatingly dull work.

I *still* insist that you did not waste your time learning Fortran, but a procedural language is not always the optimum way to get work done with a computer.

Text processing and formatting

An application of computers that is currently enjoying rapid growth is that of *text processing and formatting.* When this application is the primary purpose of a small computer, the term *word processing* is commonly used.

The basic idea is to let the computer help in the entry, correction, modification, and printing of textual material such as letters, reports, scientific papers, and books. In a sense, word processing involves nothing that can't be done with an ordinary (nonelectronic) typewriter, but with word processing the job is faster and generally produces much better-looking results.

The entry and correction of the textual material is an entirely separate process from the formatting and printing of it. Indeed, although the two parts may be carried out on the same equipment, such as a typewriter-type computer terminal, entirely different equipment may be involved. One common combination involves a CRT-type terminal for the entry, correction, and modification of text,

(29) >bstat provarray finalarray

    * NOMAD2 * Basic Descriptive Statistics *

No.	Variable name	Number of obser	Mean	Standard deviation	Skewness	Kurtosis
1	PROVARRAY	17	87.176	14.271	-0.651	-0.781
2	FINALARRAY	17	79.706	16.143	-0.931	0.318

    Correlation matrix

No.	Variable name	1	2
1	PROVARRAY	1.000	
2	FINALARRAY	0.662	1.000

(30) >plot finalarray heading 'final grade' -
      >   versus provarray heading 'provisional grade before final'

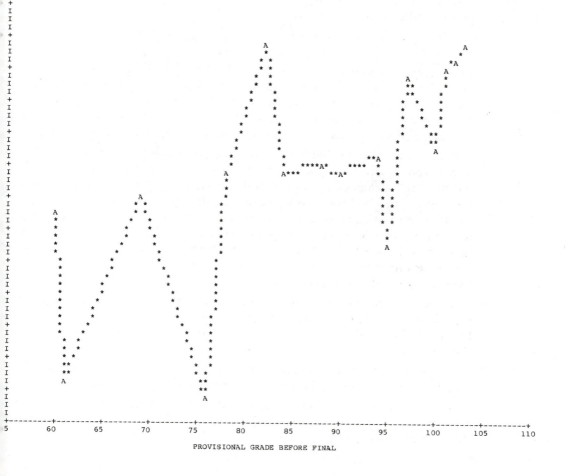

PROVISIONAL GRADE BEFORE FINAL

and a high-quality typewriter-type terminal for the production of the final copy. Alternatively, the output may be produced on a line-at-a-time printer, when speed is more important than appearance, or the output may be sent to a photocomposition device. This book was produced by entering the text with a Diablo terminal, producing high-quality drafts for review and copy editing with the same terminal, then modifying the text somewhat and sending it to a photocomposition device in the form of a reel of magnetic tape.

The approach that is currently most flexible and most efficient involves a CRT terminal in which the entry and editing of the text takes advantage of the fact that the full screen is visible to the operator. This is accordingly called a *full screen editor.* Unfortunately, it is difficult to demonstrate the operation of a full screen editor in a book. Accordingly, the demonstration that follows is based on a *line editor,* another widely used technique, in which the editor deals with one line at a time and always has one line that is considered the *current line.*

There are many systems for processing and formatting textual material on computers. The one demonstrated here is called Script. Like WATFIV, it was developed at the University of Waterloo. It has many similarities with the dozens of other such systems, and some differences. Since it is widely available and widely used, the demonstration that follows will be of direct interest to many readers; for others, the concepts will be valid but the details will differ.

The end product of the demonstration is a letter, the nature of which will become evident as we proceed.

1. At one second after 9:48 on a certain morning, the National CSS timesharing system prints a greater than sign (>) to indicate that it is ready to receive a command. I say that I want to edit a file named "ch10rawl" having a filetype of "script." ("ch10rawl" reminds me, when I look at a listing of all my files, of the letter in Chapter 10 to Nick Rawlings, a friend at National CSS with whom I have worked on a number of projects.) The system discovers that a file with that name does not exist, so it assumes I want to begin by entering text.

2. In Script, any line that begins with a period is a *control word.* ".hy on" means that I want Script to hyphenate long words at the ends of lines in the final output. ".nj" means that I do not want the right margin *justified,* i.e., I do not want spaces inserted between words so that the right ends of lines are aligned. (As they are in this book, for example.) ".nc" means "no concatenate," i.e., do not run the lines of input text together when the output is produced. This is so that the address, which is four lines of input text in this case, will be printed as four lines rather than being run together on one line. Next I say to space three lines, print the date centered, and space five more lines.

3. ".im" means "imbed," that is, get the file named "address" and insert it into this file before formatting. In fact, the file named "address" doesn't exist yet—but it will, by the time I run the letter off. Now I say to space one line before the salutation.

4. ".co" means *concatenate.* From now on, the lines of input text will be joined together, then broken up into lines of the desired length before printing.

5. ".pp" means to start a new paragraph. Unless I specify differently, which I haven't, this means to insert one blank line and indent three spaces to mark the start of a new paragraph.

(1) 09.48.01 >edit ch10rawl script
NEW FILE.
INPUT:

(2) >.hy on
>.nj
>.nc
>.sp 3
>.ce April 7, 1983
>.sp 5

(3) >.im address
>.sp
>Dear Nick:

(4) >.co

(5) >.pp

6. I now type in the text of the letter. I completely ignore the question of where to end the lines so that the letter will "look good." In fact, at this stage I don't even know how long the lines will be in the final letter, because I don't know whether it will fit on one page or not. I make a number of errors in this process. Those that I notice before entering a line by striking the carriage return key, I correct immediately. The printout is a log of the terminal input and output, and since these kinds of errors never actually enter the log file, they are not shown here.

7. The files named "regards" does already exist. It is shown later, at paragraph 14. This is one of a number of files that I need so routinely that I have set them up in my file system. This is immediately followed by the copy list. We shall see below how the file named "regards" handles the spacing and concatenation questions. After this I hit return before entering any text on a line; this tells the editor that I want to get out of data entry mode.

8. Nervous that I could lose all my text if the telephone connection were broken, or if any number of other mishaps occurred, I say to save the file. This editor is set up to assume that I want to return to data entry mode after saving; since that is not true in this case, I enter another *null line,* that is, one with nothing in it.

9. Now I have some errors to correct. I move to the top of the file in preparation for the first correction. All the commands here have abbreviations, which I would use if this were not a demonstration. I would normally type just "t" for "top," "s" for "save," etc.

10. I seem to have crossed wires in my brain, such that I very often type "tion" as "tin." I may as well correct all such instances of that error to begin with. I accordingly say to change the string consisting of the letters "tin" followed by a space, to the string "tion" followed by a space. I enclose the two strings in slashes, and add an asterisk to mean that I want the change made to the current line and all following. Since I am positioned at the top of the file, that means to do it to the entire file. The system shows me all the changed lines, of which there turn out to be two. If by chance there had been any lines in which there was a correct word ending in "tin," I would have had to change them back, of course. The EOF: means "end of file."

11. I notice that "fourth" has been mistyped as "fourther." I say to move up to the line containing the erroneous word. I could also have said "up 30," the line I want being 30 lines up from the current position (I think). But it's silly for me to have to count such things, so I use the *context editing* feature of Script to say, instead, "go up to the line containing the following string." Having gotten there, I make the necessary change.

12. Seeing no other errors at the moment, I file the text. The system remembers the name I used when I said to edit the file, and files it under that name and filetype.

13. Now I need to put Nick's address into the file named "address," which I said I wanted embedded in the letter. I accordingly say to edit such a file, enter the text, and file it.

(6)
```
>This is to thank you most sincerely for your help
>on my book, especially the NOMAD2 material.
>.pp
>I was delighted to end the presentatin with NOMAD.
>The book is basically a Fortran book,
>but I don't think engineers in the next decade
>will actually do much programming,
>in Fortran or any other third generatin language.
>My last chapter is therefore a horizon-broadening
>sampler of the many nonprocedural approaches to
>using computers,
>ending with the material on NOMAD,
>as a leading example of a fourther generation language.
>.pp
>It is a real pleasure to address this letter to you,
>as an old friend,
>as a founder of National CSS,
>and as one of the implementors and
>tireless expositors of NOMAD.
>I am sending copies to some of the many other people,
>now or formerly at CSS,
>to whom I also owe large debts of gratitude for help over
>the years.
>.pp
>Incidentally,
>the photocomposition approach is looking good.
>I have been able to run off drafts of the manuscript for
>reviewers and a final copy for editing at Wiley,
>and now I am about to convert it to magentic tape
>for submission to the firm that will do the computer-based
>photocomposition.
>This method should save time,
>eliminate errors caused by re-keyboarding, and
>reduce the anxiety of proofreading--a task that I actively despise.
>.pp
>Thanks again.
```

(7)
```
>.im regards
>cc: Bill Clarke
       Alan Codkind
>      Hal Feinleib
>      Chris Grejtak
>      Larry Smith
>      May Stinson
       Scot Stoney
       Rob Vonderhaar
>
EDIT:
```

(8)
```
>save
INPUT:
>
EDIT:
```

(9)
```
>top
>change /tin /tion / *
```

(10)
```
I was delighted to end the presentation with NOMAD.
in Fortran or any other third generation language.
EOF:
```

(11)
```
>up /fourther/
as a leading example of a fourther generation language.
>change /fourther/fourth/
as a leading example of a fourth generation language.
```

(12)
```
>file
```

(13)
```
10.00.54 >edit address script
NEW FILE.
INPUT:
>Mr. Nicholas A. Rawlings
>National CSS, Inc.
>187 Danbury Road
>Wilton, CT 06897
>
EDIT:
>file
```

14. I also said that I wanted a file named "regards" embedded in the letter. This one already exists, since I use it frequently in writing letters. Let's see what it is. The first line says to indent 25 places (the plus sign means "to the right"). Then there is a command to space one line. Since this command is on the same line as the command to indent, the two are separated by a semicolon. Now comes the complimentary close, a command to space four lines, and my name. Next I say to move the indentation back to where it was, space one line, and turn off all formatting (".nf"). This last is mostly so that the names in the copy list that follows in the body of the text will not be concatenated.

15. Now it's time to see how the letter looks. I still don't know whether it will fit on one page, there are likely more typing errors, and I want to read it over to see if I like it. I tell CSS to run off the letter, using the command "script" and naming the file. On the assumption that I might want to remove the terminal paper and insert a sheet of my letterhead, Script issues an appropriate comment and waits. Since this is only a draft, I go ahead and print it on the terminal paper.

⑭ 10.01.31 >print regards script

```
    .in +25;.sp
    With best regards,
    .sp 4
    Daniel D. McCracken
    .in -25;.sp;.nf
    DDM:css
    .sp
```

⑮ 10.01.40 >script ch10rawl
   SCRIPT/MASTER 3.5 (03/20/80)
   Load paper; hit return:
   >

                        April 7, 1983

Mr. Nicholas A. Rawlings
National CSS, Inc.
187 Danbury Road
Wilton, CT 06897

Dear Nick:

   This is to thank you most sincerely for your help on my
book, especially the NOMAD2 material.

   I was delighted to end the presentation with NOMAD. The
book is basically a Fortran book, but I don't think engineers in the next decade will actually do much programming,
in Fortran or any other third generation language. My last
chapter is therefore a horizon-broadening sampler of the
many non-procedural approaches to using computers, ending
with the material on NOMAD, as a leading example of a fourth
generation language.

   It is a real pleasure to address this letter to you, as
an old friend, as a founder of National CSS, and as one of
the implementors and tireless expositors of NOMAD. I am
sending copies to some of the many other people, now or formerly at CSS, to whom I also owe large debts of gratitude
for help over the years.

   Incidentally, the photocomposition approach is looking
good. I have been able to run off drafts of the manuscript
for reviewers and a final copy for editing at Wiley, and now
I am about to convert it to magentic tape for submission to
the firm that will do the computer-based photocomposition.
This method should save time, eliminate errors caused by
re-keyboarding, and reduce the anxiety of proofreading--a
task that I actively despise.

   Thanks again.

                    With best regards,

                    Daniel D. McCracken

DDM:css

                                        PAGE 2

cc: Bill Clarke
    Alan Codkind
    Hal Feinleib
    Chris Grejtak
    Larry Smith
    May Stinson
    Scot Stoney
    Rob Vonderhaar

16. Well, it certainly didn't fit on one page, did it? When I say to edit the file, I forget to enter the filetype; people do this so often, apparently, that the CSS systems programmers provided a prompt for the filetype rather than simply rejecting the command.

17. The first thing I want to do is to insert some Script commands to change the layout of the page, so I give the editor command "input" and do it. The first new command (".ll") makes the line length 65 characters, instead of the default of 60. The next says to add three spaces to the normal position of the left margin, which will center the text on my letterhead. (I checked this by laying a sheet of letterhead over the terminal printout of the draft.) The final new command (".pl") sets the page length to 70, overriding the default of 66. The idea here is that I intend to try to get the letter on one page, which will put the copy list near the bottom of the page. I don't want Script to decide to go to a new page just to type the last one or two names.

18. The second sentence of the second paragraph starts in the general form, "The book is a book." I call myself a writer? Let's change the second "book" to "text." Asking to locate the line containing the string "book," I find an earlier instance than the one I want. I repeat the command with "again," which saves retyping it. Having found the one I want, I use the "change" command, this time abbreviated "c."

19. The parallelism in the second paragraph isn't all it could be. I'm trying to contrast Fortran, as a procedural third generation language, with NOMAD, as a nonprocedural fourth generation language. But the text as drafted doesn't note that Fortran is a procedural language. I don't need to tell Nick this, of course, or the people on the copy list—but writers have their pride. Let's insert the word "procedural" in parentheses after the term "third generation."

20. "Magentic"? Missed that one in reading over the raw text. Well, I said I hated proofreading.

21. I really don't care for the paragraph structure here. The third paragraph properly ought to be last, and the "Thanks again" paragraph could then be the last sentence of the final paragraph—thus also saving two lines. I accordingly find the "Thanks again" line, and say with the "putd" command to put it into a temporary buffer maintained by the CSS editor and delete it from this file. The current line becomes the one after the one deleted. I move up one line to the ".pp" command and delete it; this line will no longer be a separate paragraph.

22. Now I move up to the line containing the string "years." which is the last line of what will soon be the final paragraph. I say to "get" the contents of the buffer, which is the "Thanks again" line that I just put there.

(16) 10.03.52 >edit ch10rawl
PLEASE ENTER FILETYPE
>script
EDIT:
>input

(17) INPUT:
>.ll 65
>.ad +3
>.pl 70
>
EDIT:

(18) >locate /book,/
on my book, especially the NOMAD2 material.
>again
The book is basically a Fortran book,
>c /book,/text,/
The book is basically a Fortran text,

(19) >locate /third
in Fortran or any other third generation language.
>c /language/(procedural) language/
in Fortran or any other third generation (procedural) language.

(20) >locate /magen
and now I am about to convert it to magentic tape
>clocate /magne
and now I am about to convert it to magnetic tape
(21) >locate /Thanks
Thanks again.
>putd
>up
.pp
>delete

(22) >up /years.
the years.
>get
Thanks again.

23. Now I move up to the line that begins what will soon be the final paragraph, and then up one more line. This positions me at the ".pp" command. Now I say "put and delete," but do not specify the number of lines. Instead, I say that the range of the command is the current line, down to but not including the line containing the named string. This is the entire paragraph that I want to move. The "putd" command positions me at the line after the last one deleted, which is seen to be what I expected when I print it. Now I locate the line at the end of the paragraph about photocompostion, then move to the next line using the "next" command, just to assure myself that I am where I think I am and that there is not an extraneous ".pp" still there. Things look OK, so I want to move back up to the previous line. There is no real need to see it again, however, so I place a period after the command. This inhibits printing the line. Now I say "get," which retrieves the entire paragraph; the editor prints the last line of the buffer.

The process of moving blocks of text is much easier with a full screen editor. In one system, it is necessary only to place markers at the beginning and end of the block to be moved and another in the line after which it should be inserted, and press one key.

24. Finally, to get the letter down to one page, I'm going to have to give up on putting each name of the copy list on a separate line. I locate the start of the list, delete all eight lines, enter input mode, re-type them as two lines, get out of input mode, and file the letter as now corrected.

(23)
```
>up /It is
It is a real pleasure to address this letter to you,
>up
.pp
>putd /.pp
>print
.pp
>locate /despise
reduce the anxiety of proofreading--a task that I actively despise.
>next
.im regards
>up.
>get
Thanks again.
```

(24)
```
>locate /cc:
cc: Bill Clarke
>de 8
EOF:
>input
INPUT:
>cc: Bill Clarke, Alan Codkind, Hal Feinleib, Chris Grejtak,
>    Larry Smith, May Stinson, Scot Stoney, Rob Vonderhaar
>
EDIT:
>file
```

25. Now I say to format and print the file, and this time I use my letterhead.

26. To print the envelop, I need only say "script env," since I have a file of that name that is shown here. The file merely positions the address at the proper place, and specifies a page length of 40 so that when the printing is finished the system won't waste the time to space to the end of a default-length page of 66 lines.

You have seen perhaps 10% of the capability of Script. Some of the features not illustrated here:

■ Page numbering, with or without other heading information at the top or bottom of each page.

■ Footnotes, with Script deciding when to break off regular text in order to have enough room for the footnote.

■ Various controls over hyphenation, including the generation of a dictionary of words to be hyphenated according to the user's wishes.

■ The ability to print text in two columns.

■ Macro instructions, with which Script commands may be grouped into small files that are invoked with a single command named by the user. This command can call for operands to be substituted into the file, in a manner closely resembling the substitution of arguments in Fortran subroutines.

■ Various page formatting options, such as centering a block of text, a technique called the hanging indent, or placing a box around a block of text.

■ The creation of indexes. The terms to be included are marked in the raw text; Script creates the index after it has formatted the material into pages.

■ What amounts to a small programming language, with which the user can take a much greater degree of control over many matters of formatting that otherwise are handled in some standard manner or which cannot be done at all.

Text processing and formatting are important applications of computers. It is to be hoped that within not too many years, most college students will have access to such facilities, for writing papers for courses. (Not just *computing* courses—English, history, etc.) Once the chosen system has been learned, the writing is faster, there is far greater ease in doing such things as moving blocks of text to improve structure and "flow," and the finished product is far neater and easier to read.

As it happens, I don't have a secretary these days. For the most part, this is no handicap whatsoever. When I can produce a letter-perfect one-page letter in 20 minutes, why should I go through the process of dictating a letter, waiting for a draft, marking up or dictating the changes, waiting for the final draft, and then agonizing whether the likely mistakes or the second thoughts about content and wording are serious enough to justify asking for another draft? Studies indicate that the average business letter is typed four or five times before it is acceptable, and costs upwards of $10.

Word processing is here to stay. It is powerful in the hands of secretaries, too, of course, but you should be taking advantage of it yourself, right now.

(25) 10.05.08 >script ch10rawl
       SCRIPT/MASTER 3.5 (03/20/80)
       Load paper; hit return:
       >

(26) 10.07.35 >print env script

       .sp 2
       .nf
       .ad +33
       .im address
       .pl 40

       EOF:
       >

DANIEL D. McCRACKEN
160 CABRINI BLVD.
NEW YORK, NY 10033

April 7, 1983

Mr. Nicholas A. Rawlings
National CSS, Inc.
187 Danbury Road
Wilton, CT 06897

Dear Nick:

   This is to thank you most sincerely for your help on my book, especially the NOMAD2 material.

   I was delighted to end the presentation with NOMAD. The book is basically a Fortran text, but I don't think engineers in the next decade will actually do much programming, in Fortran or any other third generation (procedural) language. My last chapter therefore a horizon-broadening sampler of the many non-procedural approaches to using computers, ending with the material on NOMAD as a leading example of a fourth generation language.

   Incidentally, the photocomposition approach is looking good. I have been able to run off drafts of the manuscript for reviewers and a final copy for editing at Wiley, and now I am about to convert it to magnetic tape for submission to the firm that will do the computer-based photocomposition. This method should save time, eliminate errors caused by re-keyboarding, and reduce the anxiety of proofreading--a task that I actively despise.

   It is a real pleasure to address this letter to you, as an old friend, as a founder of National CSS, and as one of the implementors and tireless expositors of NOMAD. I am sending copies to some of the many other people, now or formerly at CSS, to whom also owe large debts of gratitude for help over the years. Thanks again.

                    With best regards,

                    Daniel D. McCracken

DDM:css

cc: Bill Clarke, Alan Codkind, Hal Feinleib, Chris Grejtak,
    Larry Smith, May Stinson, Scot Stoney, Rob Vonderhaar

APPENDIX

FORTRAN INTRINSIC FUNCTIONS

The following table lists all of the functions that are required to be supplied as part of a Fortran system. In the text these were called *built-in* or *supplied* functions; the more formal term is *intrinsic function*.

The *generic names* shown in the table permit the same function name to be used with more than one type of argument. When a generic name is used to reference an intrinsic function, the type of the result is the same as the type of the argument, except for the intrinsic functions that perform type conversion, nearest integer, and absolute value with a complex argument. Generic names have not been used in this book.

This material is reproduced with permission from American National Standard ANSI X3.9-1978, *Programming Language FORTRAN*, copyright 1978 by the American National Standards Institute. Copies of this standard may be purchased from the American National Standards Institute at 1430 Broadway, New York, N.Y. 10018.

Intrinsic Function	Definition	Number of Arguments	Generic Name	Specific Name	Type of Argument	Type of Function
Type conversion	Conversion to integer int (*a*) See Note 1	1	INT	—	Integer	Integer
				INT	Real	Integer
				IFIX	Real	Integer
				IDINT	Double	Integer
				—	Complex	Integer
	Conversion to real See Note 2	1	REAL	REAL	Integer	Real
				FLOAT	Integer	Real
				—	Real	Real
				SNGL	Double	Real
				—	Complex	Real
	Conversion to double See Note 3	1	DBLE	—	Integer	Double
				—	Real	Double
				—	Double	Double
				—	Complex	Double

Intrinsic Function	Definition	Number of Arguments	Generic Name	Specific Name	Type of Argument	Type of Function
	Conversion to complex See Note 4	1 or 2	CMPLX	— — — —	Integer Real Double Complex	Complex Complex Complex Complex
	Conversion to integer See Note 5	1		ICHAR	Character	Integer
	Conversion to character See Note 5	1		CHAR	Integer	Character
Truncation	int (a) See Note 1	1	AINT	AINT DINT	Real Double	Real Double
Nearest whole number	int$(a + .5)$ if $a \geq 0$ int$(a - .5)$ if $a < 0$	1	ANINT	ANINT DNINT	Real Double	Real Double
Nearest integer	int$(a + .5)$ if $a \geq 0$ int$(a - .5)$ if $a < 0$	1	NINT	NINT IDNINT	Real Double	Integer Integer
Absolute value	$\lvert a \rvert$ See Note 6 $(ar^2 + ai^2)^{1/2}$	1	ABS	IABS ABS DABS CABS	Integer Real Double Complex	Integer Real Double Real
Remaindering	$a_1 -$ int$(a_1/a_2)*a_2$ See Note 1	2	MOD	MOD AMOD DMOD	Integer Real Double	Integer Real Double
Transfer of sign	$\lvert a_1 \rvert$ if $a_2 \geq 0$ $-\lvert a_1 \rvert$ if $a_2 < 0$	2	SIGN	ISIGN SIGN DSIGN	Integer Real Double	Integer Real Double
Positive difference	$a_1 - a_2$ if $a_1 > a_2$ 0 if $a_1 \leq a_2$	2	DIM	IDIM DIM DDIM	Integer Real Double	Integer Real Double
Double precision product	a_1*a_2	2		DPROD	Real	Double
Choosing largest value	max$(a_1, a_2, \ldots)$	≥ 2	MAX	MAX0 AMAX1 DMAX1	Integer Real Double	Integer Real Double
				AMAX0 MAX1	Integer Real	Real Integer

Intrinsic Function	Definition	Number of Arguments	Generic Name	Specific Name	Type of Argument	Type of Function
Choosing smallest value	$min(a_1, a_2, \ldots)$	≥ 2	MIN	MIN0	Integer	Integer
				AMIN1	Real	Real
				DMIN1	Double	Double
				AMIN0	Integer	Real
				MIN1	Real	Integer
Length	Length of character entity	1		LEN	Character	Integer
Index of a substring	Location of substring a_2 in string a_1 See Note 10	2		INDEX	Character	Integer
Imaginary part of complex argument	ai See Note 6	1		AIMAG	Complex	Real
Conjugate of a complex argument	$(ar, -ai)$ See Note 6	1		CONJG	Complex	Complex
Square Root	$(a)^{1/a}$	1	SORT	SORT	Real	Real
				DSORT	Double	Double
				CSORT	Complex	Complex
Exponential	$e**a$	1	EXP	EXP	Real	Real
				DEXP	Double	Double
				CEXP	Complex	Complex
Natural logarithm	$log(a)$	1	LOG	ALOG	Real	Real
				DLOG	Double	Double
				CLOG	Complex	Complex
Common logarithm	$log_{10}(a)$	1	LOG10	ALOG10	Real	Real
				DLOG10	Double	Double
Sine	$sin(a)$	1	SIN	SIN	Real	Real
				DSIN	Double	Double
				CSIN	Complex	Complex
Cosine	$cos(a)$	1	COS	COS	Real	Real
				DCOS	Double	Double
				CCOS	Complex	Complex
Tangent	$tan(a)$	1	TAN	TAN	Real	Real
				DTAN	Double	Double
Arcsine	$arcsin(a)$	1	ASIN	ASIN	Real	Real
				DASIN	Double	Double

Intrinsic Function	Definition	Number of Arguments	Generic Name	Specific Name	Type of Argument	Type of Function
Arccosine	$\arccos(a)$	1	ACOS	ACOS	Real	Real
				DACOS	Double	Double
Arctangent	$\arctan(a)$	1	ATAN	ATAN	Real	Real
				DATAN	Double	Double
	$\arctan(a_1/a_2)$	2	ATAN2	ATAN2	Real	Real
				DATAN2	Double	Double
Hyperbolic sine	$\sinh(a)$	1	SINH	SINH	Real	Real
				DSINH	Double	Double
Hyperbolic cosine	$\cosh(a)$	1	COSH	COSH	Real	Real
				DCOSH	Double	Double
Hyperbolic tangent	$\tanh(a)$	1	TANH	TANH	Real	Real
				DTANH	Double	Double
Lexically greater than or equal	$a_1 \geq a_2$ See Note 12	2		LGE	Character	Logical
Lexically greater than	$a_1 > a_2$ See Note 12	2		LGT	Character	Logical
Lexically less than or equal	$a_1 \leq a_2$ See Note 12	2		LLE	Character	Logical
Lexically less than	$a_1 < a_2$ See Note 12	2		LLT	Character	Logical

Notes:

   1. For a of type integer, $\text{int}(a) = a$. For a of type real or double precision, there are two cases: if $|a| < 1$, $\text{int}(a) = 0$; if $|a| \geq 1$, $\text{int}(a)$ is the integer whose magnitude is the largest integer that does not exceed the magnitude of a and whose sign is the same as the sign of a. For example,

   $$\text{int}(-3.7) = -3$$

   For a of type complex, $\text{int}(a)$ is the value obtained by applying the above rule to the real part of a.

   For a of type real, IFIX(a) is the same as INT(a).

   2. For a of type real, REAL(a) is a. For a of type integer or double precision, REAL(a) is as much precision of the significant part of a as a real datum can contain. For a of type complex, REAL(a) is the real part of a.

   For a of type integer, FLOAT(a) is the same as REAL(a).

   3. For a of type double precision, DBLE(a) is a. For a of type integer or real, DBLE(a) is as much precision of the significant part of a as a double precision datum can contain. For a of type complex, DBLE(a) is as much precision of the significant part of the real part of a as a double precision datum can contain.

   4. CMPLX may have one or two arguments. If there is one argument, it may be of type integer, real, double precision, or complex. If there are two arguments, they must both be of the same type and may be of type integer, real, or double precision.

For a of type complex, CMPLX(a) is a. For a of type integer, real, or double precision, CMPLX(a) is the complex value whose real part is REAL(a) and whose imaginary part is zero.

CMPLX(a_1, a_2) is the complex value whose real part is REAL(a_1) and whose imaginary part is REAL(a_2).

5. ICHAR provides a means of converting from a character to an integer, based on the position of the character in the processor collating sequence. The first character in the collating sequence corresponds to position 0 and the last to position $n - 1$, where n is the number of characters in the collating sequence.

The value of ICHAR(a) is an integer in the range: $0 \leq$ ICHAR(a) $\leq n - 1$, where a is an argument of type character of length one. The value of a must be a character capable of representation in the processor. The position of that character in the collating sequence is the value of ICHAR.

For any characters c_1 and c_2 capable of representation in the processor, (c_1 .LE. c_2) is true if and only if (ICHAR(c_1) .LE. ICHAR(c_2)) is true, and (c_1 .EQ. c_2) is true if and only if (ICHAR(c_1) .EQ. ICHAR(c_2)) is true.

CHAR(i) returns the character in the ith position of the processor collating sequence. The value is of type character of length one. i must be an integer expression whose value must be in the range $0 \leq i \leq n - 1$.

ICHAR(CHAR(i)) = i for $0 \leq i \leq n - 1$.
CHAR(ICHAR(c)) = c for any character c capable of representation in the processor.

6. A complex value is expressed as an ordered pair of reals, (ar, ai), where ar is the real part and ai is the imaginary part.

7. All angles are expressed in radians.

8. The result of a function of type complex is the principal value.

9. All arguments in an intrinsic function reference must be of the same type.

10. INDEX(a_1, a_2) returns an integer value indicating the starting position within the character string a_1 of a substring identical to string a_2. If a_2 occurs more than once in a_1, the starting position of the first occurence is returned.

If a_2 does not occur in a_1, the value zero is returned. Note that zero is returned if LEN(a_1) < LEN(a_2).

11. The value of the argument of the LEN function need not be defined at the time the function reference is executed.

12. LGE(a_1, a_2) returns the value true if $a_1 = a_2$ or if a_1 follows a_2 in the collating sequence described in American National Standard Code for Information Interchange, ANSI X3.4-1977 (ASCII), and otherwise returns the value false.

LGT(a_1, a_2) returns the value true if a_1 follows a_2 in the collating sequence described in ANSI X3.4-1977 (ASCII), and otherwise returns the value false.

LLE(a_1, a_2) returns the value true if $a_1 = a_2$ or if a_1 precedes a_2 in the collating sequence described in ANSI X3.4-1977 (ASCII), and otherwise returns the value false.

LLT(a_1, a_2) returns the value true if a_1 precedes a_2 in the collating sequence described in ANSI X3.4-1977 (ASCII), and otherwise returns the value false.

If the operands for LGE, LGT, LLE, and LLT are of unequal length, the shorter operand is considered as if it were extended on the right with blanks to the length of the longer operand.

If either of the character entities being compared contains a character that is not in the ASCII character set, the result is processor-dependent.

Restrictions on range of arguments and results

Restrictions on the range of arguments and results for intrinsic functions when referenced by their specific names are as follows:

1. *Remaindering.* The result for MOD, AMOD, and DMOD is undefined when the value of the second argument is zero.

2. *Transfer of Sign.* If the value of the first argument of ISIGN, SIGN, or DSIGN is zero, the result is zero, which is neither positive or negative (4.1.3).

3. *Square Root.* The value of the argument of SORT and DSORT must be greater than or equal to zero. The result of CSORT is the principal value with the real part greater than or equal to zero. When the real part of the result is zero, the imaginary part is greater than or equal to zero.

4. *Logarithms.* The value of the argument of ALOG, DLOG, ALOG10, and DLOG10 must be greater than zero. The value of the argument of CLOG must not be (0.,0.). The range of the imaginary part of the result of CLOG is: $-\pi <$ imaginary part $\leq \pi$. The imaginary part of the result is π only when the real part of the argument is less than zero and the imaginary part of the argument is zero.

5. *Sine, Cosine, and Tangent.* The absolute value of the argument of SIN, DSIN, COS, DCOS, TAN, and DTAN is not restricted to be less than 2π.

6. *Arcsine.* The absolute value of the argument of ASIN and DASIN must be less than or equal to one. The range of the result is: $-\pi/2 \leq$ result $\leq \pi/2$.

7. *Arccosine.* The absolute value of the argument of ACOS and DACOS must be less than or equal to one. The range of the result is: $0 \leq$ result $\leq \pi$.

8. *Arctangent.* The range of the result for ATAN and DATAN is: $-\pi/2 \leq$ result $\leq \pi/2$. If the value of the first argument of ATAN2 or DATAN2 is positive, the result is positive. If the value of the first argument is zero, the result is zero if the second argument is positive and π if the second argument is negative. If the value of the first argument is negative, the result is negative. If the value of the second argument is zero, the absolute value of the second argument is zero, the absolute value of the result is $\pi/2$. The arguments must not both have the value zero. The range of the result for ATAN2 and DATAN2 is: $-\pi <$ results $\leq \pi$.

The above restrictions on arguments and results also apply to the intrinsic functions when referenced by their generic names.

ANSWERS TO STARRED EXERCISES

Chapter two, page 19

1. For each of these, there are a great many possible correct answers; the following are some of the obvious possibilities.

```
256.0
2.56
-43000.0    or  -4.3E4
1E12        or  1.0E12
4.92E-7
-10.        or  -10.0
1E-16
```

3.
87,654.3	comma not permitted
+987	must have decimal point or exponent
9.2E+98	exponent too large in most systems
7E-9.3	decimal point not allowed in exponent

5. Yes.

6.
+234.	decimal point not permitted
23,400	comma not permitted
1E12	exponent not permitted

Chapter two, page 21

1. Real: G, GAMMA, BTO7TH, ZCUBED, B6700, DELTA, EPSLON, ALEPH, DAYAN, CHAYA, A1P4, AONEP4, ALGOL, ADA, PASCAL. Integer: I, IJK, J79, LARGE, IBM370, LAMBDA, NUN.

   Unacceptable:

GAMMA421	too many characters
J79-1	- not permitted
R(2)19	parentheses not permitted
ZSQUARED	too many characters
12AT7	does not begin with a letter
2N173	does not begin with a letter
CDC6600	too many characters
S/370	/ not permitted
EPSILON	too many characters
L'CHAYIM	' not permitted
A1.4	decimal point not permitted
A1POINT4	too many characters
FORTRAN	too many characters
PL/I	/ not permitted

Chapter two, page 26

1. a. $X + Y**3$
   d. $A + B / C$
   f. $A + B / (C + D)$
   h. $((A + B) / (C + D))**2 + X**2$
   j. $1.0 + X + X**2/2.0 + X**3/6.0$
   k. $(X/Y)**(G-1.0)$

2. For parts, b, e, and g, the expressions shown in the exercise are syntactically legal but do not represent the mathematical expressions shown. The expression shown for part j is syntactically illegal, for lack of two asterisks and a right parenthesis.
   b. $(X + 2.0) / (Y + 4.0)$
   e. $((X+A+3.141593)/(2.*Z))**2$
   g. $-2.0**(R-1.0)$
   j. $A+X*(B+X*(C+D*X))$

Chapter two, page 30

1. a. 13.0 real

   b. 0.0 real

   e. 4 integer

   f. 4.0 real

   k. 1.333332 (approximately) real

   n. 8.000000 (possibly not exact) real

   o. 5 integer

3. a. `DELTA = BETA + 2.0`

   c. `C = (A**2 + B**2)**0.5`
       (or use the square root function introduced later in the chapter)

   d. `R = 2.0**0.5`

   g. `Y = (2.0*X)**2 + (X/2.0)**0.5`

   h. `G = G + 2.0`

Chapter two, page 39

1. a. `AREA = 2.0 * P * R * SIN(3.141593/P)`

   c. `ARC = 2.0 * SQRT(Y**2 + 4.0*X**2/3.0)`

   e. `S = - COS(X)**(P+1.0) / (P + 1.0)`

   f. `G = 0.5 * ALOG((1.0 + SIN(X))/(1.0 - SIN(X)))`

   i. `E = X * ATAN(X/A) - A/2.0 * ALOG(A**2 + X**2)`

   l. `Q = (2.0/(3.1416*X))**0.5 * SIN(X)`

   n. `Y = (2.0*3.141593)**0.5 * X**(X+1.0) * EXP(-X)`

4.  ```
    S = (A + B + C) / 2.0
    AREA = SQRT(S*(S-A)*(S-B)*(S-C))
    ```

Chapter two, page 42

1.  ```
    REAL A, B, C, F
    READ *, A, B, C
    F = (1.0 + A) / (1.0 + B/(C + 6.0))
    PRINT *, A, B, C, F
    END
    ```

```
4.  REAL A, B, C, X1, X2
    READ *, A, B, C
    X1 = (-B + SQRT(B**2 - 4.0*A*C)) / (2.0*A)
    X2 = (-B - SQRT(B**2 - 4.0*A*C)) / (2.0*A)
    PRINT *, A, B, C, X1, X2
    END
```

```
6.  REAL A, E, H, P, X
    READ *, A, E, H, P
    X = (E*H*P) / (SIN(A) * (H**4/16.0 + H**2*P**2))
    PRINT *, A, E, H, P, X
    END
```

```
7.  REAL A, X, S, Y, Z
    READ *, A, X, S
    Y = SQRT(X**2 - A**2)
    Z = X * S / 2.0 - A**2 / 2.0 * ALOG(ABS(X + S))
    PRINT *, A, X, S, Y, Z
    END
```

Chapter three, page 50

1. a. The program reads values for the variables named A and B from the standard input device, usually a terminal or a card reader. If the value of A is greater than that of B, it prints the THE FIRST NUMBER IS LARGER; if the value of A is less than *or equal to* B; it prints THE SECOND NUMBER IS LARGER.

   c. The program reads values for three integer variables named I, J, and K. If the sum of the squares of I and J is equal to the square of K, it prints THE NUMBERS DEFINE A RIGHT TRIANGLE; otherwise, it prints THE NUMBERS DO NOT DEFINE A RIGHT TRIANGLE. Note that it prints the latter message for combinations like (3, 5, 4) and (5, 4, 3), which *do* define right triangles.

   e. The program reads values for two real variables U and V. It then reverses the sign of a negative value for either variable and prints the message shown and the sum of the two (possibly modified) variables. This sum will be the sum of the absolute values of the numbers as read. A simpler way to accomplish the same result would be to use the single statement:

   ```
   PRINT *, ' THE SUM OF THE ABSOLUTE VALUES IS ', ABS(U) + ABS(V)
   ```

2. a. The equal sign is not a valid relational operator; presumably .EQ. was intended.

   c. The line beginning with IF will not be recognized as an IF statement, for lack of parentheses around the relational expression.

4. a.

```
READ values for X, Y, and R

IF X**2 + Y**2 < R**2 point is in the circle


* Exercise 4a, page 53
      REAL X, Y, R

      READ *, X, Y, R

      IF (X**2 + Y**2 .LT. R**2) THEN
          PRINT *, ' THE POINT ', X, Y
          PRINT *, ' LIES WITHIN A CIRCLE OF RADIUS ', R
      ELSE
          PRINT *, ' THE POINT ', X, Y
          PRINT *, ' DOES NOT LIE WITHIN A CIRCLE OF RADIUS ', R
      END IF

      END
```

The exercise does not really specify what is to be done after determining whether the point lies within the circle. Many other ways of reporting the result of the test would be acceptable.

c.

```
READ earnings

IF earnings are over $100 THEN
    tax is 0.02 * (earnings - 100)
ELSE
    tax is zero
END IF

PRINT earnings and tax


* Exercise 4c, page 53
      REAL EARN, TAX

      READ *, EARN

      IF ( EARN .GT. 100.00 ) THEN
          TAX = 0.02 * (EARN - 100.0)
      ELSE
          TAX = 0.0
      END IF

      PRINT *, EARN, TAX

      END
```

e.

```
READ X, Y, R, S, D

IF ( SQRT((X - R)**2 + (Y - S)**2) .LE. D )THEN
     the two points are within D of each other


*  Exercise 4e, page 53
        REAL X, Y, R, S, D

        READ *, X, Y, R, S, D

        IF ( SQRT((X - R)**2 + (Y - S)**2) .LE. D ) THEN
            PRINT *, ' POINTS ARE WITHIN ', D, ' OF EACH OTHER'
        END IF

        END
```

Problem statement was not explicit about what to report; other answers would be acceptable.

Chapter three, page 62

1. a. The program prints the sum of the integers from 1 to 10, which is 55.

   c. The program reads the value of an integer N, then prints the squares of the odd integers not greater than N.

   e. The program prints the angles from 0 to 180 degrees, in steps of 10 degrees, each with its sine. The angles in degrees are converted to radians within the expression for the argument of the sine function.

2. a. Statement 20 is defined but not used; the GO TO 10 refers to a statement number that does not exist. I is never initialized.

   c. The program is syntactically correct, but no value is given to X before entering the loop. WATFIV would report this fact when first attempting to execute the IF statement and would not proceed. Programs from most other compilers would execute, starting X at whatever value happened to be in the location assigned to it.

3. Many ways are possible; the simplest would be to increment TENRAT by 5 instead of 1.

5.

```
* Exercise 5, page 65

      REAL A

      READ *, A

      IF ( A .EQ. 0.0 ) THEN
         PRINT *, ' FORMULA NOT VALID FOR A = 0'
         STOP
      END IF
      IF ( A .EQ. 1.0 ) THEN
         PRINT *, ' FORMULA NOT VALID FOR A = 1'
         STOP
      END IF

      IF ( A .LT. 1.0 ) THEN
         PRINT *, ' INTEGRAL = ', 3.141593/A * ALOG(1.0 + A)
      ELSE
         PRINT *, ' INTEGRAL = ', 3.141593*ALOG(1.0 + 1.0/A)
      END IF

      END
```

The natural logarithm is needed, so we write ALOG, not ALOG10.

7.

```
MU = 15
LAMBDA = 0
WHILE LAMBDA <= 14
   Compute and print average queue length
   LAMBDA = LAMBDA + 1
END WHILE


* Queue length calculation for Exercise 7, page 65
* Variables:
*    LAMBDA:  Mean arrival rate of customers at service window
*    MU:      Mean service rate
*    N:       Average queue length

      REAL MU, LAMBDA, N

      MU = 15

      LAMBDA = 0
***** WHILE LAMBDA <= 14
   10 CONTINUE
      IF ( LAMBDA .LE. 14 ) THEN

         N = (LAMBDA/MU) / (1.0 - LAMBDA/MU)
         PRINT *, LAMBDA, N
         LAMBDA = LAMBDA + 1

      GO TO 10
      END IF
***** END WHILE

      END
```

9.

```
READ N

I = 1
SUM = 0

WHILE  I <= N
    Add I to SUM
    Add 1 to I
END WHILE

PRINT SUM and formula value
```

```
* Sum of first N integers, for Exercise 9, page 66
* Variables:
*      N:        Limit in summation
*      I:        Summation index variable
*      SUM:      At completion, = sum of first N integers

       INTEGER N, I, SUM

       READ *, N

       I = 1
       SUM = 0

*****  WHILE I <= N
  10   CONTINUE
       IF ( I .LE. N ) THEN

           SUM = SUM + I
           I = I + 1

       GO TO 10
       END IF
*****  END WHILE

       PRINT *, ' SUM OF 1ST ', N, ' INTEGERS BY ADDITION = ', SUM
       PRINT *, ' SUM OF 1ST ', N, ' INTEGERS BY FORMULA  = ', N*(N+1)/

       END
```

11.

```
X = -300
WHILE X <= +300
   N = 1
   SERIES = 0
   WHILE N <= 99
      SERIES = SERIES + SIN(NX)/N
      Add 2 to N
   END WHILE
   PRINT X, (4/PI) * SERIES
   Add 20 to X
END WHILE
```

```
* Fourier series for square wave, Exercise 11, page 66
* Variables:
*      X:       Angle, in degrees
*      N:       Variable in series terms
*      SERIES: Value of series

       REAL X, N, SERIES

       X = -300.00

*****  WHILE X <= +300
   10  CONTINUE
       IF ( X .LE. 300.0 ) THEN

           SERIES = 0.0
           N = 1

*****      WHILE N <= 99
   20      CONTINUE
           IF ( N .LE. 99 ) THEN

               SERIES = SERIES + SIN(N * X / (180.0/3.141593)) / N
               N = N + 2

           GO TO 20
           END IF
*****      END WHILE

           PRINT *, X, (4.0/3.141593) * SERIES
           X = X + 20.0

       GO TO 10
       END IF
*****  END WHILE

       END
```

13. The output:

```
   3    13
   4    17
   5    22
   6    28
   7    35
   8    43
```

Chapter three, page 79

1. a. Same truth value: the extra parentheses in the second line surround an operator (.AND.) of high precedence.

  c. Same truth value: the extra parentheses are around .AND.s.

  e. Same truth value: in the second expression the .NOT. has been *distributed* over the subexpressions inside the parentheses that appear in the first expression. When this is done, and .AND.s are changed to .OR.s and vice versa, the truth value does not change. (This is a simple example of *De Morgan's law,* a part of Boolean algebra.)

  g. Same truth value, namely, false. 1.0 is never equal to 2.0 and, since that test is ANDed with other tests in the first expression, that expression is always false regardless of the values of the variables. In the second expression, A cannot at the same time be greater than B and *not* greater than B, so ANDing those two relations also produces false.

2. a. No periods around "AND". The compiler diagnostic might be quite different, however, since it would simply see BANDC as a variable name.

  c. No parentheses around logical expression.

3. a. Remove the .NOT.. As written, the expression returns .TRUE. for N greater than or equal to 18.

  c. Either remove the .NOT. or change the .NE. to .EQ.. As written, the expression is true only for J = 18.

4.

```
*  Exercise 4, page 80
*  Can A, B, and C be the sides of a triangle?

      REAL A, B, C

      READ *, A, B, C

      IF  (       (A .GE. B+C)
     $      .OR. (B .GE. A+C)
     $      .OR. (C .GE. A+B)  ) THEN
          PRINT *, ' NO'
      ELSE
          PRINT *, ' YES'
      END IF

      END
```

6. Change the three statements that initialize, test, and modify X. All the changes are toward the end of the program.

```
* Start X at right end of range, then work to left
      X = RIGHT

***** WHILE X >= LEFT
   10  CONTINUE
       IF ( X .GE. LEFT ) THEN

           POLY = 2.0*X**4 - 15.0*X**3 - 2.0*X**2 + 120.0*X - 130.0
           PRINT *, X, POLY
           X = X - DELTAX

       GO TO 10
       END IF
***** END WHILE

       END
```

9.

a.

```
D = 1
SERIES = 0
WHILE D <=  11
    SERIES = SERIES +1/D**6
    D = D + 1
END WHILE
PRINT series value, formula value
```

```
* Exercise 9a, page 81
* A series that sums to PI**6/945

       REAL D, SERIES

       D = 1.0
       SERIES = 0.0

***** WHILE D <= 11
   10  CONTINUE
       IF ( D .LE. 11 ) THEN

           SERIES = SERIES + 1.0/D**6
           D = D + 1

       GO TO 10
       END IF
***** END WHILE

       PRINT *, SERIES, 3.141593**6/945.0

       END
```

c.

```
D = 2
SUM = 0
WHILE D <= 34
    SUM = SUM + 1/((D-1)**2 * (D+1)**2)
    D = D + 2
END WHILE
PRINT series value, formula value



* Exercise 9c, page 81
* A series that sums to (PI**2 - 8)/16

      REAL SUM, D

      SUM = 0.0
      D = 2.0


***** WHILE D <= 34
   10 CONTINUE
      IF ( D .LE. 34.0 ) THEN

          SUM = SUM + 1.0 / ((D-1.0)**2 * (D+1.0)**2)
          D = D + 2.0

      GO TO 10
      END IF
***** END WHILE

      PRINT *, SUM, (3.141593**2 - 8.0) / 16.0

      END
```

17.

```
N = 1
SUM = 0
WHILE N <= 10
    READ a value
    Add it to SUM
    Add 1 to N
END WHILE
PRINT SUM/10
```

```
* Exercise 17, page 85
* Finding the average of 10 numbers read as data

         REAL SUM, X
         INTEGER N

         SUM = 0.0
         N = 1

*****  WHILE N <= 10
   10    CONTINUE
         IF ( N .LE. 10 ) THEN

            READ *, X
            SUM = SUM + X
            N = N + 1

         GO TO 10
         END IF
*****  END WHILE

         PRINT *, ' THE AVERAGE IS: ', SUM/10.0

         END
```

Chapter four, page 92

1. a. The result would be the same as before. The way shown in the exercise is very slightly faster, since it avoids computing the expression for X twice, but seldom will such small speed differences be important.

   b. The result would be the same as before, but careful analysis is necessary to be sure. In STEPER = 0.0, the real constant would be converted to integer; in the relational expression in the IF statement, the comparison of an integer and a real causes no problems; the mixing of real and integer quantities in the PRINT statement is legal and causes no problems; the incrementing of STEPER would work. However, such a simple change as incrementing STEPER by 0.5 instead of 1.0, would cause a program that had been running "correctly" to go into an infinite loop! Mixed-mode arithmetic should be used only when there is a specific reason to do so.

   c. The function would be evaluated for X values from LEFT to RIGHT in steps of 0.01*DELTAX. There would 100 times as much output.

   d. The program would work as before, except that there would be one final line giving the value of the polynomial at X = 7.5. This would be printed, of course, immediately following the last line of the table of values, with an inappropriate heading. The result would be confusing.

   e. The results would be completely unchanged. The variable named STEPER in the main program would not be affected in any way by the existence of a variable of the same name in the subprogram.

   f. The compiler would produce a diagnostic error: a Fortran FUNCTION is not permitted to call itself.

   g. The program would operate as before.

2. The FUNCTION MID returns the arithmetic mean ("midpoint") of its two arguments. The first PRINT would produce 2.5. With A having been given the value 2.0, the second PRINT would produce 3.5. The mean of 2.0 and 16.0 is 9.0, the output of the third print. The fourth PRINT has the FUNCTION MID as one of its arguments; this invocation would return 3.0, which would then be sent to the FUNCTION along with 7.0. The mean of 3.0 and 7.0 is 5.0. The last PRINT has an argument with an undefined value: Y is still 16.0, but no value has ever been given to X. WATFIV would flag the error at execution time. My compiler raised no objection and, at execution, simply used whatever value happened to be in the location assigned to X, which evidently was zero. Unless you have a system that explicitly initializes all variables to zero, this last PRINT would give different results on different runnings, depending on what had been left in the location by previous computer operations.

4.

```
* Exercise 4, page 93
      REAL FUNCTION Y(X)
      REAL X
      IF ( X .LT. 0.0 ) THEN
         Y = 1.0 + SQRT(1.0 + X**2)
      END IF
      IF ( X .EQ. 0.0 ) THEN
         Y = 0.0
      END IF
      IF ( X .GT. 0.0 ) THEN
         Y = 1.0 - SQRT(1.0 + X**2)
      END IF

      RETURN

      END
```

The "main program" has to be fragmentary, since there were no specifications for obtaining values for the variables or what to do with the results.

```
REAL F, G, H, A, Z, X1, X2, X
REAL Y

F = 2.0 + Y(A + Z)
G = (Y(X1) + Y(X2)) / 2.0
H = Y(COS(2.0*3.14159*X)) + SQRT(1.0 + Y(2.0*3.14159*X))
```

6.

```
*  Exercise 6, page 94
      REAL X, ESINX

      X = 0.0

*****  WHILE X <= 0.1
   10  CONTINUE
      IF ( X .LE. 0.1 ) THEN

          PRINT *, X, ESINX(X), EXP(SIN(X))
          X = X + 0.01

      GO TO 10
      END IF
*****  END WHILE

      END

      REAL FUNCTION ESINX(X)
      REAL X
      ESINX = 1.0 + X + X**2/2.0 - X**4/8.0 - X**5/15.0
      RETURN
      END

0.000000000E+00    1.00000000        1.00000000
0.100000016E-01    1.01004886        1.01004982
0.200000032E-01    1.02019882        1.02019977
0.300000049E-01    1.03044796        1.03044986
0.400000065E-01    1.04079818        1.04080009
0.500000082E-01    1.05124759        1.05124950
0.600000098E-01    1.06179714        1.06179809
0.699999928E-01    1.07244491        1.07244682
0.799999833E-01    1.08319282        1.08319473
0.899999737E-01    1.09403896        1.09404087
0.999999642E-01    1.10498428        1.10498714
```

Chapter four, page 102

1. The SUBROUTINE returns, as the output in the third argument, twice the sum of the first two arguments. Twice 1.0 plus 2.0 is 6.0, which is now the value of X. Two times 6.0 plus 6.0 is 24.0, which becomes the value of G. Two times 24.0 plus 24.0 is 96.0, which is sent back as the new value of G. Unusual, and from a documentation standpoint, probably unwise—but legal.

   X is now given the value 2.0, and the two input arguments are therefore 4.0 and 0.0; twice their sum is 8.0, which is now the value of X. The next CALL is trouble: the value of X is sent as the two input arguments, but the output is sent to the value of a constant, which should never be an output argument. To this compiler, however, that is not a syntactic error; the result is to change the value in the location that the compiler set up to hold the constant 2.0, to 32.0—the result of doubling 8.0 plus 8.0. When we now print X, it is still 8.0. The last CALL now looks like it should produce 10.0, but where there should be a constant 2.0, there is now a 32.0, so the last PRINT produces twice 32.0 plus 3.0, or 70.0.

Moral: Triple-check the actual arguments in your FUNCTION and SUB-ROUTINE invocations to be sure that "there is agreement in number, order, and type." "Type" usually refers to real versus integer, and so forth, but this example shows that it also means whether a given argument is input or output (or both).

3. The program prints two lines. The first three numbers on each line are the values read by the READ in the main program. The last number on the first line is the sum of the three inputs, and the last number on the second line is the product of the inputs. The point is simply to demonstrate that a subprogram may invoke other subprograms.

4. The main program reads three numbers and sends them to ADDEM—but the SUBROUTINE never gives a value to its fourth argument. In the PRINT statement, therefore, TOTAL is an undefined variable. My system printed zero, but that is not dependable.

6. Absolutely not. Every program, whether it is a main program or a subprogram, and independent of whether it is compiled with others, must have an END as its final statement. (In many compilers, that means that a blank line after the final END will give strange diagnostics.)

7. It is illegal because a Fortran FUNCTION is not permitted to call itself. In Pascal, PL/I, APL, LISP, and many other modern programming languages this *is* permitted; such a FUNCTION is said to be *recursive*.

    If recursion were permitted in Fortran, this function would compute the factorial of its argument. If the argument is 1, 1 is returned as the function value; otherwise, the argument is multiplied by the value of the same function with an argument one less. If the original argument was 2, this gives 2 times 1, which is 2! If the original argument was 3, the function invokes itself twice, and so on.

    The factorial is the most common illustration of recursion but, in practice, it would be a silly way to compute a factorial; a simple loop is faster and easier to understand. Recursion comes into prominent—and, indeed, essential—use in somewhat different kinds of applications, such as the analysis of arithmetic expressions by compilers.

8. This is also illegal in Fortran, but permitted in some other languages.

9. a. If A, B, and C are all zero the first time, the GO TO 20 would return to the reading loop. For the first set of values that are not all zero, the program would sort them and then stop.

    b. The statement TEMP = X would cause X to be truncated to an integer and subsequently returned as the value of Y. If the program were tested using only integral values for the input, this mistake would not be noticed.

    c. The program operation would be unchanged.

    d. The values would be sorted into descending order instead of ascending.

    e. The program would stop if *any one or more* of the input values was zero instead of requiring that all three be zero.

    f. Same, answer as e.

11. The function is negative at 1.0 and 1.5, positive at 2.0. With LEFT = 1.0 and RIGHT = 2.0, the root finder "sees" only the sign change between 1.5 and 2.0. Running the program with LEFT = 1.0 and RIGHT = 1.2, say, produces the root at 1.1.

13.

```
* Exercise 13, page 105
* A main program that calls a SUBROUTINE named SORT3 that sorts
*    three numbers into ascending sequence
* Utilizes a SUBROUTINE named SWAP

* Main program reads sets of three data values until all are zero
* For each nonzero set, it prints values before and after
*       sorting into ascending sequence
        REAL A, B, C

***** REPEAT until all of values are zero
   10   CONTINUE

        READ *, A, B, C
        IF ( A .EQ. 0.0 .AND. B .EQ. 0.0 .AND. C .EQ. 0.0 ) GO TO 20
        PRINT *, ' BEFORE: ', A, B, C
        CALL SORT3 (A, B, C)
        PRINT *, ' AFTER:  ', A, B, C

        GO TO 10
***** END REPEAT

   20   CONTINUE

        END

* A SUBROUTINE that sorts three numbers into ascending sequence
        SUBROUTINE SORT3 (D, E, F)
        REAL D, E, F

        IF ( D .GT. E ) CALL SWAP (D, E)
        IF ( D .GT. F ) CALL SWAP (D, F)
        IF ( E. GT. F ) CALL SWAP (E, F)

        RETURN

        END

* A SUBROUTINE that exchanges two values sent to it by calling program

        SUBROUTINE SWAP (X, Y)

        REAL X, Y, TEMP

        TEMP = X
        X = Y
        Y = TEMP

        RETURN

        END
```

Chapter six, page 147

1. a.

```
    0
    1
   10
 -587
90061
12345
```

b.

```
   0.00
   1.00
  16.87
-568.12
   0.04
  12.34
```

c.

```
   0.
   1.
  17.
-568.
   0.
  12.
```

d.

```
 0.0000E+00
 0.1000E+01
 0.1687E+02
-0.5681E+03
 0.4000E-01
 0.1234E+02
```

3. a.

```
EXP= 12   P= 407.8   T=  32.9
```

b.

```
EXPERIMENT NUMBER =   12   PRESSURE =   407.8 TEMPERATURE =   32.9
```

c. The first line of the following output is at the top of a new page.

```
EXPERIMENT NUMBER =   12
```

```
 PRESSURE =   0.408E+03   TEMPERATURE =   0.329E+02
```

5. a.

```
        WRITE (6, 100) K, A, B
100     FORMAT (' ', I3, F10.3, F9.2)
```

b.

```
        WRITE (6, 200) K, A, B
200     FORMAT (' ', 'K =', I3, '   A=', F6.2, '   B=', F7.3)
```

c.

```
        WRITE (6, 300) K, A, B
300     FORMAT (' ', 'READING NO:', I3, '   1ST:', F8.3, '   2ND:', F8.3)
```

d.

```
        WRITE (6, 400) A, B, K
400     FORMAT (' ', F8.3, '    FACTOR 1'/
     $          ' ', F8.3, '    FACTOR 2'/
     $          ' ', I8, '   READING')
```

7. a.

```
100     FORMAT (' ', 2I7, 2F8.2)
```

b.

```
100     FORMAT (' ', 2I7, 2E12.3)
```

c.

```
100     FORMAT (' ', 'I=', I7, ' J=', I7, ' R=', F6.1, ' S=', F6.1)
```

d.

```
100     FORMAT (' ', 'I = ', I6/
     $          ' ', 'J = ', I6/
     $          ' ', 'R = ', F6.1/
     $          ' ', 'S = ', F6.1)
```

9.

```
        READ 100, BOS, PHL, SFO, ATL
100     FORMAT (4F8.2)
```

It is not necessary to provide separate format codes for the four fields. Since only one of them does *not* contain a decimal point, we can write a format code for that one and let the actual decimal point in the input override for the other cases.

11. a. There should be a four-digit integer, a five-digit integer, a ten-digit number that will be treated as a whole number if it does not contain an actual decimal point, and a five-digit number that will be treated as having two decimal places if it does not contain an actual decimal point. Any of the numbers may have leading blanks without affecting the result.

b. There should be two three-digit integers followed, eight positions later, by two eight-digit numbers that will be assigned to real variables. The values for the real variables will be taken to be whole numbers unless they contain decimal points. It makes no difference what, if anything, is in the eight positions skipped over.

c. The first 20 positions are skipped, then a one-digit integer is read, then another 10 positions are skipped, and two ten-digit real values are read. The values for the real variables will be taken to be whole numbers unless they contain decimal points.

d. There should be two five-digit integers on one record followed by another record with two ten-digit values for real variables. The "records" could be two cards, two lines of input from a terminal, or other possibilities.

13.

```
      PRINT 100, A, B, X, Z
100   FORMAT (' ', 2F12.4, 2E20.6)
```

15.

```
100  FORMAT ('1', 'TOTAL LABOR', F8.2/
$              ' ', 'TOTAL PARTS', F8.2/
$              ' ', 'ACCESSORIES', F8.2/
$              ' ', 'OIL, GREASE', F8.2/
$              '0', 'SUBTOTAL   ', F8.2/
$              '0', 'SALES TAX  ', F8.2/
$              '0', 'GRAND TOTAL', F8.2)
```

17.

```
* Exercise 17, page 151
      INTEGER SSN1, SSN2, SSN3

10    CONTINUE
      READ (5, 100, END=20) SSN1, SSN2, SSN3
100   FORMAT (I3, I2, I4)
      WRITE (6, 200) SSN1, SSN2, SSN3
200   FORMAT (' ', I3, '-', I2, '-', I4)
      GO TO 10

20    CONTINUE

      END
```

Chapter seven, page 158

1. a. It prints the odd integers from 1 to 11, together with their cubes.

   c. It reads input records (up to a maximum of 10,000) until detecting end of file. As it reads each record, it adds the first number on the record to a sum; the sum of all such numbers is printed when the end of the input is detected.

2. a. There is no statement number in the DO statement.

   b. The index of a DO must not be redefined within the range of the DO.

   c. The DOs are not properly nested: the inner DO is not completely contained within the range of the outer DO. To obtain the presumed intention, the statement numbers in the DOs must be reversed, or else the statement numbers on the CONTINUEs must be reversed.

   d. The parameters of a DO must not be altered within the range of the DO.

3. The problem specification did not say how to present the output; you are expected to make reasonable assumptions of your own from now on, unless otherwise instructed. The variable C could have been real, but then the output would have been printed with a terminal decimal point, which is slightly less attractive.

```
* Exercise 3, page 160
      INTEGER C
      REAL F

      PRINT *, ' TEMPERATURE CONVERSION TABLE'
      PRINT *, ' '
      PRINT *, ' C      F'
      DO 10 C = 0, 100
          PRINT 100, C, 1.8*C + 32.0
100       FORMAT (' ', I3, F6.1)
10    CONTINUE

      END
```

5.

```
* Exercise 5, page 160
      REAL N

      DO 10 N = 1.0, 20.0
          PRINT 100, N, SQRT(N)
100       FORMAT (' ', F3.0, F12.5)
10    CONTINUE

      END
```

7. This task can reasonably be done with or without a DO. If a DO is used, its only function is to provide an (essentially) infinite loop that is terminated by the END=20 clause in the READ.

```
*  Exercise 7, page 160
*  Version using DO
      REAL X
      INTEGER I

      DO 10 I = 1, 10000
          READ (*, 100, END=20) X
100       FORMAT (F8.0)
          IF ( X .GT. 100.0 ) THEN
              PRINT *, X
          END IF
10    CONTINUE

20    CONTINUE

      END


*  Exercise 7, page 160
*  Version using REPEAT loop
      REAL X

***** REPEAT until end of file
10    CONTINUE

          READ (*, 100, END=20) X
100       FORMAT (F8.0)
          IF ( X .GT. 100.0 ) THEN
              PRINT *, X
          END IF

      GO TO 10
***** END REPEAT

20    CONTINUE

      END
```

9.

```
*  Exercise 9, page 160
*  The DO index in an integer variable, which is converted to real
*     for use in the formula evaluation.  This avoids accumulation
*     of roundoff errors in adding 0.01 repeatedly.
                XN
      INTEGER NX
      REAL X, Y

      DO 10 XN = 100, 300
          X = XN / 100.0
          Y = 41.298 * SQRT(1.0 + X**2) + X**(1.0/3.0) * EXP(X)
          PRINT 100, X, Y
100       FORMAT (' ', 2PE20.6)
10    CONTINUE

      END
```

11.

```
* Exercise 11, page 160
* Can be written with or without DO; REPEAT used here.
      INTEGER ERRORS, I
      REAL X

      ERRORS = 0
      I = 1

***** REPEAT until zero sentinel
   10  CONTINUE

          READ 100, X
  100     FORMAT (F8.0)
          IF ( X .EQ. 0.0 ) GO TO 20
          IF ( X .GT. 0.0 ) THEN
              PRINT *, I, X
          ELSE
              ERRORS = ERRORS + 1
          END IF
          I = I + 1

      GO TO 10
***** END REPEAT

   20  CONTINUE

      PRINT *, ' THERE WERE ', ERRORS, '  ERRORS'

      END
```

13. The following program was run on a computer with the IBM System/370 architecture; the results will be different on machines with different architectures. No errors were indicated for the integer version but, for values of N larger than 13, the results were obvious nonsense. This relates to the way integer multiplication is done on this machine. For the real case, the results were correct through N = 56, after which the program blew up on an exponent overflow exception. This is because for N = 57, the result would have been too large to represent as a real number in this machine.

```
* Exercise 13, page 161
* A program to determine maximum factorials
      INTEGER N, NFACT
      REAL RN, RNFACT

      DO 10 N = 1, 20
          WRITE (6, 100) N, NFACT(N)
  100     FORMAT (' ', I3, I20)
   10  CONTINUE

      DO 20 RN = 1.0, 60.0
          WRITE (6, 200) RN, RNFACT(RN)
  200     FORMAT (' ', F3.0, E16.7)
   20  CONTINUE

      END
```

```
      INTEGER FUNCTION NFACT(N)
      INTEGER N, I
      NFACT = 1
      DO 10 I = 2, N
          NFACT = I * NFACT
10    CONTINUE
      RETURN
      END

      REAL FUNCTION RNFACT(RN)
      REAL RN, I
      RNFACT = 1.0
      DO 10 I = 2.0, RN
          RNFACT = I * RNFACT
10    CONTINUE
      RETURN
      END
```

Chapter seven, page 164

1. a. It reads 50 numbers into the 50 elements of an array named X. Then it forms the 49 *first differences* of the X array. (If "first difference" is a new term to you, consider the program to be an implicit definition.)

3. a.

```
* Exercise 3a, page 165
      REAL A(10), PROD
      PROD = A(1) * A(2)
```

b.

```
* Exercise 3b, page 165
      REAL A(10)
      A(3) = (A(1) + A(3) + A(5)) / 3.0
```

c.

```
* Exercise 3c, page 165
      REAL A(10)
      IF ( A(10) .LT. 0.0 ) THEN
          A(10) = -A(10)
      END IF
```

d.

```
* Exercise 3d, page 165
      REAL A(10)
      INTEGER I
      DO 10 I = 1, 10
          A(I) = 2.0 * A(I)
10    CONTINUE
```

5.

```
*  Exercise 5, page 165
       REAL A(30), B(30)
       REAL SUM, D
       INTEGER I
       SUM = 0.0
       DO 10 I = 1, 30
           SUM = SUM + (A(I) - B(I))**2
   10  CONTINUE
       D = SQRT(SUM)
```

7.

```
*  Exercise 7, page 166
       REAL R(40), S(40), T(40)
       INTEGER I, M
       DO 10 I = 1, M
           T(I) = R(I) + S(I)
   10  CONTINUE
```

9.

```
*  Exercise 9, page 166
       REAL A(15), B(15)
       INTEGER I, FLAG
       FLAG = 0
       DO 10 I = 1, 15
           IF ( A(I) .LE. B(I) ) THEN
               FLAG = 1
           END IF
   10  CONTINUE

       IF ( FLAG .EQ. 0 ) THEN
           PRINT *, ' OK'
       ELSE
           PRINT *, ' NO WAY'
       END IF
```

11.

```
*  Exercise 11, page 166
       INTEGER M(20), I
       DO 10 I = 1, 20
           M(I) = I*M(I)
   10  CONTINUE
```

13.

```
*  Exercise 13, page 166
       REAL B(50), BIGB
       INTEGER I, NBIGB
       BIGB = B(1)
       NBIGB = 1
       DO 10 I = 2, 50
           IF ( B(I) .GT. BIGB ) THEN
               BIGB = B(I)
               NBIGB = I
           END IF
   10  CONTINUE
```

14.

```
* Exercise 14, page 166
      REAL A(15, 15), X(15), B(15)
      INTEGER I, J
      DO 20 I = 1, 15
         B(I) = 0.0
         DO 10 J = 1, 15
            B(I) = B(I) + A(I, J) * X(J)
10       CONTINUE
20    CONTINUE
```

Chapter seven, page 174

1. The grades would be sorted but, at the end, the correspondence between grades and IDs would have been lost. We would know what the high grade was but not who got it.

3. Replace the PRINT 500 statement with these statements.

```
IF ( MOD(N, 2) .EQ. 0 ) THEN
   MEDIAN = (GRADE(N/2) + GRADE(N/2 + 1)) / 2
ELSE
   MEDIAN = GRADE((N+1)/2)
END IF
PRINT 500, MEDIAN
```

5.

```
* Exercise 5, page 175
      INTEGER CLASS(20), I
      REAL WEIGHT(20)
      INTEGER C
      REAL W

* Initialize arrays
      DO 10 I = 1, 20
         CLASS(I) = 0
         WEIGHT(I) = 0.0
10    CONTINUE

***** REPEAT until detecting sentinel
20    CONTINUE

         READ 100, C, W
100      FORMAT (I2, F6.0)
         IF ( C .EQ. 0 ) GO TO 30
         CLASS(C) = CLASS(C) + 1
         WEIGHT(C) = WEIGHT(C) + W

      GO TO 20
***** END REPEAT

30    CONTINUE

* Print headings
      PRINT *, ' CLASS     WEIGHT'

* Print averages
      DO 40 I = 1, 20
         IF ( CLASS(I) .EQ. 0 ) THEN
            PRINT 200, I, 0.0
         ELSE
            PRINT 200, I, WEIGHT(I) / CLASS(I)
         END IF
200      FORMAT (' ', I4, F12.4)
40    CONTINUE

      END
```

Chapter seven, page 182

1.

```
* Exercise 1, page 182
******************************************************************
* A SUBROUTINE to read the data and count errors
* This version applies the sufficient test for convergence
******************************************************************
        SUBROUTINE READER (A, N, EPS, MAXIT, ERRORS)

        REAL A(50, 51), EPS, BIGGST, TEMP
        INTEGER N, I, J, MAXIT, ERRORS
        REAL SUM

* Read the parameters
        READ *, N, MAXIT, EPS, BIGGST

* Initialize the error count
        ERRORS = 0
* Read the elements of the array, with checking.

***** REPEAT until end of file
   10   CONTINUE
            READ (*, *, END = 20) I, J, TEMP
            IF (        (I .LT. 1)
     $            .OR. (I .GT. N)
     $            .OR. (J .LT. 1)
     $            .OR. (J .GT. N+1)
     $            .OR. (ABS(TEMP) .GT. BIGGST) ) THEN
                ERRORS = ERRORS + 1
                PRINT 100, I, J, TEMP
  100           FORMAT (' ', 'ERROR IN ELEMENT WITH I = ', I2, ' J = ', I2,
     $                       ' VALUE = ', F10.6)
            ELSE
                A(I, J) = TEMP
            END IF
        GO TO 10
***** END REPEAT

   20   CONTINUE

* Apply the sufficient test to each row
        DO 50 I = 1, N
            SUM = 0.0
            DO 30 J = 1, I-1
                SUM = SUM + ABS(A(I,J))
   30       CONTINUE
            DO 40 J = I+1, N
                SUM = SUM + ABS(A(I,J))
   40       CONTINUE
            IF ( ABS(A(I,I)) .LE. SUM ) THEN
                PRINT *, ' CONVERGENCE CRITERION FAILS FOR ROW ', I
                ERRORS = ERRORS + 1
            END IF
   50   CONTINUE

        RETURN

        END
```

3.

```
Exercise 3, page 183
* Call the solution subroutine, ITRATE, up to MAXIT times.
      DO 30 ITER = 1, MAXIT
         CALL ITRATE (A, X, N, RESID)
*        check for convergence:
         IF ( RESID .LT. EPS ) GO TO 40
   30 CONTINUE

*     if next statement ever reached, there was no convergence:
      PRINT 300, MAXIT
  300 FORMAT ('0', 'NO CONVERGENCE IN', I5, '  ITERATIONS')

   40 CONTINUE

* Print solution
      PRINT 200, (I, X(I), I = 1, N)
  200 FORMAT (' ', 6(I3, F12.6))

      END
```

4. In the second version of the program, the addresses are derived as follows. The element in the first row and first column is in position 1 of the linear array. Now suppose we are trying to determine the location in the linear array of the element in row I and column J. Consider that row 1 has 1 element, row 2 has 2 elements, and so forth; row I-1 has I-1 elements. The sum of the series from 1 to I-1 is I*(I-1)/2. That number is also the element number of the element in row I-1 and column I-1, that is, the element preceding the first element in row I. The element in row I and column J is therefore in position I*(I-1)/2 + J.

Naturally, the main program and the READER subroutine would have to be changed to comply with the changes in the data storage, but this is not difficult once the addressing scheme is understood.

```
* Exercise 4a, page 183
*********************************************************************
* A SUBROUTINE to solve a lower diagonal system of equations
*********************************************************************
* This version stores the coefficients in a 100x100 array.
* The constant terms are in a 100 element array named B.

      SUBROUTINE SOLVE1 ( A, B, X, N )
      REAL A(100, 100), B(100), X(100), SUM
      INTEGER I, J, N

      DO 20 I = 1, N
         SUM = 0.0
         DO 10 J = 1, I-1
            SUM = SUM + A(I, J) * X(J)
   10    CONTINUE
         X(I) = (B(I) - SUM) / A(I, I)
   20 CONTINUE

      RETURN

      END
```

```
* Exercise 4b, page 183
***************************************************************
* A SUBROUTINE to solve a lower diagonal system of equations
***************************************************************
* This version stores the coefficients in a one dimensional array
*     having 5050 elements instead of 10,000.
* The conversion from a row number I and a column number J,
*     to an element number in this linear array is handled by
*     subscript computations.
* The constant terms are in a 100 element array named B.

      SUBROUTINE SOLVE2 ( A, B, X, N )
      REAL A(5050), B(100), X(100), SUM
      INTEGER I, J, N

      DO 20 I = 1, N
         SUM = 0.0
         DO 10 J = 1, I-1
            SUM = SUM + A(I*(I-1)/2 + J) * X(J)
   10    CONTINUE
         X(I) = (B(I) - SUM) / A(I*(I+1)/2)
   20 CONTINUE

      RETURN

      END
```

Chapter eight, page 192

1. Each of the numbers appearing in the system of equations can be represented exactly, but significance is lost in forming the products. The data was chosen to represent two lines that almost coincide, so that a small difference in the computations moves the point of intersection a good distance. If the lines intersected at any angle in the vicinity of a right angle, the difference in the results would be negligible.

Chapter eight, page 201

1. The only relational operators permitted in comparing two complex quantities are .EQ. and .NE.; there is no operator between the constant and the variable in the assignment statement for Z3.

2. Yes.

4.

```
* Exercise 4, page 203
      REAL Y
      COMPLEX Z

      DO 10 Y = -5.0, 5.0
         Z = CMPLX(1.0, Y)
         WRITE (6, 100) Z, CEXP(Z)
  100    FORMAT (' ', 2F5.0, 2F12.5)
   10 CONTINUE

      END
```

10. I'm afraid I don't see how a computer could help in this situation. If you do, I would be delighted to hear from you; write care of the publisher (John Wiley & Sons, Inc., 605 Third Avenue, New York, NY 10158).

Chapter eight, page 208

3. The first value printed is the value read for TRUE, whether that is .TRUE. or .FALSE.. The second value printed is always .TRUE., whether the value read for TRUE was .TRUE. or .FALSE.. The programmer should have to write weekly essays on successive paragraphs from *Finnegans Wake* for the rest of his or her life.

Chapter eight, page 218

1. With D not having been declared or given a value in the program, the expression for E concatenates a character variable with a real variable, which is illegal. A similar comment goes for the mixture A // 12345. The statement A = A // '12345' is legal but, after the expression is evaluated, it will be truncated back to the five leftmost characters, which is just the value of A used in the concatenation.

6. It will operate correctly. One might think that the program would get confused, because with exactly 50 records and the READ being embedded in a loop controlled by the statement DO 10 K = 1,50, the value of K upon exit from the loop would be either wrong or undefined. However, it turns out that the Fortran 77 standard specifies, first, that incrementing the DO index is done before the testing for completion, and second, that upon exit from a DO loop the index has its last defined value. Under the conditions stated in the exercise, the value of K on exit from the DO would be 51 and the value given to N would be 50.

   The Fortran standard prior to this one did not specify what value the DO index had to have on exit from the DO, and a program like this could operate unreliably.

INDEX